Conservation and
evolution

Conservation and evolution

O. H. FRANKEL AND MICHAEL E. SOULÉ

CAMBRIDGE UNIVERSITY PRESS

CAMBRIDGE

LONDON   NEW YORK   NEW ROCHELLE

MELBOURNE   SYDNEY

Published by the Press Syndicate of the University of Cambridge
The Pitt Building, Trumpington Street, Cambridge CB2 1RP
32 East 57th Street, New York, NY 10022, USA
296 Beaconsfield Parade, Middle Park, Melbourne 3206, Australia

First published 1981

Phototypeset in V.I.P. Times by
Western Printing Services Ltd, Bristol

Printed in Great Britain at the
University Press, Cambridge

British Library Cataloguing in Publication Data
Frankel, *Sir*, Otto Herzberg
Conservation and evolution.
1. Nature conservation
2. Ecological genetics
I. Title   II. Soulé, Michael
Ellman
639'.9   QH75   80-40528
ISBN 0 521 23275 9 hard covers
ISBN 0 521 29889 X paperback

Contents

Preface

In this book we attempt to bring together the genetic principles for the conservation of all forms of life, wild or domesticated, lions or lizards, oaks or orchids, cattle or ducks, rice or potatoes. The unifying factor underlying survival and adaptation, in time and in space, is genetic diversity; and the nature, distribution and preservation of genetic diversity is the central theme of this book.

Clearly, the subject itself is of immense diversity, and we attempt to present the underlying biological principles, irrespective of the diversity of breeding systems, life tables or ecological conditions and associations. Yet while we stress the essential unity of conservation genetics, the intensity of the human impact separates the domesticates from the mainstream of organisms which survive and reproduce within more or less natural communities of which they form a part. To a greater or lesser degree man provides an environment for domesticates in which to live, reproduce and die according to human precepts. He assembles and activates genetic diversity and is the arbiter in the evolutionary process. Yet the key to the system is the availability of the genetic building materials, the genetic resources, which alone make it possible to continue and advance the adaptive process on which evolutionary success, and indeed man's survival, depend.

Wild biota depend on their own genetic resources for survival and adaptation. Man can reduce or extend these resources, largely by manipulating the living space available to a species, but the scope for direct participation is limited indeed. Yet the principles derived from population and quantitative genetics govern sampling in domesticates and the minimal population size in nature reserves. Indeed, there are many cross-links. They are most patent in the preservation of the wild plants used by man in forestry or in pastures or range lands, and in the breeding of crops.

At first sight it may appear that this book really consists of two books, one dealing with wild biota (chapters 2–5), the other with domesticates (chapters 8–10), with the zoological and botanical gardens (chapters 6 and 7) in between. The difference is not restricted to the subject, wild versus domesticated, but extends to the treatment. In the former, no comprehensive statement had been available. The issue of a genetic component of nature conservation had been raised (Frankel, 1970b) but not fully examined. On the other hand genetic conservation of domesticated plants – though not of animals – had been the subject of two books (Frankel and Bennett, 1970b), and Frankel and Hawkes, 1975) and while we raise and discuss many aspects which previously had not been considered, this book consolidates concepts and principles which had been examined previously. Hence the second part contains a good deal of detailed discussion of techniques and procedures which would not have been possible in the first part.

The book is the result of joint planning, and all chapters bear the marks of consultation, critical exposure, and constructive comment. However, chapters 2–6 are mainly the work of M.E.S., chapters 7–10 that of O.H.F.

We have received a great deal of generous help from colleagues and friends. M.E.S. gratefully acknowledges the following individuals for their criticisms and suggestions: Kurt Benirschke, Paul R. Ehrlich, Nathan R. Flesness, Thomas J. Foose, Michael Gilpin, Stephen J. Gould, Ernst Mayr, Mark Pomerantz, Peter H. Raven, Michael Rosenzweig, Thomas J. M. Schopf, John W. Senner, Leigh Van Valen, Robert Warner, Bruce A. Wilcox, Christopher Wills, and E. O. Wilson.

O.H.F. acknowledges with gratitude the continued interest and help from his colleagues A. H. D. Brown and D. R. Marshall. M. J. D. White gave helpful opinions on some chapters. Suggestions and critical comments from F. W. Morley, J. M. Rendel and H. N. Turner greatly assisted in the writing of Chapter 10. B. M. Bindon, J. E. Frisch, B. L. Sheldon and J. E. Vercoe provided useful references and other information for the same chapter.

O. H. FRANKEL
Division of Plant Industry, CSIRO, Canberra

M. E. SOULÉ
Institute for Transcultural Studies, Los Angeles

1

Introduction

1.1 Components of survival and adaptation

Any book with the word 'conservation' – in the sense of biological conservation – in its title starts from the premise that it is better that organisms continue to exist than that they become extinct. Yet it is extinction and not survival which provides the motive for conservation. Indeed, the conditions and processes of extinction, discussed in Chapter 2, must be an ever-present background to the exploration of the conditions of survival. This is far from being an academic consideration, in the light of the drastic increase in species extinction in many parts of the world, which has added an unprecedented urgency to the conservation of natural communities and the species they contain. The domesticated plants and animals, which directly serve our own species, are also, though less immediately, threatened. Here it is the evolutionary reservoirs, the raw materials for future adaptation, which are under threat, but the effect of losing them could be equally drastic, hence the current worldwide drive for the conservation of genetic resources.

As an intrinsic facet of evolution, extinction is inevitable. But in the evolutionary record, and in the history of crops and livestock, the emergence of new forms is interwoven with the passing out of older ones. It is implicit in the concept of organic, or progressive (Dobzhansky, 1967, p. 129) evolution that it is a continuing process, with survival, adaptation and speciation balancing extinction. It is the growing threat to evolutionary processes which renders the increase in extinction doubly formidable.

Survival and adaptation can be viewed in terms of three components: time, space and fitness; the latter signifies adaptation, genetic stability and variability (Thoday, 1953). Each of these extends over a wide range: time, from an individual life span to geological or evolutionary time; space, from the local habitat to global distribution; and fitness which may relate to the individual, the population or the species. In evolutionary history, one attempts to unravel the role of these parameters in terms of historical, geological or palaeontological records, past and present distribution, taxonomic and genetic relationships, etc. Conservation, on the other hand, must proceed in the opposite direction, since its concern is not the past, but the future. Indeed, implicitly or explicitly, every conservation project, whether a nature reserve, a zoological garden, or a 'gene bank', is defined, or definable, in terms of the three parameters. It will have a time projection for its anticipated duration, a spatial projection (in the case of gene banks a definition of the areas from which their stocks are derived), and a fitness rating in terms of community biology or of population genetics. The three parameters are as basic to conservation biology as they are for conservation planning and management.

1.2 The time scale of concern

Whether acknowledged or not, every conservation effort has at least a notional time dimension, the period for which it is expected, or hoped, that it will remain operative. This may be as short as a lifetime, e.g. the preservation of a specific tree; a limited number of years or generations, such as the preservation of a 'national monument' in the shape of a specialized or otherwise significant community or site; or it may be in perpetuity in nature reserves such as national parks or biosphere reserves.

In domesticated plants and animals one can be more specific. The objects of conservation are the ancient, highly variable breeds of crops and live-stock, now on the verge of extinction, which are needed by breeders to adapt our modern breeds to ever changing natural and social environments. For how long are they likely to be needed? Clearly, the answer depends on the dynamics of economic, social and scientific change. For example, the very existence of major animal industries may be subject to human population dynamics in the years to come. New techniques derived from molecular biology may supersede the current use of genetic resources. We cannot guess whether all the present-day crops will even be in use a hundred years hence. Considering the slow rate of social change and the delay in introducing new technologies, one might project a time perspective for gene pool conservation of between 50 and 150 years, to be reviewed in the light of technological and economic changes as they occur. Its utility is that it provides a realistic conservation target: techniques and procedures, as discussed in chapters 9 and 10, must be capable of providing safeguards for survival and genetic integrity extending over that period, to be extended if required. But there can be no hard and fast rule. Forestry species are wild species with their richest genetic resources in natural forests which can be preserved just like other natural communities Forest trees are not only long-lived, but are likely to be needed, and planted, for as long as one can foresee. Hence a time perspective of say, a thousand years seems not unreasonable for the preservation of forest communities and the genetic reservoirs they contain.

From this discussion a concept emerges which has been called 'The time scale of concern' (Frankel, 1974). Relevant as it is for an understanding and planning of genetic conservation in domesticated species, its significance is even greater for the conservation of natural communities. Here the objectives of conservation inevitably include a time scale, whether expressed or not, with implications for all phases of conservation, from site and size, to design and management. A consideration of the realities of nature reserves will illustrate the significance of the time concept. There are numerous nature reserves which serve a multiplicity of social objectives – recreational, educational, commercial (such as tourism) – in addition to being sites in which a community of plants and animals remains in being. They may be extensively traversed by roads and tracks and be visited by and accommodate large numbers of people. Reserves of this kind exist in all parts of the world and perform an important social function. They are not primarily designed for preserving the integrity of flora and fauna, and will require a great deal of management to preserve the general character of the ecosystem. Depending on size and intensity of use, sooner or later they will be transformed into man-made communities. As preserves of natural communities, such nature reserves clearly have a limited time prospect and they could not be relied upon for the preservation of the species they contain. At the other end of the scale are nature reserves which are dedicated to provide a secure and lasting habitat for the species they contain, with the minimum of interference compatible with, or essential for their survival and welfare, subject to whatever access to human use and enjoyment is conceded as a matter of policy. Such reserves implicitly have a 'for ever' time scale, though this may not be formally recognized or declared, seeing that such a declaration would be subject to cancellation or amendment by subsequent generations or governments.

It is the nature reserves of the latter type which derive an enormously enhanced significance from the rapidly increasing destruction of so far relatively undisturbed natural and semi-natural communities, especially in the humid tropics. Conversion of forests to agricultural or plantation crops, devastation by shifting cultivation, and drastically exploitative forest utilization increasingly contribute to habitat destruction, a process which is virtually inevitable in the face of acute population and economic pressures in developing countries. A high rate of destruction of rain forests is widespread throughout the tropics (Whitmore, 1980). Indeed, it is widely believed that by the end of the century there will be no undisturbed tropical lowland forest left anywhere in the world, except in whatever reserves may remain. Many other terrestrial, freshwater and marine ecosystems are similarly threatened, but the prospects for species extinction are greatest in the tropics. Increasingly, the main hope for wildlife continuing to exist and to evolve under natural or near-natural conditions, with a minimum of human interference, rests on those residual habitats which are dedicated as repositories and refuges of wild biota – the nature reserves.

1.3 Preservation or conservation

We use the term 'conservation' to denote policies and programmes for the long-term retention of natural communities under conditions which provide the potential for continuing evolution, as against 'preservation' which provides for the maintenance of individuals or groups but not for their evolutionary change. Thus, we would state that zoos and gardens may preserve, but only nature reserves can conserve.

These terms have been variously used, hence some clarification is needed. In his remarkable book, *Man's Responsibility for Nature*, Passmore (1974) defines as *preservation* 'the attempt to maintain in their present condition such areas of the earth's surface as do not yet bear the obvious marks of man's handiwork and to protect from the risk of extinction those species of living beings which man has not yet destroyed' (p. 101). If protection from the 'risk of extinction' implies an evolutionary imperative, as we insist it must (chapters 4 and 5), then Passmore's 'preservation' is essentially our 'conservation'. However, Passmore's *conservation* is defined as 'the saving of natural resources for later consumption' (p. 73), i.e. the 'saving for' rather than 'saving from' the dominant species. While Passmore's use of 'conservation' is in accord with usage by resource economists, ours accords with that commonly understood by biologists and conservationists.

Passmore devotes a chapter to preservation (as defined above) but does not discuss conservation in the sense in which the term is used in this book. He does, however, discuss at length the issue of man's obligations towards the future, in the light of attitudes through history. His own conclusions are, in his words, 'limited and confused', but he appears to be in general agreement with the economists and philosophers who conclude that 'our obligations are to *immediate* posterity, we ought to try to improve the world so that we shall be able to hand it over to our immediate successors in a better condition, and that is all' (p. 91). Nor ought any one generation be called upon to make undue sacrifices on behalf of future generations. But he recognizes different views. 'We stand now, if the more pessimistic scientists are right, in a special relationship to the future; unless we act, posterity will be helpless to do so. This imposes duties on us which would not otherwise fall to our lot.' (p. 98).

We unhesitatingly endorse this 'pessimistic' view. As we have attempted to show, nothing but incisive action by *this* generation can save a large proportion of now-living species from extinction within the next few decades. Inevitably, pressing needs of today over-ride the claims for tomorrow, especially in the hard-pressed developing countries. Yet it is gratifying to note that such claims do receive recognition. Five countries of South-East Asia have combined to collect and preserve the genetic resources of crops and fruits indigenous to their region, though well aware that much of this effort will largely benefit future generations (IBPGR, 1977a). Moreover, the same countries have set aside large areas for biosphere and other nature reserves.

1.4 The evolutionary potential

In chapters 3–5 we explore the genetic basis of conservation in terms of fitness, population size and genetic diversity. We examine, first, the requirements for the maintenance of fitness which would safeguard survival at least in the shorter term; and, second, we examine the more stringent requirements for long-term conservation, involving the maintenance of the evolutionary potential, the capacity to evolve in response to environmental change.

Why is the capacity to evolve a necessary condition for survival? Environmental change is an inescapable reality. Even if climate were forever frozen in its present state, the biotic environment for all species will continually change. For example, small organisms will continue to evolve, even if large ones cannot, changing the selection regime for their predators, hosts, symbionts and competitors. In addition, as natural habitats become fragmented due to the impact of civilization, extinctions and shifts in community structure and dominance will greatly change the conditions of life for all species. Given these inevitable processes, a conservation programme must provide an evolutionary potential.

For many organisms survival may ultimately be a function of available space. This is the reason why large animals – and to a lesser extent large trees – are more vulnerable than smaller organisms, and why special management measures may be needed for their preservation. Indeed, it is the availability of space for adequate numbers of the larger species which is crucial for long-term survival of an ecosystem. Yet, as we show in chapters 4 and 5, the sizes of most existing nature reserves are inadequate for the desirable population size of large animals. For this reason we emphasize the 'genetics of scarcity': species with large numbers look after themselves. It is only those with inevitably small numbers which present genetical conservation problems.

Let us now compare the evolutionary potential, and its realization, in the three conservation systems with which we are concerned – nature reserves, genetic resources conservation of domesticated plants and animals, and zoological and botanical gardens. In nature reserves, genetic diversity consists of the genetic variance within and between populations, with or without gene flow between them. Populations are thought to be in a dynamic state, responsive to environmental change (including the inevitable human influence), i.e. subject to natural selection.

The genetic resources of domesticates are, for the most part, in a static state. The exceptions are the wild relatives of crop plants and the wild plants used by man, such as forest and pasture species, which are located in natural communities. Some local breeds of livestock also retain a fairly dynamic state. All other genetic resources are maintained 'frozen' – which in many cases is literally true (see chapters 9 and 10). Their evolutionary potential is enormous, but it needs to be realized through recombination, mutation and selection which, of course, is in the hands of the plant and animal breeders.

In zoological and botanical gardens we face a minimal intraspecies genetic diversity which is retained with a minimum of recombination. Except for matings with outside stock to avoid inbreeding, matings of both animals and plants are restricted to the small numbers necessitated by available resources. There is little scope for evolutionary change except domestication.

These comparisons illuminate the very nature and purpose of conservation. Genetic resources of domesticates are *preserved*, not for their own sake, but because of their immediate or potential usefulness to man, be it in breeding or in some form of research. The reason for nature *conservation*, as we see it, is diametrically different. Its essence is for some forms of life to remain in existence in their natural state, to continue to evolve as have their ancestors before them throughout evolutionary time. The uses and benefits that we derive from nature reserves have already been recalled and will be more fully demonstrated in further chapters. Here we stress their unique role in maintaining the continuity of self-regulating communities with their infinitely complex adaptive balance, which no man-made system could attempt to recover, and which may not exist in such complexity anywhere else in the universe.

It could be argued that we are able to produce animals more beautiful and more congenial as companions of man than are the animals we attempt to preserve in their wild state. As examples, one need only to point to the cat or the dog – the most ancient domesticated animal, whose wild ancestor we have all but exterminated; the budgerigar or the peacock; and to the many wild plants brought into our gardens. Some of these have been given increased genetic diversity in terms of numbers of alleles per locus, as in the garden cultivars of *Phlox* in comparison with populations of wild *Phlox drummondii* (Levin, 1976).

This is undoubtedly a possibility, and one of the acknowledged uses of nature reserves is to serve as reservoirs for the continuing process of domestication which began some 10 000 years ago (see chapter 8). But domestication could save only a small minority of species, and thus would mean the extinction of the vast majority of plants and animals left out in the cold, and an end to progressive evolution.

A more futuristic alternative to conservation is the possibility that the rapidly advancing field of genetic engineering will soon permit the design and production of organisms to fit any taste. Entirely new faunas and floras could, perhaps, be created. It is very unlikely, however, that biota as diverse, complex and integrated as those on coral reefs or in tropical rain forests will ever be duplicated, if for no other reason than cost. Thus we do not foresee a 'technological fix' for the massive wave of extinctions which looms just years or decades away.

1.5 An evolutionary responsibility?

The concern for the future of evolution is often subjected to criticism on the grounds that we are incapable of planning even five or ten years ahead, so how can provisions for tens of thousands of years ahead be taken seriously? The comparisons are scarcely relevant. In conservation it is not a question of blueprints for the future. All that is attempted is to provide *conditions*, based on our best scientific insight and subject to the present-day social and economic restraints, which will make it possible for an evolutionary succession of organisms to continue, inevitably subject to the social consent of future generations. The likelihood of such continuing consent may not be judged as very great; yet the risk in terms of our own investment is negligible provided the requirements for what one may call 'evolutionary conservation' do not materially exceed the requirements for the accepted form of conservation in national parks or biosphere reserves. Indeed they do not. It is generally acknowledged that in long-term or ecological reserves the cardinal requirement is for *large size*, to provide for fluctuations in population numbers and for sufficient space to accommodate the most space-demanding 'target species'. Space is also the principal requirement in 'evolutionary' reserves, since population size and fitness (see p. 1) in practice turn into area terms. The injection of the evolutionary intent tends to reinforce the accepted ecological criteria of conservation, although it adds considerable weight to arguments for larger numbers of big reserves. We conclude that the evolutionary concept results in strengthening the principles of nature conservation as now widely accepted. The essential difference is in the deliberate application of a 'for ever' time scale of concern.

If as biologists we accept the proposition that life cannot continue without opportunities for evolution, there remains the question why we should be concerned about the continued existence of living organisms except on grounds of actual or potential use to our own species. After all, extinction has always occurred, and we have a long history of contributing to this process. In the shorter term, the answers are many and are readily found in the conservationist literature, ranging from ecological and economic to scientific and educational arguments. Besides, there is a strong appeal in the very natural demand that the elephant, the giraffe, the leopard, the tiger, or

our close relatives, the gorilla and the orangutan, be seen alive by generations after us rather than in natural history books, as we now see the
traditional dodo.

This is a very natural feeling, perhaps due to no more than childhood
memories, yet at its root there is a sense of responsibility for the unprecedented and irremediable destruction of life which is taking place in the
short space of a generation or two. Heslop-Harrison (1976) believes 'we
have something of a moral duty (to preserve) for posterity some reasonable
fraction of the richness and diversity of the plant kingdom as we ourselves
have inherited it'. There is, we believe, a growing feeling that to end all
evolution that is not induced by our own species – except that of organisms
we are as yet unable to control – is an arrogant if not a fatal step for man to
undertake, and that as biologists we bear a special responsibility. 'No longer
can we claim evolutionary innocence. We are still subject to evolutionary
processes . . . but we are also major operators. We are *not* the equivalent of
an ice age or a rise in the sea level: we are capable of prediction and of
control. We have acquired evolutionary responsibility.' (Frankel, 1970b).

But how far does this responsibility extend? Difficult choices there must
be, since it is clearly impossible to preserve representatives of *all ecosystems*
– not even in a vast and relatively empty continent like Australia (Specht,
Roe and Boughton, 1974). Less still is it possible to preserve all species.
Opportunities for ecosystem selection are shrinking, and in most parts of the
world the choice, if any is to be made, is on grounds of suitabililty. But there
is choice, and diversity of views, with regard to the preservation of species.
'How many species do we wish to preserve, and for what purpose? . . . What
is the social and economic justification for such preservation?' (Raven,
1976). And, one may add, what is the scientific justification?

In this book we focus on the plight of rare species, existing or potential –
species whose effective population sizes are small. The reason for this focus
is simply that the genetics of nature conservation is the *genetics of scarcity*.
That is, our concern is with fitness and evolutionary potential concomitant
with loss of genetic variation. Abundant and widespread species are, so long
as they stay that way, in little danger of losing this variation, and of becoming
inbred and subject to genetic drift. As geneticists, therefore, we have little to
say about the conservation of species which are now and will remain abundant. Obviously, then, the optimal conservation scenario is one in which the
principles of numerical scarcity discussed in this book are rarely applicable.
But so long as human populations continue to overrun and destroy natural
habitats, there will, perforce, be a need to consider scarcity. Of course as
conservationists, we also insist that it is better and cheaper to prevent species
from becoming endangered in the first place.

Just as prevention is usually better than therapy, so, ultimately is compromise sometimes better than adherence to purist or absolutist doctrines.

The strict conservationist view is that any interference with an ecosystem affects the system as a whole. On the other hand, it may be held that a compromise with man's alternative needs may return valuable advantages in terms of space and long-term security. The question is, which is the greater value? A road through a nature reserve may disrupt the life cycle of a butterfly and ultimately remove it from the system, with various chain reactions to follow. Yet the road may secure and sustain goodwill with unforeseeable beneficial effects for the reserve. In Passmore's words, 'Society, as much as nature, resists men's plans; it is not wax at the hands of the scientist, the planner, the legislator. To forget that fact, as a result of conservationist enthusiasm, is to provoke rather than forestall disaster.' (Passmore, 1974, p. 100).

Significant as these issues are in a tactical sense, they do not affect the core issue of conservation strategy. Provided the great majority of species we manage to include in viable habitats retain their evolutionary potential, extinctions may yet exceed survivals, but, hopefully, will not overwhelm them before mankind decides (and the choice must be a conscious one) to leave nature the space it needs to reestablish an equilibrium between extinction and speciation (chapter 4).

What future for speciation? Whatever the modes of speciation adopted by any organism, the historical record of speciation on islands (see chapter 4) does not encourage expectations for speciation of birds, mammals and trees occurring within the confines of nature reserves.

On the other hand, the evidence of evolutionary radiation in single, small oceanic islands (see e.g. Basilevsky *et al.*, 1972, on the evolution of coleopteran genera and species on St Helena, and comments by White, 1978) encourages the belief that very large nature reserves may provide opportunities for continuing speciation, at least for lower vertebrates, invertebrates and for most families of plants.

2

The process of extinction

Mankind is like a lapidary with a sack of uncut
diamonds. His resources are limited, he can
only cut one at a time; to finish the sack would
take a very long while. But meantime there are
other ways of keeping warm than shooting the
sackful into the anthracite stove and burning
them up overnight.

G. Clifford Evans, 1976

2.1 What is extinction?

Extinction is the failure of a species or a population to maintain itself
through reproduction. Extinction has occurred when either (1) the last
individual has died, or less stringently, (2) when the remaining individuals
are incapable of producing viable or fertile offspring. These definitions of
extinction are simple, operational and relatively unambiguous, but they are
also uninteresting. This is because they focus on the event of extinction
rather than its history. It is more heuristic to think of extinction as a process
(a verb) rather than as an event (a noun).

An analogy will help to explain this. Death is an event, but dying is a
process, and just as the process of dying or senescence can be said to begin at
the moment of fertilization (at least for higher animals), so might the process
of extinction begin when a nascent species becomes reproductively isolated
from its parent or sibling species. If an individual's life is, in part, the process
of becoming more and more dead, then, by analogy, species may slowly
suffer an erosion of fitness, leading ultimately to rarity and extinction. This
analogy, like many, may be misleading but might there be a grain of truth in
the dead and buried concept of species senescence (Simpson, 1953)? Some
modern versions of this idea are reviewed by Soulé (in preparation).

In viewing extinction as a biological process, one is promptly led to a
fundamental and ubiquitous conceptual branch point – the split between
randomness and determinism. In most biological phenomena, both kinds of

processes interact. In this chapter we set aside the stochastic element and review the deterministic forces that contribute to extinction.

2.2 Factors contributing to extinction

Classifications are always arbitrary constructs diverging more or less from reality (viz the classification of ecological factors according to whether they are biotic and abiotic). We nevertheless indulge for the following practical and pedagogical reasons. First, it provides a perspective on extinction to the non-specialist. Second, our classification emphasizes the human impacts on extinction rates. Third, our classification stresses the role of isolation in extinction, a point that is often neglected. Here is the classification:

1. Biotic Factors
   (a) competition
   (b) predation
   (c) parasitism and disease
2. Isolation
3. Habitat Alteration
   (a) slow geological change
   (b) climate
   (c) catastrophe
   (d) man

The reader should keep in mind that this scheme has little or no meaning in itself, rather it is merely a convenient way of ordering our discussion of interesting phenomena.

2.2.1 Biotic factors

(a) *Competition.* The first biotic factor is competition. Competition between two forms is said to occur when individuals are utilizing a common resource and when increase in the number of either decreases the fitness (e.g. growth rate, population size) of the other. Nowadays it is customary to divide competition into two types: (1) exploitation competition where both forms are using the same resource, say food or shelter, but otherwise not interacting, and (2) interference competition where one form hampers another form by poisoning, attacking, or killing and eating the competitor (see Case and Gilpin, 1974 for a recent discussion). Territoriality is a common expression of interference; so is allelopathy (Muller, 1966).

Rarely, if ever, does competition *alone* cause extinction at least on continents, and the literature on invasions contains surprisingly few examples (Elton, 1958). The Argentine fire ant (*Iridomyrmex humilis*) drove out virtually all species of North American ants in its path during its invasion of North America (Smith, 1936), but so far as is known, no native species became extinct. The European starling has driven out native American song birds from around towns and suburbia, but none of the song birds are extinct as a consequence. Elton (1958) notes that while the American grey squirrel (*Sciurus carolinensis*) has replaced the native red squirrel (*Sciurus vulgaris*) in many parts of the south of England, the latter still persists in many regions. These examples illustrate the point now widely understood; '. . . unless one species is uniformly superior to another in every habitat, it will not exterminate the other completely. That is, the usual effect of introducing a competitor is to restrict but not eliminate the previous species' (MacArthur, 1972).

One would expect that islands would offer many examples of extinction by competition. Indeed several workers have described mosaic distributional patterns of ecologically similar species on small islands. Such patterns certainly suggest the impossibility of long-term coexistence of ecological analogues where the island size is below some threshold. For example, Wilson (1961) mentions the mosaic distribution of species of ants on very small Melanesian and Polynesian islands. Soulé (1966) observed that the lizard species in the genera *Uta, Urosaurus* and *Sator* do not coexist on small islands, and Diamond (1975a) described the 'checkerboard' distribution of closely related bird species in archipelagos.

Two forms of competition could account for these mosaic distributional patterns: (1) the 'excluded' species coexisted in the past with the surviving species, but was eventually extirpated because (at least in part) of the competitive advantage of the survivor; (2) the 'excluded' species were never able to establish a colonizing foothold because the island was already occupied by the putative competitor (see Williams, 1969). These two possibilities obviously mark two ends of a 'competitive exclusion continuum', and it would serve no purpose to indulge in nit-picking distinctions.

More species can be packed together on larger islands. Descriptively, this is the well-known species–area relationship (see chapter 5). Wilson (1961), for example, points out that coexistence of competing ant species is the rule for islands slightly larger than those for which the mosaic pattern applies. The coexistence of more species on larger land masses is, in part, explained by the increase in habitat diversity with island size. But the habitat diversity explanation is just one factor. In all probability, area *per se* facilitates the coexistence of competitors. That is, on small islands the extinction of a patch is tantamount to extinction of the species, whereas on large islands where the species is patchily distributed, extinction requires the simultaneous elimination of all patches. Assuming the validity of the principle that the time course of those extinctions which are impelled by competition increases with island size, it is not surprising that there are no documented continental examples of competitive extinctions in recent history.

Perhaps the dingo, a dog introduced by the Aborigines, caused the extinction of the Tasmanian wolf (*Thylacinus cyanocephalus*) on the Australian mainland although it may have been undergoing a range contraction already due to climatic changes. It is likely that North American mammals caused the extinction of many of their South American counterparts during the great faunal interchange that began during the Pliocene about four million years ago (Fig. 2.1). But again, we are struck by the virtual absence of direct evidence of a decisive role of competition in extinction. The persistence of the native Hawaiian rat (*Rattus hawaiiensis*) alongsìde the notoriously aggressive Norway and black rats testifies to the tenacity of species in the face of competition (Svihla, 1936).

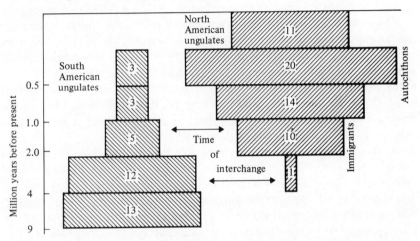

Fig. 2.1 Rapid turnover of ungulate genera in South America during the faunal interchange occurring between four and one million years ago. Numbers in squares represent number of genera. After Webb (1976).

The available information seems to support the premise that the extirpation of an *established* species on a continent by competition is a very slow process. After searching the literature we concluded that throughout the recorded history of human introductions of plants and animals, there is, with the possible exception of Australia, apparently not a single clear cut case of extinction of an indigenous form due to competition with an introduced species on a continent or in a geographic realm.

Perhaps this is because competition on continents operates in a richly diverse and fluctuating world. In the face of intense competition, an inferior species retreats to patches of habitat where it has a competitive edge and can cling to existence. The larger the area, the more likely it will be that such refugia exist in the habitat mosaic (MacArthur, 1972).

In summary, competition may reduce the density and geographic range of species, but except for extreme cases (such as on islands or when geological events or man precipitates an encounter between highly convergent but highly unequal biota) competition is just one among several strands of extinction's rope.

(b) *Predation*. Excluding the island continent of Australia, there are no cases of extinction of vertebrates on *continents* in recent history attributable to predation by animals other than man. Predation, in fact, seems to be even less efficacious than competition in causing extinction. The reasons for this are well known.

Perhaps the most important reason is environmental heterogeneity and the related factor – the size of the environment. Gause (1934) was one of the first to demonstrate these principles. He developed a simple laboratory predator–prey system with *Paramecium caudatum* as the prey and another ciliated protozoan, *Didinium nasutum*, as the predator. When both species were introduced into a simple test tube, *Didinium* ate the *Paramecium* down to the last individual. On the other hand, when one vial contained sediment in which the *Paramecium* could hide, the *Didinium* all starved before the *Paramecium* were exterminated. (These results mimic what happens on islands or in isolated patches of habitat.)

In a further elaboration of these experiments, Luckinbill (1973) achieved much longer coexistence of *Paramecium* and *Didinium* by adding a tangle of microscopic fibres to the system, thus establishing a kind of ecological cat and mouse game. The important thing for coexistence is that there is heterogeneity in space, which implies both hiding places for prey and sufficient space to allow semi-isolation between patches of favourable habitat.

Another reason that predators do not usually annihilate their prey is the phenomenon called 'predator switching'. A predator that exclusively searches for a rare prey species could easily starve. What we would expect to observe and what is typically found is that predators prey on relatively common species, and switch to other species when the original prey becomes rare or an alternative prey species becomes more abundant. In vertebrates, the underlying mechanism of predator switching is the formation of a 'search image' (Tinbergen, 1960). Though we may not be consciously aware of it, it is much easier to find an object if we have an image in our minds of what it looks like. That is, it is easier to find a frog in the grass when we are told to look for a frog than it is to find the frog when told only to search for an animal. For predators, search images develop when an important item becomes common enough to deserve attention. Further, images fade when the item becomes so scarce that the cost of searching for it outweighs the benefit. So it is unlikely that predation by itself would result in extinction if the predator has alternative prey species, and assuming the prey has the ability to disperse.

In addition to predator switching and habitat complexity, there is another mechanism that has been suggested to explain the coexistence of predators and prey. This is the idea of the 'prudent' predator (Slobodkin, 1968). A wise and highly organized predator population would elect to maximize the production of prey by cropping those individuals with the lowest growth rates and the least potential for reproduction. Such an 'optimal yield' strategy would work only if none of the predators 'cheated' by preying on the relatively inexperienced, or more succulent, young prey, and if there were no other less prudent predator species utilizing the same prey.

Seen in this light, being 'prudent' is a form of enlightened self-interest, verging on altruism. That is, instead of engaging in an unorganized scramble for the prey, prudent predators require considerable restraint, a 'restraint' that could only be born of natural selection, operating to increase the ultimate, long-term evolutionary fitness of the predator, rather than the immediate satisfaction of appetites. Such a complex behaviour could evolve by group selection or kin selection (see review by Wilson, 1975), but other explanations for the survival of prey are available. Prey can evolve ways to avoid being eaten, and the evolution of avoidance is a simpler explanation of predator–prey coexistence than is the evolution of predator prudence. The list of such predation-related self-protection adaptations is long and many books have been written on single items. The panoply includes protective coloration and cryptic behaviours, mimicry, distastefulness, poisons and apostatic polymorphism. As pointed out by Fisher (1958) nature is a cat and mouse game; that is, the more pressure a particular predator puts on a prey population, the more likely is the prey to evolve some way of escaping the predator.

At least in theory, therefore, we might expect that predators are unlikely to extirpate their prey, except locally in habitat patches. The statistics of recent extinction bear out this conclusion; only on islands is predation a direct and immediate cause of extinction. Ziswiler (1967) lists the causes of extinction in species and races of birds and mammals that have vanished during the last 300 years or so. Of the approximately seventy-seven species which are extinct, predation is a major factor in twenty-five. All of these forms were insular except for four Australian marsupial species: a bandicoot, two rat-kangaroos and a wallaby; the predators blamed are introduced foxes and feral house cats (although habitat destruction by man and sheep could have been a major contributing factor).*

These Australian examples may be the exceptions that prove the rule, which is that predation rarely is the direct cause of extinction on continents.

---

* Goodwin and Goodwin (1973) list twenty-one Australian mammals that have been extinguished; sixteen are marsupials and five are murid rodents. These Australian forms account for over one-third of all mammal extinctions since 1600. It is noteworthy that there have been no extinctions of rodents on continents outside of Australia, suggesting that the susceptibility to extinction of the Australian fauna is not unique to marsupials, but is related somehow to the isolation of its biota from continents with larger and more diverse taxa.

For predation to be effective, the prey must be unusually vulnerable because of limited and homogeneous habitat (islands) or because of long isolation from behaviourally different (superior?) forms, as in the case of Australian mammals. If predation is ever found to be the proximate cause of extinction on a large continental area, it is likely that one or a combination of other factors have so reduced the range of the species that it is essentially insular and predation merely eliminates the last patch. On the other hand, the struggle for persistence of a prey species is exacerbated by the pressures of predation, and predation must be counted among the major contributing factors in the extinction process.

By virtue of his intelligence, stamina and social organization, man is the super-hunter. No other predator has ever come close to matching his marvellous and awesome capacity to catch and kill. The two rules of predation given in the preceding section are happily violated by man. First, environmental heterogeneity does not confuse the determined tribesman; man ignores habitat boundaries and geographic barriers like no other predator. Second, scarcity of a particular prey does not extinguish his motivation because man hunts for prestige as well as for meat. Bringing down a rare or odd animal is often a status symbol. Furthermore, the economic or social value that human cultures place on commodities is often inversely proportional to their abundance.

Sadly there are many examples of humans hunting a species to extinction or to the brink of extinction. 'The extermination of such large gregarious animals as the blue buck, quagga, and Burchell's zebra probably followed such a course. Once settled in South Africa the Boer farmers shot every animal that ran before their rifles, and it is not difficult to imagine that these species were murdered to the last individual.' (Ziswiler, 1967, p. 57). A long list of other animals might be appended including the Mongolian wild horse (*Equus przewalskii*), the passenger pigeon (*Ectopistes migratorius*), many marine mammals, the great auk (*Alca impennis*), the carolina parakeet (*Conuropsis carolinensis*), the dodo (*Raphus cucullatus*) and the Caucasian wisent (*Bison bonasus caucasicus*) to mention only a few. For most of the above species, human competition for habitat may have been the ultimate cause of their demise.

Firearms have been the weapon of choice in these very recent exterminations. There is considerable debate, however, over whether primitive man with primitive weapons had the ability to cause the rather cataclysmic extinctions of large mammals that occurred throughout most of the world in the interval from about fifteen to about five thousand years ago. The abruptness with which thirty genera disappeared in North America is certainly suggestive of a 'prehistoric blitzkrieg' waged on the native megafauna by recently arrived early man from Asia (Martin, 1973; Mosimann and Martin, 1975; but see MacNeish, 1976 who believes in an earlier arrival date for man). Webb (1969), Grayson (1977) and others prefer to attribute the massive extinction to an abrupt deterioration of climate at the end of the last glaciation, though no such dramatic extinctions occurred at the ends of the earlier Pleistocene glaciations (Martin, 1967).

It has been argued that aboriginal human groups would have more sense than to exterminate their food supply. While it is true that in recorded history no species has been driven extinct by aborigines (Hester, 1967) there are accounts of the complete and wasteful annihilation of individual herds, for example narwhal and musk-ox by the Eskimo (Freuchan, 1935, pp. 180–1, 212) and buffalo by Plains Indians (Roe, 1951, pp. 334–520). These events (and untold numbers of similar calamities) reinforce the well known irony that nature's paragon of intelligence is an ecological idiot. In this light it is reasonable to keep an open mind about man's role in causing prehistoric extinctions.

(c) *Parasitism and disease.* In terms of ultimate fitness, a virulent strain is poorly adapted. By killing or debilitating a large fraction of its major host species, it establishes the conditions that could cause its own extinction. On *a priori* grounds, therefore, diseases are unlikely to be a major factor in extinction, at least under 'normal' conditions. Disease and predation are, in this respect, analogous – they both require the vitality of their cohabiting species.

A corollary of this principle is 'endemic balance' in Van der Plank's (1975) terminology. The disease is present all the time but its virulence is relatively low. This situation is apparently one that gradually emerges by a coevolutionary process during which the host becomes resistant to the disease and the pathogen becomes less virulent. One of the best known examples is the introduction of the myxoma virus into Australia in order to control rabbits (Fenner, 1971). Following the dramatic die-offs occurring in the early stages of the campaign, the surviving rabbits were those with relatively high genetic resistance to the virus, and the virus strains that tended to persist in the wild were those that produced less acute symptoms in unexposed laboratory rabbits. Whereas the disease is still a serious one for the rabbits, the system is clearly moving towards a mutual genetic accommodation or balance.

From these principles, we might deduce that epidemics* should be extremely rare unless a long-standing ecological balance is perturbed. The facts bear this out. In reviewing the impact of disease on the genetics and evolution of plants, J. R. Harlan (1976) remarked that 'Man is the direct or indirect cause of most (possibly all) of the epidemic imbalances we know about.'

* For the sake of simplicity we use this anthropocentric term instead of the more specific epiphytotic (for plants) and epizootic (for animals).

In his review, Harlan (1976) points out that epidemics in plants usually occur when (a) a susceptible host is introduced from a disease-free region into a region where the disease is indigenous (fire blight in American pears and apples, *Dothistroma* needle blight in pines (Bingham, Hoff and McDonald, 1971), and introduced land races of rice), or (b) when the disease is introduced into a region theretofore free of disease (e.g. white pine blister rust, Dutch elm disease, chestnut blight, coffee rust in the American tropics, late blight of potato, powdery mildew of potato, and maize rust in Africa and Asia).

Exactly the same generalizations apply to animals, including man. Trypanosomiasis still excludes cattle from many parts of Africa, and Europeans were repeatedly decimated by epidemics in the tropics until the discovery of quinine and the invention of vaccines against diseases such as yellow fever. There are also numerous examples of category (b) in the preceding paragraph. The endemic diseases of Europe, brought to the tropics by traders, colonists and adventurers, were highly virulent to aborigines. Entire American Indian tribes were exterminated by smallpox and measles introduced by overzealous missionaries (Aschmann, 1959; McNeill, 1976). Some authors (see Saville, 1959; Recher, 1972) have gone so far as to suggest that epidemics are the cause of the wave of extinction often accompanying the sudden contact of separately evolved biota. If this were usually the case, the fossil record should show simultaneous extinctions, but data (Webb, 1969) usually show a more gradual elimination of species.

Other forms of disturbance can cause epidemic imbalances. Extensive and contiguous plantings of single crops, as occurs in energy-intensive agriculture, provide a medium on which new disease mutants can devastate a 'population'. Natural habitats, too, can be shifted from endemic balance to a state of epidemic imbalance by human activities. Bingham *et al.* (1971) provide several examples of localized epidemic outbreaks caused by human-induced habitat changes.

A most interesting fact is that frequent contact between groups provides the necessary condition for the persistence of a class of highly virulent diseases that otherwise could not survive. For example, Hare (1967) and F. L. Black (1975) suggest that primitive, prehistoric man, living in small, relatively isolated groups, was free of diseases which are transmissible only in their acute stages, and in which the organisms disappear upon death or recovery. These diseases, including measles, smallpox, influenza and poliomyelitis, have been some of the most prominent infectious diseases of modern, urban man. Hare (1967) analysed historical documents and was unable to find convincing evidence for smallpox, measles and cholera in ancient times.

It thus appears that a species can avoid epidemics if it occurs in small, relatively isolated (perhaps territorial) groups. Many organisms have patchy distributions or, if social, exist in groups numbering less than a few hundred. One would have to invoke group selection in order to argue that such behaviour was an evolutionary strategy to avoid epidemics, but even if patchy distributions have other explanations, it is still apparent that the extinction of such a species by a pathogen is highly improbable. As far as we could determine, there are no examples of extinction by epidemics, although the American chestnut has been almost eliminated by a fungus of Oriental origin (see Bingham *et al.*, 1971).

An apparent paradox in the foregoing survey of the biological factors contributing to extinction is the insignificance of predation, disease and especially competition in stable continental biota. Except on islands, there are no known extinctions attributable solely to their effects, at least in the last 300 years. It would be grossly incorrect, however, to conclude that competition, disease and predation are unimportant. Each one of these factors imposes an energetic and a genetic load. In combination they certainly depress the life expectancy of individuals and, as pointed out elsewhere (Soulé, in prep.), these interactions often greatly reduce the population size and range of species, thus making them much more susceptible to extinction by chance, catastrophe and habitat alteration.

Our problem in weighing and ranking the role of biotic interactions in the extinction process is one of perspective and complexity. The life span of a species may be millions of years, yet our observational and analytical window on biological processes has been open for 150 years at the most. In addition, the relative importance of factors and their interactions probably changes, so that predation may be the main factor limiting population size during one period, whereas disease, climate or competition may dominate during a later period. A process taking a million years is not easily understood in the space of one scientist's career.

2.2.2 The corrosive effects of isolation

Of the seventy-seven or so species of birds and mammals that have gone extinct in recent history (Ziswiler, 1967), fifty-three were insular forms. If Australian marsupials are included, the number is fifty-eight (or about 75%). There are two reasons for the susceptibility of island forms to extinction; these are (1) patch size, and (2) erosion of defences against other species. First, an island is made up of one or just a few very similar habitat patches. In contrast, the mainland may have thousands of such habitat patches and these patches are much less uniform. On an island, therefore, an introduced predator is quite likely to find and destroy its prey in all patches. The same argument, *mutatis mutandis*, applies to competition.

Second, island animals gradually lose their competitive 'tone' and their defences against predators; the longer they are isolated, the more their defences degenerate. Deterioration of predator defences is the most easily demonstrable. Of the approximately fifty-eight island species recently extinct, about twenty-five succumbed to introduced predators including dogs, cats, rats, foxes, mongooses and mustelids (from data in Ziswiler, 1967). In contrast, none of the continental extinctions (except in Australia) can be blamed on introduced predators. Illustrative of this class of extinction, if somewhat atypical because of its simplicity, is the saga of the Stephen Island wren (*Xenicus lyalli*), Fig. 2.2, a species that had lost the power of flight. The following account is taken from Carlquist (1965). 'No chronicle of depredations of island native species by introduced animals would be complete without the story of [this bird], a native of New Zealand's Stephen Island . . . as told by the ornithologist W. R. B. Oliver:

The history of this species, so far as human contact is concerned, begins and ends with the exploits of a domestic cat. In 1894 the lighthouse keeper's cat brought in eleven specimens, which came to the hands of H. H. Travers . . . A few more captures [were] made and duly reported by the cat and then no more birds were brought in. It is evident, therefore, that the cat which discovered the species also immediately exterminated it.

Fig. 2.2 The extinct, flightless Stephen Island wren, *Xenicus lyalli*. From Carlquist (1965).

Of the fifty-eight island extinctions of full species, man's hunting is responsible for eleven. Some of these animals were easy prey for men or for any other large predator. Among the best known is the Mauritius dodo (*Raphus cucullatus*). The dodo, synonymous with extinction, was a tasty, turkey-sized pigeon (Fig. 2.3). Its wings were reduced to stumps and, like many island forms, it had no fear of man. It fell easy victim to spice traders hungry for fresh meat (Fig. 2.4). In the seventeenth century convicts from

Fig. 2.3 The extinct Mauritius dodo, *Raphus cucullatus*. From Ziswiler (1967) after Van den Broecke.

Fig. 2.4 The first Dutchmen arrive on Mauritius. Contemporary representation. From Ziswiler (1967).

the Dutch penal colony there introduced pigs which soon ran wild and added to the dodo's problems by destroying their eggs. The last dodo perished around 1681.

When predators are absent on islands, prey animals will inevitably lose defensive structures and behaviours that are no longer necessary for survival. Whereas on the mainland any individuals with genotypes that decrease fear of predators or escape abilities would be eliminated, on islands, cryptic and escape organs such as wings might not only be neutral with respect to survival, but actually deleterious since energy is needed to grow and maintain them. Becoming airborne could also result in suicidal dispersal missions. This reasoning applies equally to behaviour. On islands, alert, nervous individuals would often be distracted from the essential business of finding and eating food. The same argument is often invoked to explain the loss of vision and pigmentation in cave animals and internal parasites.

It is the consensus among evolutionists and biogeographers that island species rarely reinvade continents to establish major evolutionary lines. If this is true and we grant that island forms are amusing side shows but rarely if ever responsible for a main act in the evolutionary circus, then why all the furor and publicity about endangered species, assuming most continental species will survive? But can we assume this? First, most animal species officially designated as endangered in the *Red Data Book* published by the IUCN are now continental forms. Approximately twice as many continental as island animals are considered endangered (Goodwin and Holloway, 1972). Island forms are simply the least tolerant of man and the changes in habitat and biota that follow in his wake. Now, the tidal wave that first passed over them is beginning to inundate continental species (Myers, 1979; Soulé and Wilcox, 1980).

2.2.3 Habitat alteration and destruction

Every species has habitat requirements. Therefore, every species is only as safe as its habitat. Cosmopolitan species are no exception: ospreys require shoreline and protected nesting sites; the edible mussel, *Mytilus edulis*, occurs in all seas, but rarely below the intertidal where it is usually eaten by fish and other predators. Many large marine predators such as the thresher shark *Alopias vulpes* and the sperm whale are so widespread that it is difficult to imagine how they could be extinguished by any conceivable change in the world's climate or geology. Nevertheless, they do require salt water, and in large amounts.

Not all organisms, however, are so catholic in the habitat tastes, and any naturalist could produce a long list of organisms that require very specific conditions. From there is a short step to creating an extinction scenario for each such species by altering some critical habitat variable. The agencies responsible for habitat alteration or destruction can be grouped into four categories:

(a) slow geological change
(b) climatic change
(c) catastrophic events
(d) human disturbance

(a) *Slow geological change.* Slow geological habitat changes include those shifts in the earth's crustal plates that create and destroy seas, change major current patterns and either increase or decrease the area of certain habitats. As an example of the effect of change in area concomitant with continental drift, Schopf (1974) and Simberloff (1974) attribute the massive extinction of marine animals living on continental slopes at the Permo-Triassic boundary to the creation of the supercontinent 'Pangaea' and the concomitant loss of a large fraction of the previously available habitat space (Fig. 2.5).

(b) *Climatic change.* Climatic changes may shift the range of species latitudinally and altitudinally. In some cases such shifts are impossible, in other cases they are possible but other factors may preclude range adjustment. As an example of impossible range adjustments we can imagine what would happen to the coral reef communities in the Eastern Tropical Pacific (best developed in the Gulf of Chiriqui off Western Panama). Dana (1975) points out that conditions for reef development are at best marginal in this part of the tropics. Upwelling of cold water, freshwater runoff and high turbidity combine to restrict severely the vertical and areal distribution of hard corals. Dana also concludes that these reefs are young, probably interglacial, and that another glacial period would wipe them out. If such were to occur, it is likely that some of the associated fish (including some endemic forms) and invertebrates would also perish. There are many such geographic culs-de-sac in the world; the Gulf of California has species that would succumb from either a warming or a cooling of the climate.

Islands are the terrestrial analogues of these oceanic traps. Range adjustments for island biota are severely limited and wholesale extinctions of island faunas in the subtropics and temperate zones undoubtedly accompany every major change in climate.

Range adjustments, even on continents, may not be as common as we might imagine. The extinction of most of the large mammals during and following the last glacial retreat may or may not be an example of man's expertise in hunting, but it is naive, nonetheless, to assume that plants and animals just pick up and go when the climatic belts begin to leave them behind. There is more involved, for example, in going north than changing venue. At very high latitudes, there are very few hours of daylight in winter requiring the readjustment of biological clocks. Besides, large grazing animals require

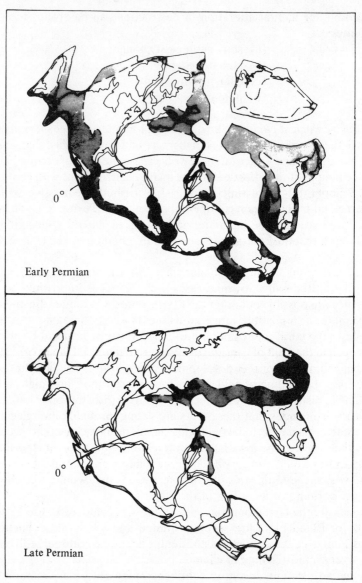

Early Permian

Late Permian

Fig. 2.5 Change in the extent of shallow marine seas (ruled pattern) from the lower
(early) to the upper (late) Permian. After Schopf (1974).

several hours of feeding every day, and the colder the climate, the more food is
required. One wonders if part of the reason for the late-Pleistocene extinction
of tundra grazers, such as mammoths, was the difficulty of getting enough to
eat in the few hours of daylight in December at high latitudes.

To leave this subject without a word about dinosaurs would be unconscionable. The dinosaurs were dwindling in numbers throughout the latter part of the Cretaceous, but there was, indeed, a rather abrupt decline at the Cretaceous–Paleocene boundary (Colbert, 1951). Van Valen and Sloan (1977) analysed the disappearance of dinosaurs at this boundary in a sequence of faunas from Montana. The extinction lasted about 100 000 years and was accompanied by a climatically induced replacement of a subtropical flora by a temperate flora. In reviewing their data and those from other studies, Van Valen and Sloan throw their support behind the hypothesis that the extinctions were caused by diffuse competition from primitive mammals better adapted to the cooler climate. They also apply their climate-competition argument to genera of marine plankton and benthic filter-feeding organisms, more than half of which became extinct at the same time. The implication is that dinosaurs would have survived the rather sudden worldwide cooling had not the mammals been undergoing their dramatic adaptive radiation.

The lesson here is that close scrutiny usually reveals that causation is complex. Possibly the interaction of climatic change and competition produces an impact which is much greater or faster than either factor alone would be. We will never know for sure.

(c) *Catastrophic events*. Not much need be said about catastrophic habitat alteration. A catastrophe is usually a localized event, and is unlikely to extinguish any but the most localized forms. (A worldwide catastrophe such as an epidemic of plague proportions, on the other hand, might be beneficial to most species by reducing the size of the human population.) Localized events such as cyclones or volcanic eruptions could cause the extinction of a localized species, but any species already so localized was on the verge of extinction anyway having been pushed there by biotic or climatic factors.

(d) *Man*. We now come to habitat alteration caused by man. No other agent of environmental change is so devastating as man, or so thorough (Ehrlich, Ehrlich and Holdren, 1977). For example, the rate of destruction of tropical forests today is approximately 47 ha/min (Myers, 1979); at this rate the tropical forests will be gone in twenty-five to fifty years. This is why virtually all primates are considered to be threatened. Fosberg (1973), in describing the rate of habitat destruction in the tropics, and the sensitivity of tropical habitats, points out that a large proportion of tropical species will probably disappear before their existence is documented by systematists.

Evidence for the overwhelming effect of habitat destruction and human interference is presented by Greenway (1967) in his exhaustive treatment of modern extinctions of birds. Fig. 2.6, adapted from Greenway, shows the inverse relationship between amount of virgin forest per human and the number of extinctions on islands in the Lesser Antilles. Except for the

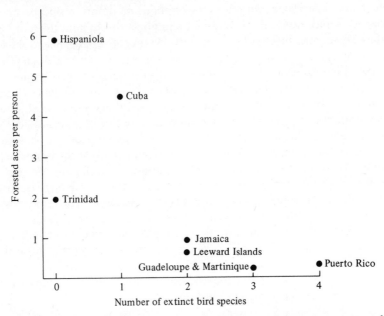

Fig. 2.6 The relationship between the number of forested acres per person and the number of bird species extinguished in historical times on islands in the West Indies. Omitted are the Windward Islands (St Lucia, St Vincent and Grenada; they average about ½ acre of forest per person, but have suffered no extinctions). After Greenway (1967).

Windward Islands, there is a clear association between human population density and faunal destruction. The Windward Islands have relatively dense human populations, but have not suffered any extinctions, apparently because the humans on these islands (St Lucia, St Vincent and Grenada) were much less assiduous in the hunting of birds than were the inhabitants of the French-speaking islands. Also, the humans of the Windward Islands are prevented from cultivating much of the lowland habitat because of the steep relief which is unique to lower elevations of these islands.

All these islands have introduced populations of rats and mongooses, notorious predators on ground-nesting birds. Yet in spite of the presence of these mammals, the ground-nesting wrens continue to survive on the Windward Islands. This underlines a point mentioned earlier: the introduction of a few competitors or predators does not necessarily presage a wave of extinction, but if the element of significant habitat alteration is added, the consequences can be grave.

There is simply no way that evolution in large plants and animals can keep up with the rate that man is modifying the planet's surface. The damming of a river may take a decade, but the evolution of fishes adapted to lakes rather than rivers may take thousands of generations. The same principle obviously applies to other simplified habitats such as plantations of sugar cane, coffee, tea or cocoa, not to mention croplands that in many parts of the tropics are replacing rain forests.

Has anything similar to man's planetwide destruction of habitats ever happened before? A qualified 'yes' to this question is probably fair, but only if we accept a vast difference in time scales. Great extinctions have occurred in the past, precipitated by geological or climatic alterations, but whereas the present crisis has a time scale of hundreds of years or less, those of the past have lasted many millions. The best understood of these crises is probably the Permo-Triassic event already mentioned. This particular mass extinction is worthy of the closest scrutiny by conservationists, because its consequences impel us to some profoundly important conclusions.

Two aspects of modern biogeography have recently been joined to produce a satisfying explanation of the Permo-Triassic crisis during which the number of marine invertebrate families was reduced by about half (Schopf, 1974). The first is plate tectonics. After an initial separation of the continents, the late Permian saw their coalescence and the formation of the single supercontinent of Pangaea. A result was the disappearance of most of the continental shelf habitat which had ringed each continent before the coalescence. Two reasons are suggested for the marked diminution of shallow seas. The first is topological – the separate pieces of a puzzle have more total periphery than the completed puzzle. Second, the joining together of continental plates has a worldwide effect on the rate of downward flow of materials (subduction) around plate margins. According to theory, any decrease in the rate of subduction reduces the rate at which oceanic ridges are created, and this, in turn, reduces the amount of undersea topography. The result is a lowering of the eustatic sea level and draining of much of what remained of the continental shelf surrounding the supercontinent. In all, the reduction left about 15% of the shallow seas.

Assuming that there was indeed a very great loss of continental shelf habitat, why would this result in such a dramatic decline in the number of kinds of animals inhabiting the remnants of shallow seas? The answer comes from a second aspect of contemporary biogeography. Schopf realized that the well-known empirical relationship between habitat area and number of species (chapter 5) would dictate that a reduction in habitat area would be followed by a decline in animal diversity, but palaeobiological methods and data cannot resolve the number of species present at remote times, only the number of families and possibly genera. In an attempt to overcome this difficulty, Simberloff (1974) in a companion paper, developed a method for approximating the number of families from the available palaeontological data and was able to demonstrate that the observed number of families is very near the expected, assuming appropriate time lags.

Two points are worth reemphasizing from this example. First, the decline in diversity following the reduction in habitat area was catastrophic. Second, the recovery of diversity (Fig. 2.7), following the breakup of Pangaea at the beginning of the Triassic, required a very long time, *ca* twenty million years for a 50% recovery in the number of invertebrate families. A large body of biogeographic data is consistent with these points (MacArthur and Wilson, 1967; Diamond and May, 1976; Soulé, Wilcox and Holtby, 1979). The conclusion is that the large terrestrial flora and fauna of this planet are just beginning a plunge into an unprecedented abyss of extinction. The main cause is simple. Most of the planet's species are tropical, and never before have the tropics and their habitats been so seriously reduced in area (Whitmore, 1980).

2.2.4 Conclusions and summary

1. In the light of the previous discussion, what conclusions can be reached about the relative weights of the various factors in causing extinctions in the

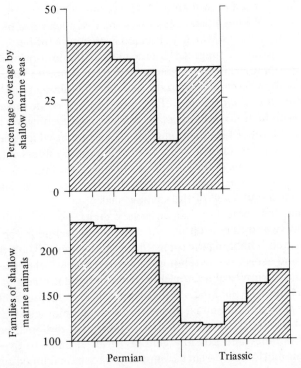

Fig. 2.7 A summary of changes in the number of families of shallow marine invertebrates and in the area of shallow marine seas during the Permo-Triassic. After Schopf (1974).

past? At the outset one point stands out above all the others: in historical time, not a single species of plant or animal is known to have become extinct except by the direct or indirect hand of man. In one way or another, therefore, all of our information about recent extinctions falls under the heading of human interference, and we have few data on the significance of the other factors (competition, predation, disease) except as they have been caused by man. This certainly distorts our understanding. Nevertheless, we will attempt to reach some tentative conclusions, though they are based on very limited data. Many of the following generalizations have been tendered by others (Hester, 1967; Ziswiler, 1967; Greenway, 1967).

2.   Island forms are far and away the most sensitive to environmental change of any kind. They cannot migrate if the climate changes, or if there is a local catastrophe such as an eruption, drought or hurricane. They are especially susceptible to introduced predators because of reduced habitat size and diversity and because in isolation there is an evolutionary erosion of fear, flight and other predator defence mechanisms. Additional reasons why isolation and small population size enhance the probability of extinction are discussed in chapters 3 and 4.

3.   Isolation on continents from (1) man or (2) behaviourally more advanced vicars (ecological analogues) is the forerunner of mass extinctions. That is, whenever a part of the world (such as South America or Australia) is cut off from a larger land mass (particularly Eurasia and Africa), the stage is set for a massive wave of extinctions once contact is reestablished, whether the mode of bridge construction is a natural geological process, such as plate tectonics, or human migration and transport of biota.

4.   Predation can reduce the density of prey and (more rarely) even annihilate local populations of prey, but if the prey species is sufficiently vagile to colonize empty patches of habitat, predation alone is unlikely to be the immediate cause of extinction, though it must increase the probability of extinction of many species by reducing the number of patches or area occupied by the prey at any given time.

5.   Competition, like predation, will tend to decrease the density and the distributional range of interacting species. The effect of competition is difficult to demonstrate directly, though we can infer its importance from such observations as (a) range contraction upon the introduction of similar species, (b) constant ratios of body size or feeding organs between sympatric congeners (Hutchinson, 1959; Schoener, 1965), (c) the replacement of one group for another ecologically analogous one in the fossil record (Simpson, 1953; Webb, 1969).

6.   Habitat alteration or destruction is today (though not always) without peer as an agent of extinction. Major extinction crises in the fossil record are generally believed to be due to either the dissappearance of most of a habitat (the Permo-Triassic crisis) or to relatively rapid changes in

climate which brought about great habitat alteration (the Cretaceous extinction of large reptiles). Throughout the history of life, there has never been as wanton nor as rapid an agent of habitat destruction as twentieth-century man.

3

Population genetics and conservation

3.1 Population size and genetic variability

3.1.1 Introduction

In the previous chapter the process of extinction was discussed in very general terms. In this chapter we turn, rather abruptly, to some very specific, down-to-earth problems. Everywhere, particularly in the tropics, habitats are being lost to a rising sea of humanity. Soon only tiny islands of natural habitat will be left, mostly as arid or cold lands unfit for agriculture, or as isolated nature reserves. Many species will perish completely, and many others will only survive because of the ministrations of man. Hence the need for conservation genetics – the genetics of scarcity.

The scarcity is in numbers. Whether our concern is the wild relatives of cultivated plants or wild animals, the conservationist is faced with the ultimate sampling problem – how to preserve genetic variability and evolutionary flexibility in the face of diminishing space and with very limited economic resources. Inevitably we are concerned with the genetics and evolution of small populations, and with establishing practical guidelines for the practising conservation biologist.

The task has its hazards, not the least of which is the heterogeneity of the biological world. No two species are genetically the same, and no generalization (for example, about minimum population size) can be universally valid. To those who insist on bludgeoning us with this hazard, our response is that Noah must have had similar critics to whom he probably remarked, 'I can either stand here in the rain arguing about precision, or I can start building. Goodbye.'

Do populations suffer a significant genetic deterioration as a consequence of a sudden or gradual decrease in numbers? The fate of thousands of species may hang on the correct resolution of this issue, and on our ability to put principles into practice. To intelligently discuss this question, it is first necessary to describe the relationship between population size and genetic variability, and second, establish the consequences of decreasing genetic variability. In this chapter we review what is known about the immediate or short-term genetic effects of small numbers on genetic variation and fitness. The relationship between population size and long-term fitness, i.e. evolutionary potential, is the subject of Chapter 4. Readers wishing to pursue these topics in greater depth should consult such standard works as Crow and Kimura (1970), Wright (1977), or less specialized treatments such as Pirchner (1969), Spiess (1977), or Wilson and Bossert (1971).

Fig. 3.1 illustrates the kinds of events with which we are concerned. The 'normal' population size is that found in nature prior to significant inroads by man. The 'crash' or reduction is shown to occur suddenly, portraying, for example, the establishment of a breeding group from a few founder individuals. The decline can also be gradual, such as when habitat destruction diminishes the occupiable territory, as is happening throughout the tropics. The bottleneck is the minimum population size as a result of a crash.

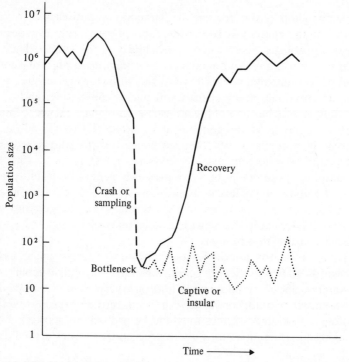

Fig. 3.1 The kinds of changes in population size relevant to conservation genetics.

Whether the population makes a significant numerical recovery or remains in a chronic state of impoverishment depends mostly on the degree of man's hospitality. For economic reasons most captively bred populations will never rebound to their former abundance, whereas some managed populations, such as whales, have the potential for complete numerical recuperation, requiring only a major revolution in resource utilization by maritime nations.

3.1.2 Effects of bottlenecks on variation and allelic diversity

A bottleneck is an observable and dramatic collapse of numbers. It can be produced by a gradual or sudden environmental change, such as a drought or flood; it can be a natural colonization (founder) event, such as when one or more individuals establish a new population in a previously unoccupied region or on an island; it can be an artificial founder event, such as the establishment of the coastal redwood (*Sequoia sempervirens*) in New Zealand or the Arabian oryx (*Oryx leucoryx*) in Arizona. Whatever the circumstances though, a bottleneck is equivalent to taking a relatively small sample of items, in this case, genes, from a large population. Because small samples rarely are completely representative of the source population from which they are drawn, a bottleneck will usually initiate an interval during which the population lacks some or most of the genetic diversity of the source population. Depending on the degree of genetic 'sampling error' or depauperateness of the bottlenecked population, it may be temporarily or permanently handicapped in ways described in this and the next chapter.

The loss of genetic variability concomitant with a bottleneck event has both qualitative and quantitative aspects. Qualitatively, specific alleles will either be lost or retained, and if lost, it is highly improbable that mutation will replace them as long as the population is small. Quantitatively, the amount of variability for specific characteristics will be reduced. In other words, the *variance* of quantitative traits is reduced by a bottleneck. For example, Fig. 3.2 shows the reduction in the variance of some trait by one-half, the amount of reduction expected when the population size of the bottleneck is a single individual.

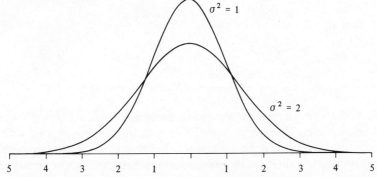

Fig. 3.2 Two 'normal' frequency distributions; the variance of the higher distribution is one-half that of the wider distribution.

Bottlenecks will usually have a greater qualitative than quantitative impact. That is, the loss of alleles, especially rare ones, is much greater than is the loss of genetic variance *per se*. We will examine first, the loss of genetic variance.

An approximation of the amount or proportion of genetic variation or heterozygosity that remains following the sudden reduction of a large population to a small one containing N individuals is

$$1 - 1/2N \qquad (3.1)$$

As shown in Table 3.1, most genetic variation is conserved unless the bottleneck is very severe. Even a sample of four or five pairs contains most of the genetic variance of the source population. Experimental data generally support these approximations. Franklin (1980) cites unpublished results of Keith Hammond showing the effects of bottlenecks of sizes 2, 20 and 100 on the loss of genetic variance in abdominal chaetae of *Drosophila*. The results (originally expressed as heritabilities) are shown in the third column of Table 3.1; they are close to the expected values. Unless the number of founders is of the order of two pairs or fewer, the bottleneck, *per se*, is not the villain in a genetic melodrama, at least with respect to genetic variance. Rather, as shown in section 3.1.3, most of the loss that ensues is attributable to events *following* the bottleneck.

TABLE 3.1   *The percentage of genetic variance remaining in founder populations*

No. of individuals in sample	Expected percentage of genetic variance remaining	Empirical results (see text)
1	50	–
2	75	74
6	91.7	–
10	95	–
20	97.5	90
50	99	–
100	99.5	–

The other way of looking at the consequences of bottlenecks is in terms of the loss of alleles. Relatively rare alleles with frequencies of, say, 0.05 or less, contribute little to genetic variance. Yet rare genes, including perhaps genes for disease resistance, may be important in special circumstances, such as during an epidemic. A gene of this kind might be neutral or close to neutral in its effect on fitness for many generations, but during an environmental crisis it could spell the difference between extinction and survival for the population.

How do rare genes fare during bottlenecks? Nei, Maruyama and Chakraborty (1975) and Denniston (1978) have shown that rare genes have a high probability of being lost during bottlenecks. The formula for estimating the number of alleles (n) remaining after a bottleneck of N individuals is

$$E(n) = m - \sum_j (1 - p_j)^{2N} \qquad (3.2)$$

where m is the number of alleles prior to the bottleneck, p is the frequency of the jth allele, and N is the effective number of individuals at the bottleneck. For example, consider a diploid species in which each locus has four alleles segregating in the population, one allele of which is common while the other three alleles are rare. Table 3.2 gives these expected values for bottlenecks of various sizes and for two sets of allele frequencies for which $m = 4$.

TABLE 3.2   *The number of alleles retained, beginning with four, in samples of sizes N calculated for two sets of allele frequencies in the source population*

No. of individuals in sample (N)	Average number of alleles retained	
	$p_1=0.70, p_2=p_3=p_4=0.10$	$p_1=0.94, p_2=p_3=p_4=0.02$
1	1.48	1.12
2	2.02	1.23
6	3.15	1.64
10	3.63	2.00
50	3.99	3.60
∞	4.00	4.00

In contrast to the relatively minor effect of bottlenecks on genetic variance, the results in Table 3.2 show that allelic diversity can suffer very seriously during founder events (see Marshall and Brown, 1975). We can only guess about the consequences of such attrition on fitness, however. In the short run, the loss of rare alleles is probably not very important, especially in benign environments. In the long run, though, such alleles might be crucial. The prudent tack would be to hedge our bets: whenever possible, maximize the size of founder groups.

3.1.3 Effects of genetic drift on variation and allelic diversity

A bottleneck is a single event of sampling error, the amount of error and the loss of variation being proportional to the number sampled. When numbers are low, a population is, in effect, going through a serious bottleneck every generation, and the effects are cumulative because the regeneration of variation by mutation is insignificant in small populations. The random changes in gene frequencies that occur due to sampling error, including the loss of alleles, is called genetic drift.

From equation 3.1 the expected proportion of variation remaining after t generations is

$$(1 - 1/2N)^t \tag{3.3}$$

Some useful results are tabulated in Table 3.3. For example, a population must number at least 100 if it is to retain more than 60% of its genetic variance for 100 generations.

TABLE 3.3   *The retention of genetic variance in small populations of constant size for t generations*

Population size (N)	Percentage genetic variance remaining after 1, 5, 10 and 100 generations			
	1	5	10	100
2	75	24	6	<< 1
6	91.7	65	42	<< 1
10	95	77	60	< 1
20	97.5	88	78	8
50	99	95	90	36
100	99.5	97.5	95	60

Consider the case of a gravid female who establishes a population on an island. In this situation the population will grow until competition or space begin to act as brakes. Assume that the population reproduces annually and that the population size triples every generation. The sequence of sizes over a period of ten years is therefore 2, 6, 18, 54, 162, 486, 1458, 4374, 13 122, and 39 366. In order to estimate the genetic variance remaining in this case, the harmonic mean of the ten values of N (= 13.16) is substituted in equation 3.3 (see section 3.1.5). We see that the amount of variation retained is 67.9%, or ten times the amount retained if the population size had remained at 2. This should partially allay the fears of those concerned that a *single* bottleneck must extract most of the genetic variation in an island population. Again, we wish to emphasize that a bottleneck will not, by itself, erode much of the genetic variance. The crucial issue is whether the population remains small or grows to a relatively large size. It is perennial low numbers that erode genetic variation.

Another relevant factor in the conservation of variation is the *rate* of population growth. As shown by Nei *et al.* (1975) the proportion of heterozygosity (equivalent to genetic variance for practical purposes) retained subsequent to a bottleneck is negligible if the growth rate, r, is less than 0.1. With $r > 1.0$, however, the post-bottleneck loss of variation is relatively insignificant.

The same conclusions are apparent when considering the number of alleles that survive during an interval of substantial genetic drift. The question becomes one of estimating the number of alleles out of the original m that are retained after t generations. The rather complex mathematics discussed by Denniston (1978) are not reproduced here. Table 3.4 gives the theoretical results for a constant population size of six individuals. Note that after sixteen or twenty generations most populations will have only a single allele remaining at each locus, regardless of how many alleles were present to begin with. Obviously, the leakage of alleles is less in a larger population. However, the prognosis for rare alleles is poor, even at moderate values of N, unless some form of selection elevates the fitness of individuals carrying such genes.

3.1.4 Effective population size: unequal sex ratio in dioecious species

So far in this chapter we have made the assumption that the number of males

TABLE 3.4 *The expected number of alleles after t generations in a population of six individuals given three different starting frequencies*

No. of generations	Number of alleles when:		
	$m=2, p_1=p_2=1/2$	$m=4, p_j=1/4$	$m=12, p_j=1/12$
0	2.00	4.00	12.00
1	1.99	3.87	7.78
2	1.99	3.55	5.88
4	1.91	2.94	4.08
8	1.67	2.18	2.64
16	1.34	1.52	1.68
20	1.24	1.36	1.44
56	1.01	1.02	1.02
∞	1.00	1.00	1.00

From Denniston (1978).

and females contributing to each subsequent generation is the same. This permitted us to sidestep a major complexity in genetic calculations, namely, the problem of the genetically effective population size, N_e. Kimura and Crow (1963) should be consulted for a comprehensive treatment of this subject. N_e is not necessarily the same as the actual number of breeding individuals. Unless the sexes are equal, N_e is less than N.

The reason for this can be seen intuitively. Consider a herd of zebra comprised of a male and nine mares. All the offspring in such a group will be half-sibs or full sibs. Now, in a population comprised of five males and five females, the progeny will, on the average, be much less closely related. Clearly the chance of an allele becoming lost is greater in the former population. That is, the amount of genetic drift in the herd with the skewed sex ratio is higher than the amount in the herd in which the sexes are equal. To be precise, the formula for N_e when considering the sex ratio is

$$N_e = \frac{4N_m N_f}{N_m + N_f} \tag{3.4}$$

where N_m and N_f are the number of breeding males and females, respectively. In the zebra example, N_e for the skewed herd is 3.6. In other words, the sampling error in a population of 3.6 individuals with an equal number of males and females is equal to the sampling error in a population of 10 individuals with a 9:1 sex ratio. Thus, N_e is the size of an ideal population that is subject to the same degree of genetic drift as a particular real population. In this definition 'ideal' means a randomly breeding population with a 1:1 sex ratio, and in which the number of progeny per family are randomly (Poisson) distributed.

The zebra example is really not far-fetched. In zebras, like other equids, the reproductive group is a harem, and a single male may sexually monopolize as many as 6 mares (Klingel, 1969). If the average size of a harem is 5 mares, a herd of 100 individuals will consist of, say, 60 females (12 harems of 5 each), 12 stallions, each with a harem, and 28 bachelor males. From the above formula, the effective size of the herd is $4(12 \times 60)/(12 + 60) = 40$. In other words, the amount of random genetic drift in these 100 adult zebra is equal to that in an 'ideal' population made up of 20 males and 20 females mating randomly.

3.1.5 Effective population size: population fluctuation

Real populations fluctuate in size. Much of the discipline of ecology deals with the causes and consequences of this phenomenon. In insects, fluctuations of several orders of magnitude are common, especially in temperate-zone species. Vertebrates tend to fluctuate less violently, but changes in food abundance, weather and pathogens often account for large swings in numbers. Even in the tropics, long thought to be synonymous with stability, there are dramatic changes in the numbers of animals (Gilbert, 1980; Foster, 1980).

When populations decline or 'crash', the survivors are the progenitors of all future generations, and any deviation in the genetic make-up of these progenitors from the gene pool of the original population will be reflected in future generations. More particularly, if the progenitors contain but a sample of the kinds of genes that existed in the original population, future generations will have a corresponding deficit in genetic diversity. In more quantitative terms, the effective size of a population when the number per generation varies over time is the harmonic mean of the effective number of each generation, or

$$\frac{1}{N_e} = \frac{1}{t} \left(\frac{1}{N_1} + \frac{1}{N_2} + \ \ldots \ + \frac{1}{N_t} \right) \tag{3.5}$$

In section 3.1.3 this formula was used to calculate the effective size of an exponentially growing population. A more typical situation for the practising wildlife manager would be a herd of large animals confined to an area of finite size. For example, say that we wish to maintain a stock with an effective size of at least 100, but that we can expect the population size to decline to 25 on the average of once in 10 generations. In order to maintain an effective size greater than 100 the population must be allowed to grow to a larger size during the 'good' years. A simple calculation will show that this larger size, or carrying capacity, K, is 150. If the minimum size over the 10-generation interval is 15 individuals, K must be 270. K reaches infinity if the minimum size is 10. It follows that the maintenance of a reasonably large effective size will require space or facilities for more individuals than might have been anticipated.

3.1.6 Effective population size: progeny distribution

One of the characteristics of a genetically ideal population is that the number of progeny is randomly distributed among the families; i.e. a Poisson distribution should describe the frequencies of offspring number. Such a distribution probably holds for organisms with non-overlapping generations, but large organisms as a rule have overlapping generations. Hill (1977) points out that 'with overlapping generations, even if there are no fertility differences among survivors and there is random death of breeding individuals, the inbreeding rate will be higher than [expected] since the distribution of lifetime family size is not Poisson. With an exponential distribution of deaths the rate can be nearly *three times as high* as in the simple formula (Felsenstein, 1971)' (emphasis added).

The effect of a non-random distribution of offspring among families on N_e is easily estimated. Let $k_1, k_2, \ldots$ represent the number of gametes contributed by different individuals to the next generation, and let $\bar{k}$ and V_k be the mean and variance, respectively, of the k's. Then

$$N_e = \frac{N\bar{k}}{\left(\frac{N}{N-1}\right)\frac{V_k}{\bar{k}}(1+F) + (1-F)} \tag{3.6}$$

where N is the actual number of parents and F is the inbreeding coefficient (Kimura and Crow, 1963), the latter defined in section 3.2.2. In an infinitely large (ideal) population both $\bar{k}$ and V_k equal 2.0, and F is a very small number. A few 'thought experiments' with formula 3.6 should convince you that N_e will be less than N when some families have no offspring and others have many. At the other extreme, where all families have exactly the same number of offspring $N_e = 2(N-1)$. Thus in a closely managed population, such as in a zoo or possibly a remnant herd of a large mammal in a nature reserve, it is possible for the effective size to be twice the number of breeding individuals. This principle is probably the most powerful weapon in the hands of captive breeders. It should be noted, however, that the exploitation of this method forestalls the operation of natural selection and would rarely if ever be appropriate in populations numbering several hundred or more individuals.

One way of reducing V_k and thus increasing N_e is by culling offspring from larger families, thereby levelling out the genetic contributions of parents. Culling of excess offspring is often necessary in managed populations, so it might as well be practised with this genetic purpose in mind. The following example from Denniston (1978) demonstrates the efficacy of culling to increase N_e. Removal of offspring is done as follows: if one individual is culled, it is taken from the largest family; if two or more individuals are culled, individuals are taken from the largest family until it is reduced to the size of the second largest family, then individuals are removed alternatively from both families until they are reduced to the size of the third largest family, and so on. Denniston's population is monogamous and consists of fifteen families contributing 0, 0, 1, 1, 1, 1, 1, 6, 6, 7, 7, 7, 7, 8 and 9 progeny, respectively. Here, the value of $\bar{k}$ is 4.13, and V_k is 11.05. The impact of culling in this manner is shown in Fig. 3.3. Note that half of the offspring can be removed without decreasing N_e. In fact, if there is significant inbreeding in the population, culling in this manner actually increases N_e. The mathematically inclined reader should refer to James (1962) and Latter (1959).

3.1.7 *Effective population size: close management of breeding*

In situations where individuals can be identified and where breeding can be manipulated (such as in zoos), it is possible to minimize the effects of genetic drift and inbreeding. Animal breeders have developed several such breeding schemes; they include *maximum avoidance of inbreeding, circular half-sib mating* and *double first cousin mating*. These schemes require precise pedigree information which in turn permits the calculation of inbreeding coefficients (Wright, 1977).

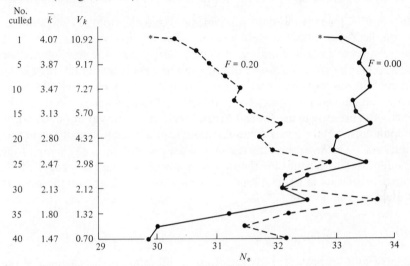

No. culled	$\bar{k}$	V_k
1	4.07	10.92
5	3.87	9.17
10	3.47	7.27
15	3.13	5.70
20	2.80	4.32
25	2.47	2.98
30	2.13	2.12
35	1.80	1.32
40	1.47	0.70

Fig. 3.3 The effect on effective population size of culling progeny from the largest of fifteen families contributing 0,0,1,1,1,1,1,6,6,7,7,7,7,8 and 9 offspring, respectively. Asterisks indicate the N_e values for the unculled population.

The impact of such schemes on the effective size of a breeding group varies according to the scheme. The effect of maximum avoidance of inbreeding is essentially the same as that produced by equalizing progeny number among families – there is a doubling of the effective population size compared to random mating. Circular half-sib mating is less effective in the early generations but surpasses maximum avoidance later on (Kimura and Ohta, 1971). In practice, however, we feel that these schemes rarely will be implemented. The reason is that specific matings would be dictated by the pedigree rather than by social position, but most zoo breeders are reluctant to separate established and productive breeding pairs and to disrupt social hierarchies by shifting animals from one group to another. In fact, such manipulations can often result in infanticide and other forms of mayhem (Kleiman, 1980).

In any case, there is virtually no advantage in using these schemes over the equalization of progeny number among families. The doubling of N_e in the maximum avoidance system, for example, is almost entirely attributable to each parent contributing the same number of progeny to the next generation of parents. The breeder is well advised, therefore, to use common sense in the management of breeding, and to emphasize the practice of culling offspring (as described in the preceding section) for the minimization of genetic attrition.

3.2 Genetic variation in natural populations: data, models and hypotheses

In section 3.1 we reviewed the principles and statistical tools which permit us to predict the consequences for genetic variability in those situations where population size is low. In this section we attempt to go one step further – to determine the effect on *fitness* of reduction in genetic variation, the fundamental question of conservation genetics. Before fully engaging this topic, however, one must be somewhat familiar with three related subjects: (1) the methods currently in use for estimating levels of genetic variation in natural populations; (2) the generalizations about genetic variation in different taxa and how these might be affected by population structure; (3) some of the models used to account for the existence, persistence and heterogeneity of genetic variation in populations. Readers already familiar with these topics should skip to section 3.2.3.

3.2.1 Estimation of genetic variation in natural populations

Several authors have recently reviewed the literature on biochemical variation in natural populations (Powell, 1975; Selander, 1976; Soulé, 1976; Nevo, 1978; Wright, 1978). Each author tends to champion a particular hobbyhorse, but on some points the data speak for themselves, and there is general agreement. But before discussing these generalizations, it is necessary to take a short excursion on methodology.

There are many ways to express genetic variability at the gene product level (Wright, 1978). With respect to a local population, one can consider the proportion of the observed loci which have more than one variant (allelozyme or allozyme), the percentage polymorphism, P. A common convention is to consider as polymorphic only those loci at which the commonest variant (allele, loosely speaking) has a frequency less than 0.95. Also at the level of the local population, one can estimate the observed heterozygosity, H, the percentage of observed genotypes at which the average individual is heterozygous. The two measures are highly correlated as shown in Table 3.5.

When considering many populations or the species as a whole, however, the correlation may break down completely. For example, local populations of an inbreeding plant may each be fixed for a unique constellation of alleles, thus giving, for the species as a whole, a high estimate for P, but zero H. The same phenomenon, though less extreme, is seen in highly subdivided populations of outbreeding species of plants and animals. Thus in inbred, sub-

TABLE 3.5   *Estimates of genetic variation in natural populations based on electrophoretic data*

Taxonomic group	No. of species	P Mean	P SD	H Mean	H SD	r(P, H)
Mammalia	46	0.147	0.098	0.036	0.025	0.838[b]
Aves	7	0.150	0.111	0.043	0.036	0.900[a]
Reptilia	17	0.219	0.129	0.047	0.003	0.605[a]
Bony fishes	51	0.151	0.098	0.051	0.034	0.883[b]
Plants	15	0.259	0.166	0.071	0.071	0.206
Insecta (exc. *Drosophila*)	23	0.329	0.203	0.074	0.081	0.680[b]
Amphibia	13	0.269	0.134	0.079	0.042	0.785[a]
Invertebrata (exc. Insecta)	27	0.399	0.275	0.100	0.074	0.788[b]
Drosophila	43	0.431	0.130	0.140	0.053	0.637[b]

From Nevo (1978).
P = proportion of loci polymorphic per population.
H = proportion of loci heterozygous per individual.
$r(P, H)$ = correlation between P and H over all species.
[a] $p < 0.01$.
[b] $p < 0.001$.

divided or poly-typic species, the greatest fraction of the observed variation may manifest itself among rather than within individuals.

Which of these simple measures is best for conservation genetics? Limiting ourselves to diploid, outbreeding species (and most large animals and tropical plants belong to this category), the important variable is the amount of genetic variation within the group of individuals that is actually or potentially the target of a conservation programme.* Such a group will most often be (1) a natural geographic remnant of a once widespread species, or (2) a synthetic mixture of individuals from two or more such remnants. For the present purpose, the difference between these two kinds of groups (natural and synthetic) can be ignored because it will disappear after a single generation of breeding (although certain genetic problems, such as inbreeding depression, might appear with greater frequency in the natural group, whereas other problems, such as genetic incompatibility, may be more common in synthetic groups). The simplest measure of actual genetic variability in such circumstances is observable heterozygosity. In the absence of dominance and epistasis, heterozygosity is the same as additive genetic variance, the selectable component of total genetic variance.

Table 3.5 from Nevo (1978) summarizes the general levels of elec-

* Note discussion in Chapter 1, p. 8, on the genetics of scarcity.

trophoretic variation in natural populations. The taxonomic groups are ranked in order of increasing *H*. The data for plants and birds could be misleading because of the small sample size and because most of the plant species so far studied have been inbreeders (Brown, 1978). In addition to the plants, hermaphroditic snails (Selander and Kaufman, 1973a) and non-flying Orthoptera (Nevo, 1978) also have very low levels of heterozygosity, and the inclusion of these groups in the table has reduced the *H* values for the invertebrates and the insects, respectively. There is no way to avoid this problem when averaging together species which differ among themselves in breeding system, population size, vagility and history.

3.2.2 Models for the maintenance of genetic variation in natural populations

In chapters 3 and 4 we have generally and implicitly assumed that deleterious recessive genes are the cause of the decrease in fitness that accompanies inbreeding. The term applied to this category of genetic disability is *mutational load*. Actually inbreeding depression can also be the result of over-dominance, the superiority of the heterozygote over both of the homozygotes. Genetic load resulting from over-dominance is referred to as *segregational load*.

The consensus today is that little or no overdominance exists at individual gene loci, or even for individual morphological or reproductive traits (Eberhart, 1977). Comstock (1977) concurs, and in reviewing the state of the art of quantitative genetics states that 'Studies of genetic variance components indicate that overdominance is not a major feature in the genetics of single quantitative traits, not excepting such highly heterotic ones as grain yield of maize.' Falconer (1977) comes to the same conclusion based on his selection experiments in mice. The only well-documented exception in all the scores of traits that have been genetically defined, whether simple or complex, is sickle-cell anaemia, but even this textbook example turns out to be a case of 'conditional' heterosis, since the heterozygote advantage disappears with the removal of the selective factor, falciparum malaria (cf. Berger, 1976).

One way of rescuing single gene overdominance was originally suggested by Levene (1953). Levene showed how a polymorphism could be maintained in a heterogeneous environment even though the heterozygote was nowhere superior to both homozygotes. The necessary condition is that the relative superiority of the homozygotes changes from one site to another. An analogous model could be applied to tissues within individuals or to the different life history stages. That is, if *AA* is superior to *aa* in the adult stage (or the gut epithelium), while *aa* is better in the larva (or brain), then the heterozygote, *Aa*, would be the most efficient overall (Fig. 3.4). This model requires no environmental heterogeneity.

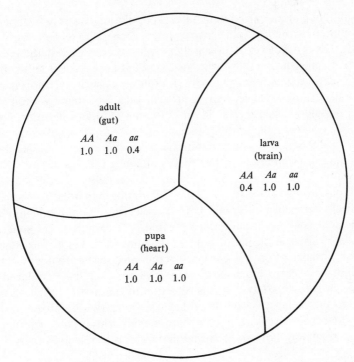

Fig. 3.4 A model of the maintenance of a structural gene (protein) polymorphism due to life cycle or tissue heterogeneity. The numbers below the genotypes are hypothetical biochemical efficiencies. Averaged over all stages or tissues, the heterozygote is superior to either homozygote, yet there is no heterosis *per se* in any single biochemical environment.

Such models, stemming from Levene and others, bear the name *marginal overdominance*, and they have attracted considerable attention and extension in recent years (Hartl and Cook, 1973; Karlin and Lieberman, 1975; Gillespie and Langley, 1974; Hedrick, 1974; Gillespie, 1977). Here, however, we have an example of theory outdistancing facts; there are simply no sophisticated tests of marginal overdominance. Nevertheless, the genetically knowledgeable conservationist should be aware that the downfall of pure (or unconditional) overdominance need not imply that the masking of deleterious recessives is the only advantage of heterozygosity. The genetic polymorphisms existing in natural populations might enhance fitness, even in the absence of deleterious recessives, assuming the reality of marginal overdominance. In summary, environmental variation in space or time (including somatic space and ontogenetic time) could account for the persistence of some fraction of allelic polymorphisms.

On the surface, this conclusion might appear to contradict our belief that the amount of genetic variation in natural populations is strongly dependent on population structure, particularly population size (see section 4.2.2). Actually it does not. Even if most polymorphisms are selected, the selection coefficients must be small, on the average, and genetic drift will still be a major force in determining the overall number of such polymorphisms.

Quite a different conceptual approach, one favoured by ecologically minded population geneticists, invokes the metaphysical concept of niche width as a principal cause of differences in genetic variation between populations. This theory posits that heterozygosity can enhance the ecological amplitude of a population, either by (1) allowing the production of more kinds of individuals, or by (2) enhancing the flexibility of each individual. These ideas, known collectively as niche-variation hypotheses, (see review by Hedrick, Ginevan and Ewing, 1976) predict that a population inhabiting a wide niche can more successfully exploit its resources by generating phenotypic variation of one or both of the above two kinds. For example, when the variance of prey size is large, a predator species that is variable in size might be able to maintain a greater population size compared to a species that is less variable; this is selection of the first (1) type. Another version of the niche-variation hypothesis assumes that heterozygous individuals can cope with environmental extremes and variation more efficiently than can relatively homozygous individuals; this is selection of type (2), above.

Marginal overdominance of the niche-variation hypotheses are mutually compatible; more precisely, the former provides a mechanism that could account for the latter. That is, a 'wide niche' is a mosaic of selection regimes permitting marginal overdominance to be expressed, whereas a 'narrow niche' could be thought of as a single regime in which selection coefficients of genotypes are fixed.

These ideas have spurred considerable debate. One of us (M.E.S.) has argued that the niche-variation hypotheses, while logical and intuitively appealing, are based on a simplistic view of genetic organization, and that all the available data can actually be explained more parsimoniously by an alternative hypothesis, namely that the differences in average heterozygosity among populations are explicable in terms of population structure and history as well as differences in the strength of directional selection (Soulé, 1976). Others base their models on predictability and reliability of food categories in the environment (Valentine, 1976). Both of these latter schools are in agreement that natural selection has a role in determining the differences in genetic variability among populations and species; they also agree that the simpler theories of the late 1960s and early 1970s (e.g. Selander and Kaufman, 1973b; Nevo, Kim, Shaw and Thaeler, 1974) do not account for the patterns of variation that have emerged in recent years.

Still another model that can account for the selective maintenance of heterozygosity is frequency-dependent selection; Lewontin (1974) gives a comprehensive discussion. While there is some laboratory evidence for a negative correlation between an allele's frequency and its contribution to fitness (an allele improves relative fitness when rare, decreases it when common), the relevance of this in nature is unknown.

Yet another class of models, a class that is in some ways quite realistic, makes an important distinction regarding the kinds of genetic load (Wallace, 1970). Wallace points out that some genotypes are unfit in all environments and at all population densities (density-independent fitness); i.e. 'hard' selection will tend to eliminate such genotypes at all places and times. Other kinds of genotypes will prosper when competition or population density is low, but will suffer when conditions are less favourable (density-dependent fitness); he refers to such conditional selection as 'soft selection'. It is hard to see how such a distinction will be of much interest to practical conservationists, however; as is discussed in chapter 5, the survival of *many* species of large vertebrate or plant is going to require intensive management, including density management, and such phenomena as hard versus soft selection, or unconditional versus marginal overdominance, will probably not be applicable (or detectable) in practice.

Finally, there are the 'neutralists' (Nei, 1975, reviews the thinking of this school) who espouse a modern version of the 'classical' position in population genetics (see Lewontin (1974) for a thorough discussion). The neutralists believe that the heterozygosity we observe is mostly irrelevant to evolution and fitness because the alleles are physiologically and biochemically indistinguishable. Ohta and Kimura (1975) have proposed a model which is classical in spirit, but can explain the persistence of much allelic diversity in terms of an equilibrium between mutations producing slightly deleterious alleles and their elimination by stabilizing selection. Wright (1978) offers another option, taking into account variation in space; he also assumes complex dominance and epistasis interactions.

3.2.3 Heterozygosity and fitness in natural populations

Immediate fitness is all those phenomena that contribute to survival (viability) and reproduction. There are two categories of genetic phenomena which can have a direct effect on fitness, independent of their effects on particular characteristics; these are *heterosis* and *inbreeding depression*. Both of these are related to N_e and to the level of genetic variability or heterozygosity, and in many situations they might even be considered opposite sides of the same coin. Nevertheless, we shall treat them separately (for reasons of convenience and didacticism), while acknowledging the somewhat arbitrary nature of this dichotomy. We begin with heterosis.

Among geneticists and breeders there is a *heterozygosity consensus*; this is the belief based on extensive laboratory and farm experience that fitness (viability, vigour, fecundity, fertility, etc.) is enhanced by heterozygosity, and that any decrease in genetic variation will be paralleled by a diminution of fitness. Enhancement of fitness due to increased heterozygosity is called heterosis, and it is virtually universal in outbreeding domesticated plants and animals.

The question we address here is whether a decrease of heterozygosity in a *natural* population will lead to a corresponding decrease in fitness. If it does, some species could become moribund or extinct for no other reason than loss of heterozygosity, *per se*. Another way of phrasing this question is this: how much of fitness depends on being heterozygous at a significant proportion of an individual's loci? There are many possible functional relationships between fitness and heterozygosity in natural populations. Some are illustrated in Fig. 3.5. Model 1 portrays the 'null hypothesis', that of no effect of heterozygosity *per se* on fitness. (The intercept for this curve is arbitrary.) Model 2 portrays a linear relationship – the more loci are heterozygous, the more fit the individual, and each additional heterozygous locus, on the average, adds a constant increment of fitness. Model 3 portrays an asymptotic hypothesis model: each additional heterozygous locus confers less benefit.

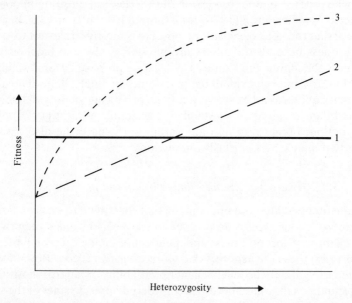

Fig. 3.5 Three possible relationships between genetic variation and fitness. The intercept is shown to be greater than zero in order to indicate that viable populations can exist without any heterozygosity. The intercept of curve '1' is arbitrary.

The central theoretical problem of conservation genetics is to decide which among these models provides the closest fit to reality. One approach is to compare the heterozygosities and fitnesses of individuals within populations. Until recently it was impossible to assay the relative differences in heterozygosity among individuals, unless, that is, their pedigrees were known. Since about 1965, however, electrophoretic techniques have permitted us to rank individuals according to their electrophoretic variation (Lewontin, 1974). The most promising strategy for demonstrating the superiority of relatively heterozygous individuals is the 'longitudinal' sampling of a cohort, i.e. repeated sampling of a year class in order to test the hypothesis that individuals which are relatively heterozygous have a higher than average probability of surviving a long time. The next best thing is a 'vertical' (one-time) sampling of the age classes, although this method introduces the possibility of error because of heterogeneity due to birth or development during different seasons or years. Data from both kinds of studies are slowly accumulating.

A caveat before proceeding with a discussion of the results: there have now been hundreds of electrophoretic surveys of natural populations and only a small fraction of these report any significant excess of heterozygotes. By emphasizing the latter studies and ignoring the rest there is obviously a danger of bias. On the other hand, there are two reasons why this possible bias is not so serious. First, most studies use relatively small sample sizes per population, so statistical significance for departure from Hardy–Weinberg frequencies would not be expected unless the heterozygote excess was quite large. Second, few studies were designed to sample age classes, either horizontally or vertically, and the chance of finding a heterozygote advantage is very much decreased when age classes are pooled.

One final technical point: on *a priori* grounds we might expect never to observe an excess of heterozygotes when many loci are sampled in a series of age classes. This is because, as Sved, Reed and Bodmer (1967) have shown, the variance of the number of heterozygous loci per individual in a randomly mating population is very low. For example, in a population in which there are 10 000 polymorphic loci, each segregating for two equally frequent alleles, the average individual will be heterozygous for 5000 with a standard deviation of fifty; this means that 95% of the individuals will have between 4900 and 5100 heterozygous loci. With such a tiny differential in heterozygosity, natural selection might be hard pressed to distinguish between the extremes. Later (p. 53) we suggest that structured populations, which are probably the rule for many terrestrial species of plants and animals, have much higher heterozygosity variances than would be the case for the ideal, randomly mating population, in turn allowing much more latitude for natural selection to distinguish individuals based on their total heterozygosity.

3.2.4 Intrapopulation studies

In some investigations only one or two polymorphic loci have been examined. One of the first and most widely cited of such studies was a survey of transferrin groups in species of tunas by Fujino and Kang (1968). They divided a sample of 790 skipjack tuna from Hawaii into five age (size) classes in order to test for differential fitness of the transferrin classes with length of survival. They claimed to show that the smallest fish (31–40 cm) had a deficiency of homozygotes and an excess of heterozygotes. There are several ways to look at these data, however, and it is by no means established that overdominance (heterosis) exists. In any case, the putative heterozygote advantage decreases with the age of the fish, so there is no evidence for the superiority of heterozygotes throughout the life of the adult fish.

Chaisson, Serunian and Schopf (1976) reported that ribbed mussels (*Modiolus demissus*) living in the Wild Harbor salt marsh, Cape Cod, Massachusetts, apparently have a higher probability of surviving to adulthood if they are heterozygous at the tetrazolium oxidase locus. At two localities in the marsh, the large mussels showed considerable excesses of heterozygotes, but the young mussels at the same sites did not. Koehn, Turano and Mitton (1973) earlier came to the same conclusion using the same material; they found that newly settled mussels have a deficiency of heterozygotes, whereas large individuals have no deficiency or have a slight excess. Analogous results were obtained by Tracey, Bellet and Graven (1975) and by Koehn, Milkman and Mitton (1976) in studies on another genus of mussel. An unresolved issue in these studies is why the populations are deficient of heterozygotes just after settling.

Similar trends to those just described for mussels have been reported by other investigators employing only one or two polymorphic loci. These include Tinkle and Selander (1973) using the lizard *Sceloporus graciosus*, Watt (1977) using *Colias* butterflies, and Converse and Williams (1978) who found that heterozygotes in man for the *HLA-B* locus live longer than do homozygotes.

A clever departure from the usual protocol of such studies was carried out by Singh and Zouros (1978) on the American oyster *Crassostrea virginica*. They noted a great heterogeneity in shell size of oysters of the same age, and asked whether the faster growing individuals were more heterozygous than slower growing ones in the same cohort. Electrophoresis was performed on 372 one-year-old oysters, half of which were the smallest (slow growing), weighing less than 1 g, and half of which were the largest (fast growing) in their sample weighing more than 4 g. Five loci were found to have substantial levels of polymorphism; they were *Lap-2*, *Pgi*, *Pgm*, *Est-3* and *Got-1*. All except *Got-1* were deficient in heterozygotes for the small class. Singh and Zouros divided the large class of oysters into three subgroups: 4–6 g,

6–8 g, and 8 g and over. The general result was a correlation between growth rate and heterozygosity (Fig. 3.6); in the largest class, none of the four loci showed a heterozygote deficiency. For *Got–1*, the pattern was the opposite with an excess of heterozygotes in the small class and a deficiency in the largest. The authors attribute the *Got–1* result to locus specific effects on fitness. The overall slower growth in relatively homozygous individuals, they think, is probably related to genetic load i.e. homozygosity for deleterious genes: on the average the slower growing individuals are more inbred and thus suffer from a mild degree of inbreeding depression (see section 3.3.3). Regardless, this study provides evidence for the superior viability or vigour of relatively heterozygous individuals.

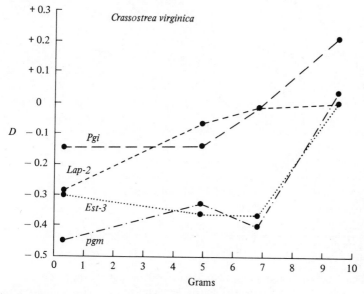

Fig. 3.6 Deviations from expected heterozygosity at four loci in the American oyster. The ordinate is the deviation: observed heterozygosity (H_o) minus expected heterozygosity (H_e) divided by H_e. The abscissa is weight in grams. After Singh and Zouros (1978).

One of the most convincing studies of this kind was performed on the perennial herb *Liatris cylindracea* by Schaal and Levin (1976). *Liatris* is a self-incompatible member of the Compositae occurring locally throughout the dry prairies of the American midwest. The perennating organ is a corm which can be aged by counting the rings of pigmented cells deposited annually in this population. The average age of individuals at the study site was nineteen years, and the range was one to forty-four years. The sample was divided into six age classes and heterozygosity estimates were obtained

by electrophoretic surveys of corm tissue, using fourteen polymorphic loci.
Survival is clearly related to heterozygosity as shown in Fig. 3.7. As in Fig.
3.6, the ordinate is deviation from expected heterozygosity. The youngest
plants are excessively homozygous, probably because of inbreeding, but this
excess decays with age, apparently due to the disproportionate mortality of
homozygous individuals.

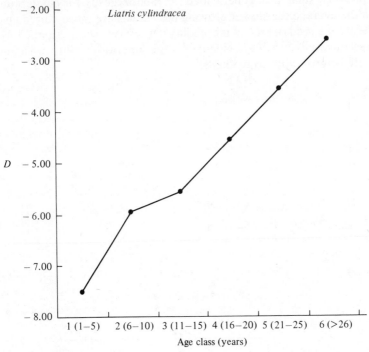

Fig. 3.7 The decrease in heterozygote deficiency with age in a natural population of
Liatris cylindracea. Ordinate is deviation from expectation assuming random mating.
Data from Schaal and Levin (1976).

Schaal and Levin exploited the relative ease of manipulation in plants,
performing greenhouse tests for relationships between heterozygosity and
(1) age at sexual maturity, (2) reproductive output and (3) vegetative
output. All these experiments confirmed the association between
heterozygosity and fitness. For example, *Liatris* seedlings were grown in the
greenhouse for two years and were then divided into those that had flowered
(in the field, flowering usually commences at between five and ten years)
and those that had not reached maturity. The average heterozygosity of
those that had flowered was 8.1%, while that of the non-flowering plants
was 5.2% ($P<0.001$). Overall, this study quite clearly demonstrates natural
selection favouring excessively heterozygous individuals.

Further evidence for the superiority of relatively heterozygous individuals in natural populations comes from studies (Mitton, 1978; W. F. Eanes, 1978) in which both external morphological as well as electrophoretic measurements are made on the same individuals. Both Mitton, whose organism was the killifish, *Fundulus heteroclitus*, and Eanes, whose organism was the monarch butterfly, *Danaus plexippus*, reach the conclusion that individuals which are heterozygous at randomly chosen polymorphic loci are less variable for the morphological traits than are homozygous individuals. Eanes and Mitton view their results in light of laboratory studies showing that homozygotes are less well developmentally buffered than heterozygotes. Even if this interpretation is incorrect, the fact remains that phenotypically extreme individuals are less likely to survive and reproduce than typical individuals (Cavalli-Sforza and Bodmer, 1971; Fox, 1975; Franklin, 1980 refers to several examples), and that fitness is therefore inversely related to an individual's deviation from the average character state for quantitative traits.

At this point we step aside from our catalogue of relevant studies and briefly discuss the issue raised on p. 49 regarding the effect of population structure on the theoretically narrow range of heterozygosities occurring in an ideal population. Recall that in a large, panmictic population, all individuals should have about the same level of heterozygosity. Actually such an ideal population is one of the extremes on a continuum of possible population types; near another extreme are highly structured populations in which small and semi-isolated subpopulations exchange individuals at a very low rate. In the latter case there will be an excess (compared to random breeding) of homozygotes, and the proportion of the genome that is homozygous will vary considerably among individuals. In addition, rather large arrays of linked alleles can be maintained in non-random or non-equilibrium combinations for several generations (Wills, 1978). In other words, structured populations will produce a much greater range of individual heterozygosities. Consequently, the chances are good of finding evidence of heterozygote superiority when assaying a random sample of loci in such populations because even neutral alleles will often be linked to genes that confer an advantage to the heterozygote. As a result, we can expect to observe heterozygote advantage much more easily in a structured population, such as *Liatris*, than in a more vagile species.

It is important, however, to appreciate that heterozygotes may be superior even in species where such fitness differences are undetectable. That is, we may not always be able to resolve the slight differences in heterozygosity existing among individuals. In such species another approach is required. One such approach is to compare the fitnesses of individuals from populations differing in average heterozygosity.

3.2.5 Interpopulation studies

Just such a study was performed by Garten (1976) on the relationship between aggressive behaviour and heterozygosity in the oldfield mouse, *Peromyscus polionotus*. Garten obtained mice from populations differing in mean electrophoretic heterozygosity and tested their aggressiveness by staging paired encounters in small arenas. He found very significant correlations between various components of aggression and heterozygosity (Fig. 3.8). As in all such studies, the danger of spurious correlation is ever present. For example, the correlation between mean body size and mean heterozygosity among populations was also very high, so that the aggressive superiority could be the result of the actual differences in strength among individuals, or, perhaps, it could have resulted from the intimidation of the smaller individuals by the larger. Other, more subtle spurious correlations cannot be ruled out.

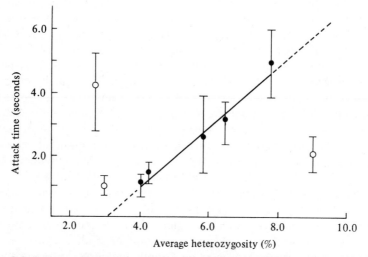

Fig. 3.8 Relationship between mean accumulated attack time and average percentage heterozygosity of male oldfield mice from five mainland (dots) and three island (circles) localities. The regression line is based on mainland samples only. After Garten (1976).

Another interpopulation study was recently reported by Soulé (1979) who reanalysed data from earlier studies on lizard populations in order to test the relationship between mean heterozygosity and fitness. Fitness was defined in terms of developmental stability or homeostasis as measured by asymmetry of bilateral morphological traits. The assumption is that asymmetry reflects accidents or 'noise' during development, and that genetically superior individuals will be less asymmetrical than genetically inferior ones.

Fig. 3.9 suggests that there may be a negative correlation between heterozygosity (estimated by electrophoresis of proteins) and asymmetry. If future studies of this kind verify a correlation between developmental stability and heterozygosity, we will be led to conclude that individuals from populations rich in genetic variation are more fit in this regard than individuals from genetically depauperate populations.

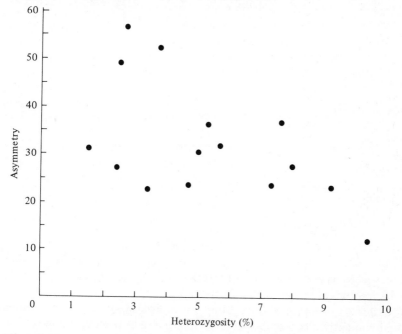

Fig. 3.9 The relationship between fluctuating asymmetry (sum of ranks) and percentage heterozygosity in one mainland and fourteen island populations of the sideblotched lizard. From Soulé (1979).

The studies discussed so far point to the ubiquity of heterozygote superiority. On the other hand, it is possible to carry this conclusion too far. Just as cake is edible without icing, so might populations be viable without any heterozygosity, at least in the short run. Nor is the mere existence of allelic variation in a population evidence for its contribution to immediate fitness. Icing can be made with salt instead of sugar, and allelic variation can be neutral or even deleterious. In any case, populations are known in which allelic variation is nil.

Some predominantly inbreeding plants, for example, have relatively high levels of heterozygosity, while others seem to have none. Four such species were included by Brown (1978) in a recent survey of genetic structure in plants. The heterozygosity data are given in Table 3.6. The important thing

TABLE 3.6   *Genetic variability in a sample of predominantly inbreeding plants*

Species and region	Number of populations	No. of loci scored	No. of loci polymorphic	H_e		
				Mean	Min	Max
Oenothera biennis Cook County	16	20	1	8	0	50
Oenothera biennis S. Illinois	28	20	4	22	0	50
Avena barbata California	16	5	5	7	0	48
Hordeum spontaneum Israel	28	28	25	11	0	20
Lycopersicon pim-pinellifolium Ecuador & Peru	43	11	11	14	0	27

H_e is the mean expected panmictic heterozygosity.
From Table 1 of Brown (1978).

to note is the range of the values. Some populations in each species had no heterozygosity; some had 50%. Except for the obvious conclusion that heterozygosity is not a necessary condition for survival in selfing plants, it may be imprudent to generalize further.

Yet, the very presence of segregating polymorphisms in some populations of predominantly inbreeding species is something to wonder at, and it has been interpreted as prima-facie evidence for heterozygote superiority. For example, Clegg, Allard and Kahler (1972) studied experimental populations of barley (*Hordeum vulgare* L.); these lines are over 99% self-fertilizing. Later they (Clegg and Allard, 1972) found that some populations maintain substantial heterozygosity for tightly linked, co-adapted gene complexes. As Clegg and his coauthors point out, the fitness of the heterozygotes in these populations must be twice as high as that of the homozygotes to maintain these polymorphisms in the face of such intense inbreeding.

Rick, Fobes and Holle (1977) also observed twice the expected level of heterozygosity in polymorphic populations of *Lycopersicon pimpinellifolium*, a close relative of the tomato. Inbreeding in these populations ranged from about 45% to 100% per generation. Add to this the evidence for heterozygote advantage in other predominantly self-pollinated grasses (Jain and Marshall, 1967), and one cannot dismiss the possibility that heterosis is common in inbreeding species.

Thus, in inbreeding plants we see (Brown, 1978) an entire spectrum of phenomena relating to heterozygosity. On the one hand there are completely homozygous populations; on the other there are populations with large and significant excesses of heterozygosity. In the middle are populations with some heterozygosity, but with no evidence, as yet, for heterozygote superiority. If generalizations are applicable to such species, they will not be widely accepted until many more studies of the Schaal and Levin model are performed.

3.2.6 The significance of monomorphic populations

Selfing plants are not alone in providing examples of negligible variation in proteins. Animals, too, occasionally lack such polymorphisms. For example, an elephant seal population with no detectable electrophoretic heterozygosity (Bonnell and Selander, 1974) is growing very rapidly, following near extinction from hunting. Other animals which are virtually lacking in detectable electrophoretic variation are an isolated gopher species (Selander, Kaufman, Baker and Williams, 1975 and references therein), lizards on tiny islands (Soulé, 1980), and a facultatively, self-fertilizing land snail (Selander and Kaufman, 1973a). Indeed, it is very likely that most of these populations are evolutionary dead-ends, but at least it is clear that survival in nature, albeit short-term, is possible despite virtual homozygosity.

Is the appearance of virtual total allelic impoverishment in the above plants and animals to be taken at face value? Admittedly, there could be considerable cryptic genetic variation in these populations, but even granting this, it still appears certain that some species (and some populations) are able to sustain themselves even though they contain much less than the average level of heterozygosity.

3.2.7 Conclusions

In this section we have asked whether there is empirical support for the hypothesis of a positive effect of heterozygosity on fitness in natural populations. Many of the foregoing studies would not, by themselves, be particularly convincing. Collectively, however, and in the absence of contradictory evidence, the judicious conclusion is that fitness in natural populations is a positive function of heterozygosity. Tentatively, then, we can rule out Model 1 in Fig. 3.5. Essentially this means that any loss of genetic variation, at least in outbreeding populations, is tantamount to erosion of immediate fitness. Further, there is no logical basis for assuming that there is a 'safe' level of fitness detriment. Given, that is, the general applicability of either Model 2 or Model 3 in Fig. 3.5, it is evident that each increment of genetic simplification costs the population an increment (not necessarily constant) of welfare.

The study by Rick *et al.* (1977) on genetic variation in *Lycopersicon* provides a tantalizing clue favouring Model 3 versus Model 2. As shown in Fig. 3.10, the excess of observed heterozygotes over the expected appears to be related to percentage cross-pollination. Heterozygosity of these populations and cross-pollination are highly correlated ($r = 0.81$; $P<0.01$), so the negative association of heterozygote excess with outbreeding suggests that the locus-specific enhancement of fitness increases as the overall heterozygosity decreases among populations. This may be the first evidence from natural populations for an asymptotic relationship between genetic variation and fitness. Earlier experimental work by Crow and his colleagues (Temin, Meyer, Dawson and Crow, 1969; Crow, 1970) provides additional support for the asymptotic model. They found that as *Drosophila* were inbred, their viability drops off slowly at first, and more rapidly as inbreeding reached higher levels.

It would be premature to base a conservation strategy on such a subtlety, but if Model 3 turns out to be a general rule in population genetics, it would mean that the absence of obvious inbreeding effects on the viability or fecundity of a managed population during the first few generations of captive breeding or intensive management is no guarantee that such immunity would persist at higher levels of inbreeding.

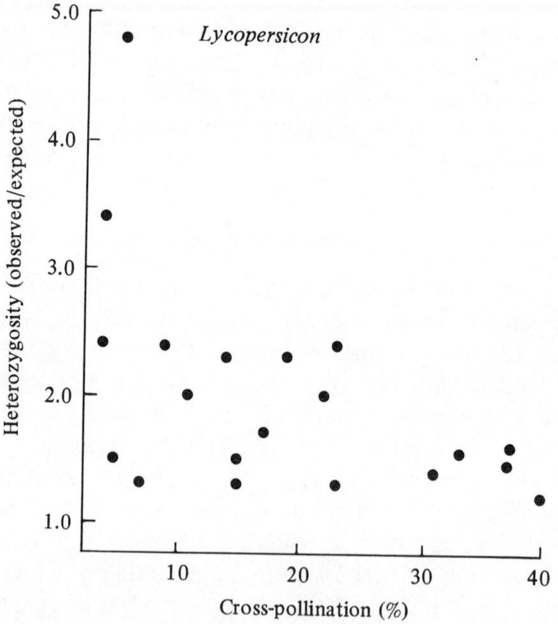

Fig. 3.10 The inverse relationship between cross-pollination and heterozygote excess in *Lycopersicon*. After Rick, Fobes and Holle (1977).

Finally, the correlation of fitness with heterozygosity in natural populations need not signal the existence of single gene overdominance. An alternative is dominance i.e. recessive deleterious genes; this is mentioned (p. 51) in the discussion of the work of Singh and Zouros (1978). One of the major tasks of empirical population genetics is to discover the mechanisms underlying the purported correlation.

3.3 Inbreeding depression

3.3.1 Inbreeding defined

We begin our discussion of inbreeding with an experiment performed on Poland China swine almost forty years ago (McPhee, Russel and Zeller, 1931). The experiment was designed to establish the effects of sib (brother–sister) mating. As it turned out, the experiment was rather short-lived, lasting only two generations. The reason for its discontinuation at this point was a precipitous drop in the fitness of the inbred line. As shown in Table 3.7 the mean number of pigs per litter dropped from 7.15 in the general herd to 4.26 in the second generation of inbreeding. This decline in fecundity was accompanied by a drop in survivorship of pigs by more than half. When the values for fecundity and survivorship are multiplied, one obtains the number of surviving offspring per litter. In the general herd this number is four; it is only one in the F_2 inbreds. Concomitant with this 75% loss in productivity was an equally serious change in the ratio of males to females; it changed from 1.1:1.0 in the general herd to 1.6:1.0 in the F_2 inbreds, thus further aggravating the decline in fecundity and survivorship.

The implications of results like these for the conservation of large organisms are profound, not only for the breeders of endangered species in zoos, but also for the managers of wildlife reserves. But before pursuing this topic we must first define inbreeding in a more quantitative fashion, and explain

TABLE 3.7   *Vital statistics of a herd of Poland China swine and the progeny of two generations of sib mating*

| | No. | Inbreeding coefficient | | Size of litter | Percentage born alive | Percentage raised to 70 days | Sex ratio |
		Dam	Litter				
General herd	694	0+	0+	7.15	97.0	58.1	109.7
F_1 inbred	189	0.09	0.33	6.75	93.7	41.2	126.1
F_2 inbred	64	0.33	0.42	4.26	90.6	26.6	156.0

From McPhee, Russel and Zeller (1931) after Wright (1977).

some of the theories that attempt to account for its effects. Those familiar with inbreeding genetics might wish to skip to section 3.3.4.

One of the most important points to grasp about inbreeding is that it is a *relative* concept. For example, John may be homozygous for deleterious genes at many more loci than Mary, but Mary, strictly speaking, may be more inbred. This is because the formulation of inbreeding is in terms of the proportion of homozygous loci in an individual relative to that proportion in the general population. That is, John may be a member of a tribe having a strong taboo against incest and inbreeding, and yet everyone in the tribe can trace his ancestry back to a single matriarch and her husbands. Mary, on the other hand, might be a typical American of mixed ancestry, as well as being the daughter of first cousins. In genetic terms, John is more homozygous, but Mary is more inbred.

Another way of saying this is that an individual who is a product of inbreeding within a very heterozygous base population can be more heterozygous than a non-inbred individual from a genetically homogeneous population. We will return to this point in the discussion of the relationship between inbreeding and genetic load.

Inbreeding means mating of close relatives, individuals, that is, who are likely to share some of their genes because they have one or more ancestors in common. The most widely used measure of inbreeding, or consanguinity, is the inbreeding coefficient of Wright (1921), denoted by F. This coefficient is most easily understood as the probability that the two alleles of a particular locus in an individual are identical by descent. For example, consider individual E in Fig. 3.11; she is the product of a brother–sister (sib) mating, so she has only one set of grandparents. When considering only one gene (we use gene and locus interchangeably), the total number of copies present in her grandparents was four. We don't mean that the two grandparents had four biochemically distinct alleles; these four copies may have been molecularly identical, but for our purposes they are still considered distinct. Now we ask what is the probability (this probability is called F) that E is homozygous for any one of these four copies present in the grandparents. The answer is 1/4. This can be seen in the following way. The probability that A transmitted a copy of one of his two alleles, say a', to C is 1/2. The probability that another copy of a' was transmitted to D is the same, 1/2. Therefore the probability that C and D both received one copy of the identical gene a' is 1/4. Given that both C and D received one copy of a', the probability that E is homozygous for a' is 1/4; therefore, the probability that E is homozygous $a'a'$, *and* that both of the a' alleles came from A is $1/4 \times 1/4 = 1/16$. Because A and B together had four alleles of the gene in question, and because the probability is 1/16 that E will be homozygous for a *particular* one, the probability that she is homozygous by descent for *any* of the four copies is $1/16 + 1/16 + 1/16 + 1/16 = 1/4$. Precise methods for calculating inbreeding

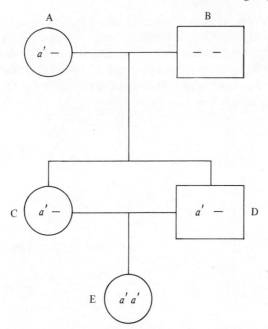

Fig. 3.11 A pedigree showing identity of a gene by descent. See text.

coefficients can be found in Cavalli-Sforza and Bodmer (1971), Crow and Kimura (1970), Pirchner (1969) and Falconer (1960).

Another way to interpret F is in terms of the relative amount of heterozygosity. In the previous example we showed that an individual who is the offspring of a sib mating has 25% chance of being homozygous by descent at a given polymorphic locus. This is equivalent to saying that she will have alleles identical by descent at 25% of all of the loci which are not already fixed (that is, were polymorphic) in the base population. Putting it slightly differently, she will be, on the average, 75% as heterozygous as the average individual in the base or source population. F, then, is a direct estimate of the genetic variability *relative* to the variability in the non-inbred population.

3.3.2 The rate of inbreeding

Next, we examine a question of critical importance in conservation genetics – the *rate of inbreeding*. The rate at which genes become fixed (alternatively, the rate at which heterozygosity is lost) depends on the breeding system. The closer the relationship between parents, the higher the rate of fixation. The most intense form of inbreeding is selfing or self-fertilization. In theory, half of the remaining heterozygous loci become fixed every generation in a selfing population. Hence, the per generation inbreeding is $F = 0.5$. Sib mating or offspring–parent mating gives a per generation loss of heterozygosity of 0.25, as we saw above. The inbreeding coefficients for sib mating as well as for other systems are given for five generations in Fig. 3.12. Note that the per generation change in the inbreeding coefficient, ΔF, is a constant for any breeding system.

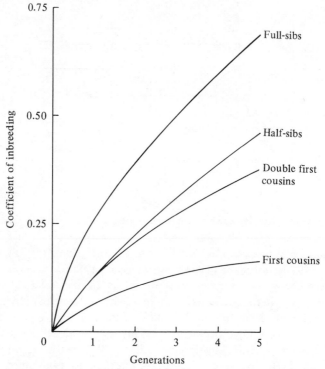

Fig. 3.12 Increase in homozygosity during inbreeding. After Underwood (1979).

3.3.3 Inbreeding depression: theory

Aside from its effect on heterozygosity, phenotypic changes are often observed in inbred lines. For simple Mendelian traits, the changes are random. Sewell Wright (1977) recounts his ability to distinguish his inbred lines of guinea pigs due to unique colour patterns and other traits that had become fixed in each line. More is said on this topic in section 3.5.5.

Inbreeding also causes a change or shift in the means of some genetically determined quantitative characters. But this shift, unlike genetic drift, is directional; *it is always towards the direction of the phenotype expressed by (homozygous) recessive alleles.* This is where the term *inbreeding depression* comes from. One might ask why this is necessarily so – why do recessive alleles when homozygous, produce inferior or 'depressed' phenotypes? The answer has to do with exposure. Dominant alleles are always exposed, or subject to natural selection; hence, a deleterious dominant is readily eliminated in a population. On the other hand, a deleterious recessive can persist at low frequency indefinitely, since it is rarely 'seen' in the homozygous state. The effect of inbreeding on a trait, in the presence of recessives, can be stated more precisely: if the variation for a particular phenotypic character has any degree of dominance or overdominance (heterozygote superiority), then there will be a shift in the average expression of the character towards the homozygous recessive phenotype, as the following example shows.

In an infinitely large, randomly breeding population, the equilibrium (Hardy–Weinberg) genotype frequencies, considering a single locus, are

$$p^2 + 2pq + q^2 = 1$$

where p is the frequency of allele a_1, the dominant allele, and q is the frequency of a_2, the recessive allele. Upon inbreeding, these Hardy–Weinberg frequencies change to

$$(p^2 + pqF) + (2pq - 2pqF) + (q^2 + pqF) = 1$$

Note that the proportion of homozygotes increases at the expense of the heterozygotes. Now, the reason the mean shifts towards the recessive (a_2) side is illustrated in Fig. 3.13. Assume a population is subdivided into a number of lines, each of which is inbred at a level F. The frequencies of the two alleles a_1 and a_2 within a line are p and q, respectively, and the average frequencies for the whole population (all lines together) are $\bar{p}$ and $\bar{q}$. For the sake of simplicity we have set $p = q = 0.5$ and we have set d, the genotypic value (average phenotype) of the heterozygote, equal to the genotypic value of the dominant, or 1.0; the respective value of the recessive is -1.0. This is merely a mathematical way of expressing complete dominance. Upon inbreeding, the change in the mean is $-2d\bar{p}\bar{q}F$, and the mean genotypic value of the average completely inbred ($F = 1.0$) population is (Falconer, 1960)

$$\begin{aligned} M_F &= M_0 - 2dpqF \\ &= 0.5 - 2(0.25) \\ &= 0 \end{aligned}$$

where M_0 is the mean genotypic value of the randomly breeding population and M_F is the value for the average inbred population. Of course, a completely inbred population will be composed of individuals that either are all a_1a_1 or all a_2a_2; 50% of lines will be monomorphic for one genotype and 50% will be monomorphic for the alternative genotype. It is by averaging together all the inbred lines that we obtain $M_F = 0$. As long as there is any

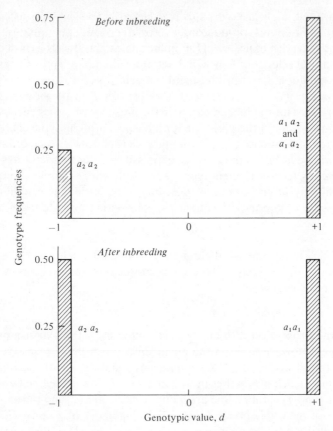

Fig. 3.13 Inbreeding depression as the result of dominance. The genotypic value of the dominant allele is 1; that of the recessive is -1. Before inbreeding, the mean genotypic value for the population is 0.5, as indicated by the symbol M_0. After inbreeding (below), all lines are homozygous, half for allele a_1 and half for allele a_2. Therefore, half of the lines have a genotypic value of -1 and half have a genotypic value of 1; the mean (M_F) is zero. After Falconer (1960, p. 113).

dominance, inbreeding always produces a phenotypic shift towards the recessive.

To counteract this tendency it is common practice for breeders to introduce 'outside' genes from relatively unrelated colonies. This procedure reduces the frequency of unfit homozygous recessives in the mixed population. Bruce (1910) explained this effect mathematically for a single locus. Assume for example, that we have two lines each having a different recessive allele with frequencies q and q', respectively. If we pick at random the same number of individuals from both lines, the overall proportion of unfit homozygous recessives is $(q^2 + q'^2) / 2$. But if we produce a *new* line by crossing individuals from one line with individuals from the other, the frequency of homozygous recessives becomes $(q/2)^2 + (q'/2)^2 + 2(q/2)(q'/2)$ which is less than or equal to $(q^2 + q'^2)/2$. This assumes (conservatively) that qq' is as unfit as qq or $q'q'$.

Crossing has an even more salubrious effect if the two lines tend to have recessives at different loci. For example, line A might have an unfit recessive, m, at locus M, and line B has unfit recessive n at locus N. Assume the frequency of m and n has the value 0.2 in each population respectively. The proportion of homozygous recessives in a new line formed by crossing equal numbers from A and B will be $2(0.1)^2 = 0.02$, in contrast to 0.04 in line A or B.

3.3.4 Inbreeding depression and fitness

In this section we delve more deeply into the actual forms taken by the deleterious effects of inbreeding. The expression of inbreeding depression is biased towards certain kinds of characteristics. That is, inbreeding depression neither affects all characters uniformly, nor are these effects random or unpredictable with respect to the traits in which they appear. In fact, we can predict quite accurately which characteristics will be depressed the most upon inbreeding.

The careful reader will have anticipated this predictability of inbreeding effects. Recall that the average phenotype of traits upon inbreeding moves away from the dominant or overdominant phenotype and towards the recessive. Traits, therefore, which have a significant amount of dominance or overdominance (non-additive genetic variance) will change the most. Furthermore, a shift towards the phenotypic expression of recessive genes is tantamount to a decline in fitness because, as already mentioned, a disproportionate number of deleterious alleles are recessive.

Now, in what kind of characters is dominance typically observed? By and large, they are traits related to reproduction; in the parlance of quantitative genetics, they are 'fitness characters' (Robertson, 1955). This means that we can expect to observe the most inbreeding depression in characters such as fecundity (total reproductive output), fertility (ability to produce viable gametes or zygotes), developmental rate or age at sexual maturity, litter size and analogous traits. In contrast, characters, the states of which are not critical to reproduction, are less affected by inbreeding. Finally, the traits which are least affected by inbreeding are those that can vary greatly without affecting the viability or reproductive contribution of individuals.

Another way of expressing the above breakdown of traits is in terms of *heritability*. A long exegesis on heritability would be inappropriate here (see Bodmer and Cavalli-Sforza, 1976 for an excellent summary). For our purposes it is sufficient to point out that heritability is a measure of the genetic determinism of a trait or of its correlation among close relatives. Such correlations are high if the genes that control a trait are acting in an additive fashion, that is, if there is relatively little dominance or overdominance. The experience of breeders has led them to generalize that reproductive traits generally have low heritability (little additive variation) while traits which are apparently less relevant to reproduction and survival have high heritabilities.

Table 3.8 summarizes the heritabilities of some traits in animals. Note that the heritabilities for coat pattern, tail length and body conformation all tend to be high (50% to 90%), whereas those for reproductive characters are usually 20% or less. The rule of thumb is that the characteristics with the lowest heritabilities are those which will be the most depressed by inbreeding, assuming a relatively large amount of dominance variation.

At the beginning of this chapter we gave an account of inbreeding depression in Poland China swine. This example is just one of several hundred in the animal and plant breeding literature. A complete review of the literature would serve no purpose here, but some additional examples are helpful in reaching some general conclusions.

Sewell Wright (1977) recently summarized his extensive and long-term studies on inbred guinea pigs. Three points are especially significant. First, out of thirty-five original lines undergoing inbreeding, only half survived for nine years, and only five were vigorous enough to warrant intensive study. Second, in these five relatively vigorous lines, the effective fecundity, compared to non-inbred controls (young raised per mating year) was only 30%.

TABLE 3.8   *Heritabilities of various traits*

Trait	Heritability	Source
Finger print ridges in humans	0.95	Holt (1961)
Amount of spotting in Friesian cattle	0.95	Falconer (1960)
Stature in human males	0.79	Osborne and De George (1959)
Femur length in mice at 3 months (male-son)	0.79	Leamy (1974)
Tail length in mice	0.6	Falconer (1960)
Length of wool in sheep	0.55	Falconer (1960)
Abdominal bristle number in *Drosophila*	0.5	Falconer (1960)
Skull length in mice at 3 months (male-son)	0.50	Leamy (1974)
Combined molar length in mice	0.30	Bader (1965)
Milk yield in cattle	0.3	Falconer (1960)
Egg production in poultry and *Drosophila*	0.2	Falconer (1960)
Litter size in pigs and mice	0.15	Falconer (1960)

Third, the overall fitness of the inbred lines compared to the controls is worst in unfavourable (nutritionally poor) environments. That is, the inbreds are less resistant to stressful conditions (Lerner, 1954; Parsons, 1971).

Bowman and Falconer (1960) studied the effects of inbreeding on litter size in mice. The decline was from 7.77 to 4.58 in five generations; expressed in terms of F, this is 0.6 of a mouse (7.7%) per 10% increase in F. (Recall that ΔF, equal to 10% is slightly less than the ΔF in a half-sib mating scheme.) Only one out of twenty inbred lines remained at the control level for litter size after generation twelve.

The difference between inbred and outbred strains of Holstein-Friesian cows was studied by Tyler, Chapman and Dickerson (1949). They found little effect of inbreeding on body dimensions (note that body dimensions are not fitness characters) but recorded a depression in milk and butterfat production of 6.2% and 5.8%, respectively, for 10%ΔF.

In gallinaceous birds, the effect of inbreeding appears to be related to the history of the species in captivity; the longer they have been domesticated, the less the inbreeding depression. Table 3.9 summarizes the results of several studies of Abplanalp and coworkers (Abplanalp, 1974). Note that the decline in fitness is least in chickens and turkeys, and most in Japanese quail and chukar partridge; the latter two species are much less domesticated than the former two. Domesticated species have a history of selection and inbreeding; i.e. they have been partially purged of their deleterious genes, so inbreeding depression concomitant with further inbreeding is relatively less severe.

Wright (1977), in reviewing the history of inbreeding in maize, reiterates that the fitness characters show the most drastic decline. Most lines

TABLE 3.9 *Effect of 25% inbreeding on the relative performance of four gallinaceous species; performance of non-inbred lines equals 100*

| Trait | Performance as percentage of non-inbred birds | | | |
	Chicken	Turkey	Japanese Quail	Chukar
Hatchability: embryo inbred	90.0	83.4	72.2	71.3
hen inbred	97.0	92.1	89.3	89.1
Fertility	99.1	98.8	79.2	71.1
Viability of females	94.3	90.7	81.5	92.1
Egg production	90.4	89.5	83.9	84.1
Total reproduction	74.4	61.6	35.9	34.1
'B' = inbreeding depression	1.183	1.938	4.098	4.303

From Abplanalp (1974).

deteriorated so rapidly upon inbreeding that they could not be saved. Those that survived to nine or ten generations of selfing deteriorated in height (27%), length of ear (28%) and, most dramatically, in yield (61%).

In sum, these studies point to the universality of inbreeding depression. More particularly, there is remarkable agreement between the inbreeding coefficient and loss of fitness, namely, a ΔF of 10% generally corresponds to a 5–10% decline for a particular reproductive trait. When considering *total reproductive performance* (rather than isolated characteristics) the decline in fitness jumps to a stunning 25% or so for species or lines that have not been extensively inbred in the past (Tables 3.7 and 3.9).

The decrease in fitness is less for most domesticated species and for previously inbred or selected lines of experimental organisms. As suggested above, this apparently less severe inbreeding depression is probably the result of the 'purging' effect of selection and inbreeding. Any inbreeding is likely to remove some of the deleterious genes from a line. For example, Roberts (results cited in Falconer, 1960) performed inbreeding experiments on mice stocks obtained by crossing inbred laboratory strains. Only two out of thirty lines were lost after three generations of sib mating. Such a high survival rate of lines is probably a consequence of the rarity of deleterious recessives in the original inbred strains.

The experience of plant breeders also leads to the conclusion that a history of slow or episodic inbreeding tends to decrease the severity of later inbreeding episodes. Many more plants than animals are prevailingly self-fertilizing, and it is possible to ask if species which are predominantly selfing have less inbreeding depression than would be expected. Young and Murray (1966) noted that the degree of inbreeding depression in domesticated plants correlated with the typical amount of self-pollination; in order of increasing cross-pollination and inbreeding depression, these plants are barley, cotton, tomato and corn. In general, self-pollinating species show less heterosis and less inbreeding depression than do out-crossing forms.

There are 'costs', however, to the purging of a stock by inbreeding. First, most attempts fail, so one must start with many lines if one hopes to pull some through the 'inbreeding crisis'. Second, a successfully purged inbred line, once obtained, will be genetically different from the outbred population in many, random, directions. Notwithstanding such costs, it is apparent that once a line has safely passed through one bout of inbreeding, it is more likely to make it through others because it will retain progressively fewer deleterious recessives.

It is clear from the previous paragraphs that inbreeding can be seen to produce two opposing effects: on the one hand it can purge some deleterious genes; on the other hand, it can fix some deleterious genes. Whether a line survives a period of inbreeding, therefore, is a matter of chance. If by chance no lethals or subvital genes are fixed, the line will survive. If, however, some such genes become fixed or reach a high frequency, then the line will probably be lost. *The critical question then is what is the ratio between lines that are purged and survive versus lines that go extinct?* The value of this ratio determines whether inbreeding is a potentially useful tool for conservation.

On this question the data speak rather loudly. Above we noted that only half of Rommel and Wright's guinea pig lines survived for nine years (approximately 11.5 generations of sib-mating). Bowman and Falconer (1960) inbred twenty lines of house mice; only one line survived after generation 12, and only ten lines survived to generation 5. Using wild mice, Lynch (1977) found that only two out of fourteen lines survived for six generations of sib-mating. The results for *Drosophila* are essentially the same; only about 10% of lines survive more than ten or twenty generations of sib-mating (Clayton, Knight, Morris and Robertson, 1957; Wallace and Madden, 1965).

Abplanalp reported that only eight out of 279 lines survived a large inbreeding experiment with white leghorn chickens. The material for the above studies was domesticated stocks, so it is probable that they were all partially purged of their deleterious recessives by population bottlenecks and selection. Thus, one could anticipate that the survivorship of inbred lines would be even less with a foundation stock fresh from nature. In summary, between 5% and 20% of lines survive after F reaches values above 0.80.

One of the myths about inbreeding is that there exists an inbreeding depression minimum, and that once a line succeeds in traversing safely this genetic purgatory, it is cleansed of deleterious genes. The basis for this misconception is the existence of apparently fit inbred domesticated lines such as white rats. Indeed, there exist apparently* homozygous lines, but it is not usually appreciated that these lines typically are severely handicapped with respect to many traits, and would certainly not survive in nature. For example, Lindstrom (1941) found that 677 *useful* lines of inbred maize had been produced by experimental stations in the US, but these were only 2.5% of all those lines that had been started. Further, the average yield of these was only 30% that of F_1 hybrids. It is probably extraordinarily rare for an inbred line to be as fit overall as an outbred population.

3.3.5 Inbreeding depression and behaviour

Up to this point we have stressed the effects of inbreeding depression on the reproductive functions. Certainly reproductive characters are usually profoundly depressed by inbreeding, but other kinds of traits can be seriously

---

* It is now accepted that inbred lines are never as homozygous as predicted from simple theory (e.g. Enfield, 1977; Eriksson, Halkka, Lokki and Saura, 1976); natural selection apparently acts through heterozygote superiority to impede fixation.

affected as well. The reason that we have emphasized reproductive depression is that breeders and experimenters find it practical and relevant to study these. But an overemphasis on such traits as fecundity could lead one to believe that if an inbred stock is relatively fecund, for example, then it should be relatively fit overall, a very misleading assumption. Fecundity and viability in the laboratory, farm or zoo is probably useless in predicting how a stock will perform in nature. This because viability and reproduction in nature depend on a vast and subtle integration of physiological and behavioural phenomena. Whereas an inbred stock might perform well in a sheltered environment, it is likely to be severely handicapped when completely on its own. White leghorn poultry, for example, are champion egg layers, but no farmer would bet on their success in a hedgerow.

To take a more concrete example, laboratory mice (*Mus musculus*) are rarely exposed to the environmental extremes encountered by their wild cousins. Hence, a deterioration of certain abilities necessary for survival in the wild might be expected in laboratory mice, and many such evolutionary changes would go unnoticed. For example, the loss of the capacity to make a warm, protective nest would, in a laboratory strain, have little or no effect on fitness, but such a change in a wild population would lead to death of young from exposure, and thus to extinction. Are behavioural traits of this type likely to be depressed by inbreeding?

They are. Lynch (1977) determined that the amount of nesting material used by mice each day (nesting score) behaves genetically in the same way as fitness characters such as litter size. That is, inbreeding decreases nesting scores within lines, and crosses between inbred lines give substantial heterosis. Fig. 3.14 shows how nesting score varies among some inbred laboratory strains in comparison with a cross between them and in comparison to a wild strain. One of the six inbred strains, *BALB*, builds much bigger nests than the wild strain (this could be as non-adaptive as building too small a nest), and two of the inbred strains build very small nests. Such aberrant behaviour, while perhaps of little significance in the laboratory, could prove disastrous in nature.

There is a rich supply of anecdotes regarding inbreeding effects. A recent example concerns the brown-eared pheasant *Crossoptilon mantchuricum*. In the West all brown-eared pheasants (descended from a male and two females captured in northern China in 1864) have been producing mostly non-fertile eggs. Eggs obtained from artificially inseminated hens are fertile, and the infertility was traced to the low libido of the cocks, according to researchers at Cambridge University (Anon., 1977). Inbreeding problems are notorious in popular breeds of dogs; hip dysplasia and ocular diseases are particularly common.

Behavioural deterioration in traits related to competition will be detrimental in nature, if not on the farm or in the field. Several studies suggest

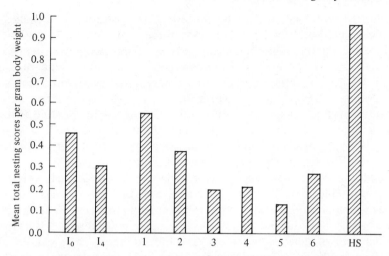

Fig. 3.14 Mean nesting scores for matings between wild caught mice (I_0), the fourth generation of full-sib matings derived from this natural population (I_4), six inbred strains (1 is *BALB/cJ*, 2 is *CBA/J*, 3 is *C3H/HeJ*, 4 is *C57BL/6J*, 5 is *DBA/1J* and 6 is *A/J*) and an eight-way cross among inbred strains (HS). After Lynch (1977).

that inbreeding severely depresses competitive ability. Latter and Robertson (1962) showed a strong effect of inbreeding on competitive ability in *Drosophila*. Mertz, Cawthorn and Park, (1976) ran competition experiments between two species of flour beetle, *Tribolium castaneum* and *T. confusum*. One of the variables in these experiments was the level of inbreeding. In a reanalysis of their results, J. W. Senner (in preparation) demonstrated a very strong depression of competitive ability, apparently attributable to inbreeding. In these examples the depression in competitive ability could be just another expression of a loss of fecundity, since any decline in reproductive fitness should reduce competitive ability.

The results of Garten (1976), discussed on p. 54 are also relevant here. Garten found that intraspecific competitiveness, as measured by aggression in oldfield mice, was positively correlated with mean heterozygosity in the population from which the mice were collected. Such studies reinforce our suspicion that *any* loss of genetic variability, whether due to natural causes (small population size, bottlenecks, directional selection) or to artificial inbreeding, is going to reduce the chances of survival in the wild.

3.3.6 Advantages of inbreeding

For certain purposes, inbred lines have clearcut advantages over non-inbred lines. Their most attractive characteristic is phenotypic uniformity under controlled conditions. In biological experimentation, the most repeatable results are obtained when the experimental material is genetically homogeneous. Much pure and applied research in physiology, developmental biology, immunology, endocrinology and pharmacology depends on inbred strains. The problem with inbred organisms, however, is that they are highly sensitive to variation in the environment, particularly nutrition and temperature. The best of both worlds, that is homeostasis in a fluctuating environment and phenotypic uniformity, is achieved by crossing two or more inbred lines with good 'combining' characteristics. The hybrid offspring are genetically identical and heterozygous, the latter tending to minimize their sensitivity to environmental heterogeneity.

Another apparent advantage of hybrids between some inbred lines is their superior productivity. The best known example is hybrid corn. This brings us to a very important question: Why isn't it possible to assemble all the best genes together in a single homozygous line, assuming heterosis results from the masking of deleterious recessives? Indeed, this is a possibility for specific agricultural or laboratory purposes, but because of the thousands upon thousands of genes involved, a project of this magnitude would be extraordinarily tedious and difficult, especially for slowly reproducing species. Even for maize, which has been bred for many decades, such an ideal strain is unlikely to replace hybrids between inbred lines for the forseeable future (Eberhart, 1977).

Whereas it is our opinion that inbreeding under most circumstances is anathema for conservation programmes, special circumstances may permit special means. In an organism, for example, that for some reason is already very homozygous, further inbreeding might do little harm and could make it easier to maintain in a domesticated or semi-domesticated state. Just such a protocol was recommended by Slatis (1960) who found that inbreeding depression in the European bison, *Bison bonasus* was nearly absent. Perhaps this is because this species has never been abundant in historical times and recently passed through a bottleneck of seventeen *related* individuals. It should be recognized, however, that the consequence of intense inbreeding, assuming the unlikely result of survival without serious depression in viability and fecundity, is the relinquishing of future adaptive options, as well as the appearance of random changes in the phenotype.

3.4 The basic rule of conservation genetics

It would appear that there is no safe amount of inbreeding for normally outbred organisms. And even changes in level of heterozygosity, such as occur normally in natural populations, might cost populations a certain amount of fitness. Of course, all species cannot be maintained in a completely outbred state, but it behoves us to minimize inbreeding and the loss of genetic variation. Notwithstanding this, inbreeding at low intensities is quite common in nature, as well as in domestic stock, so, *a priori*, it would appear that deleterious genes can be eliminated by natural selection before being fixed, given that the rate of inbreeding is low enough. Theory and experience tell us that the smaller the population, the more difficult is selection's task in the weeding out of such genes; the reason is that selection is no match for genetic drift at very small population sizes.

Is there, then, a threshold rate of inbreeding, above which fitness relentlessly declines, and below which fitness can be maintained? The answer is a qualified 'yes'. It is a qualified response because no two populations are alike: those with a large load of deleterious genes will tolerate less inbreeding than those with a lower genetic load. Incidentally, species with high levels of heterozygosity may have relatively high genetic loads (Soulé and J. W. Senner, in preparation) so there is some hope of making rough predictions of the inbreeding tolerance of different species or stocks.

The basis for our 'yes' answer regarding a maximal rate of inbreeding is an empirical rule of thumb used by animal breeders. The rule is that natural selection for performance and fertility can balance inbreeding depression if ΔF per generation is no more than about 1% (Franklin, 1980) or 2–3% (Dickerson *et al.*, 1954; Stephenson, Wyatt and Nordskog, 1953). If ΔF values are higher than this, natural selection is unable to offset the tendency for the fixation of deleterious recessives. Apparent exceptions are very rare. We prefer the lower (1%), more conservative value, for two reasons. First, domestic animals from which the rule is derived are already inbred to some degree and therefore have less genetic load than their wild progenitors. As any genetics student knows, it is virtually impossible thoroughly to eliminate a complete recessive from a population, even if it is lethal as a homozygote; nevertheless, such genes can be driven to relatively low frequencies, and in small populations, genes at low frequencies tend to be lost.

Second, domestic animals can tolerate more random phenotypic change than can stocks destined for reintroduction into unmanaged habitats. Breeders of domestic animals may safely ignore genetic and phenotypic changes in their organisms that would completely debilitate a species that had to cope with predators, inclement weather and other conditions occurring in nature. Egg ranchers for example, do not mind if their animals have genes that cause a poor body configuration, poor feather distribution, myopia, or poor predator escape behaviour. These sorts of genes often accumulate in a stock during the course of inbreeding and artificial selection.

We refer to the 1% rule as the *basic rule of conservation genetics* because it serves as the basis for calculating the irreducible minimum population size consistent with short-term preservation of fitness. An even more stringent criterion for conservation genetics is given in Chapter 4 in the context of long-term protection of adaptive potential.

How does the basic rule translate into population size? Fortunately, the relationship between ΔF and N_e is a simple one. The per generation rate of loss of heterozygosity is related to population size as given by Wright's (1931) expression

$$\Delta F = \frac{1}{2N_e} = \frac{1}{8N_m} + \frac{1}{8N_f} \tag{3.7}$$

so that the effective size must be fifty or more if the loss of heterozygosity (or increase in homozygosity) is not to surpass 1%.

As discussed in section 3.1, a census size = 50 individuals does not always mean an effective size of fifty. To take only one example, the sexes of the reproducing adults, if not equal, will require a larger census if our goal is $N_e \geqslant 50$. This is illustrated in Fig. 3.15. Note that the *minimum* number for the least numerous sex is fifteen. In other words, if a herd has less than fifteen breeding males (and between seventy and eighty females), ΔF will be above 1%, which is in the danger zone according to the rule. If a slightly more

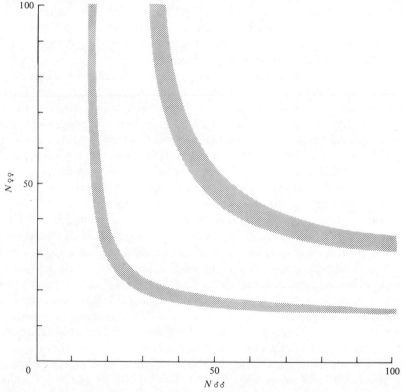

Fig. 3.15 The numbers of breeding males and breeding females needed to satisfy the 1% or $N_e = 50$ criterion (the lower left shaded region) and the 0.5% or $N_e = 100$ criterion (the upper right shaded region).

conservative criterion is employed, say ΔF of 0.5%, the equivalent minimum numbers are thirty males and ninety females. Similar effects result from population fluctuations and progeny distributions when $V_k > \bar{k}$. In the following chapters we shall apply the basic rule to various hypothetical and real conservation problems.

3.5 Mutation–selection equilibria

One final point. We have chosen to discuss the mutational origin and replacement of variation in the following chapter, so only a brief comment is made here. One might hope that the alleles that are lost in small populations could be regenerated by mutation. Such a sanguine expectation is untenable. Wright (1937) discussed this problem while examining the effects of joint pressures on the distribution of gene frequencies at single loci. The essential conclusion is that selection pressure is relatively powerless to prevent loss or fixation of alternative alleles (including new mutations) in a small population. Not only will small populations rapidly lose genetic variation, but *the beneficial alleles have roughly the same probability of being fixed as do the deleterious ones.*

3.6 Summary

1. A population bottleneck, here defined as a single generation event during which a population is severely reduced in size, has the following effects on genetic variability:
   (a) The loss of genetic variation or heterozygosity is not severe; even two individuals retain 75% of a population's genetic variance;
   (b) In contrast, the loss of alleles, especially those at low frequencies, is serious. If these alleles are ever important for survival, such as during an epidemic, then their attrition would increase the expectation of extinction;
   (c) The total amount of genetic variation that is lost following a bottleneck depends on how fast the population grows to a moderate (several hundred or more) size.
2. Chronically small population size produces random gene frequency changes and fixation or loss of alleles; such random fluctuations are referred to as genetic drift. Small populations continually 'leak' alleles and genetic variance, and a few generations of genetic drift is much more erosive of genetic variation than a bottleneck followed by a rapid recovery of numbers.
3. The impact of genetic drift is directly related to the effective population size, not the census number of individuals. The effective size, N_e, is extremely sensitive to unbalanced sex ratio among breeding adults. In

a polygynous system in which the females are monogamous (e.g. Pere David's deer, hamadryas baboon, elephant seal, zebra) N_e can be an order of magnitude smaller than the census number.

4. Large fluctuations in population size can severely depress N_e. The reason is that N_e is most strongly influenced by sampling error (drift) occurring during the 'crash' part of the sequence of generations. Mathematically, N_e is the harmonic mean, not the arithmetic mean, of the numbers in each generation.

5. If for any reason the reproductive output of some families is especially great, and/or that of other families is especially poor, N_e will be less than it would be if the number of progeny were randomly distributed among families. The converse is also true: as the variance of progeny per family approaches zero, N_e approaches twice the actual number of breeding adults.

6. It is incumbent on managers to be aware of these effects and to maximize N_e to the extent consistent with other management programme criteria.

7. The central problem of conservation genetics is the relationship between change in genetic variation and fitness. In natural populations, genetic variation can vary in space (between populations) as well as time (within populations). A review of the literature leads us to conclude that fitness within and between *natural* populations is often correlated with measures of heterozygosity. The superior viability of relatively heterozygous individuals is most apparent when sampling different age classes, especially in species with 'structured' populations. There is also some evidence that homozygous individuals are phenotypically more extreme compared to heterozygotes. Among populations of the same species, at least for non-flying vertebrates, there is evidence for a correlation between mean heterozygosity and fitness.

8. Predominantly inbreeding plants span virtually the entire range of heterozygosity levels observed in outcrossing species, though a relatively high proportion of them have no detectable allelic variation. In those populations in which heterozygosity is retained there is some prima-facie evidence for heterozygote superiority.

9. The existence of some populations of plants and animals lacking detectable genetic variation means that survival is possible without heterozygosity, at least under some circumstances. Since such populations cannot evolve, however, their extinction probability is very high.

10. There is some evidence for the asymptotic model of genetic variation and fitness, suggesting that the more genetic variation is lost, the more deleterious the losses become.

11. Inbreeding always reduces fitness in animals; the decline for reproduc-

tive traits (inbreeding depression) resulting from a 10% increase in the inbreeding coefficient is usually between 5% and 10%. For total reproductive performance, the decline may be two to five times this high. Behaviour traits and competitive ability are also depressed by inbreeding.

12. Intense inbreeding results in the loss of 90% or more of lines unless the stock has a long history of domestication or slow inbreeding.

13. The basic rule of conservation genetics, based on the experience of animal breeders, is that the maximum tolerable rate of inbreeding is 1%. This translates into an effective population size of fifty.

4

Evolutionary genetics and conservation

4.1 Introduction: theories of evolutionary change

In this chapter we discuss some of the genetic topics relevant to evolutionary change in natural populations. The reason for this discussion is our opinion that a conservation programme is inadequate if it fails to provide for continuing or future evolution of gene pools in natural habitats. The justification for our position is presented in chapter 1. There we distinguish between programmes which are 'preservationist' versus those that are 'conservationist'. While this distinction has not been sustained in other chapters, we emphasize its importance here because of the critical role of evolution in long-term survival. Awareness of this distinction by the public and by scientific consultants will force planners and policy makers to face up to and acknowledge the consequences of their decisions. Only if planners fulfill the minimum conditions necessary for evolution is the designation 'conservation' justified.

In planning for evolution it is necessary to understand the mechanisms of evolution. Contemporary evolutionists accept two major classes of long-term evolutionary change: phyletic evolution; i.e. gradual change within evolutionary lines, particularly in response to secular trends and to environmental heterogeneity in space; episodic evolutionary change giving rise to new phyletic lines. The latter depends on physical or genetic isolation and can culminate in the origin of a new species.

The data and ideas relevant to the establishment of the minimal conditions allowing continuing evolution, particularly phyletic evolution, are the subject of the rest of this chapter. The conditions necessary for speciation are treated very briefly because this subject has recently been examined elsewhere (Soulé, 1980). We do not pretend that the following treatment is exhaustive or that our conclusions are definitive. As in chapter 3 our goal is more modest. It is to provide a scientifically defensible basis for certain management decisions. In effect, this usually means employing a rough approximation of some critical parameter instead of waiting for the precise theory or information that may never arrive.

Such an approach really needs no defence. Conservationists cannot afford the luxury of methodological elegance. We are soldiers in a war and soldiers must be pragmatists. Thus it is our tenet that crude initiatives based on rough guidelines are better than the paralysis of procrastination induced in some scientists by the fear of inadequate data. To delay the implementation of conservation and management programmes until we have a definitive understanding of all the complexities of the evolutionary process is analogous to allowing cancer to go untreated until we can prevent it altogether.

More to the point, conservationists should appreciate the minimal criteria for evolution and should evaluate conservation projects with these in mind. If a programme is implemented that clearly violates these criteria, then it should be done with full comprehension of the probable consequences for long-term survival.

Finally, we also explicitly disclaim both our ability and the need to rank in importance the two classes of evolution, or even to defend this particular dichotomy. It may take decades for evolutionary theorists to arrive at a consensus about these issues, and by then it will be too late for many species; no one has yet enforced a moratorium on human population growth or on habitat destruction. Therefore we have no choice but to sidestep the arcane and emphasize the utilitarian. Besides, these two evolutionary modes blanket most of the continuum of evolutionary processes. So, by providing for these processes in most species, we probably provide for most other processes, whether we understand them or not.

4.2 Conservation biology and evolutionary genetics

4.2.1 Genetic variation and the rate of evolution

Evolutionary adaptation to a changing biological or physical environment requires, among other things, a sufficient store of genetic variation. This axiom is one of the twin pillars of evolutionary theory; the other is the concept of natural selection. The issue that most concerns conservationists is that of sufficiency: does variation ever become limiting? Are there, in other words, situations where the rate of evolution is less than optimal because variation is less than plentiful? As in the treatment of inbreeding depression, we will discuss this problem at three levels, qualitative, quantitative and practical. We begin by asking whether genetic variation is ever a limiting factor in the rate or precision of evolutionary response.

Until recently the conventional wisdom was that '. . . all natural populations contain abundant genetic variation as potential raw material for evolutionary change' (Mayr, 1963, p. 201). (Obviously Mayr did not intend to include endangered species of large organisms in his generalization, but such statements are sometimes taken literally.) This 'cornucopian' assumption was expressed as recently as 1972 by no less an authority than

Dobzhansky (1970, p. 201): 'That selection can work only with raw materials arisen ultimately by mutation is manifestly true. But it is also true that populations, particularly those of diploid outbreeding species, have stored in them a profusion of genetic variability.'* At other places in his work (e.g. p. 86), Dobzhansky pointed out that adaptation is sometimes less than optimal due to a limited supply of genetic variation, at least in laboratory populations, but the cornucopian orthodoxy still persists and flavours our thinking. Unfortunately, nearly all of our information about the genetics of natural populations came from *Drosophila*; most species have very large population sizes and this is probably a necessary condition for the maintenance of high levels of genetic variation (Soulé, 1976).

Even for *Drosophila*, however, there is evidence that small populations may respond sluggishly to natural selection. Ayala demonstrated that evolutionary rates in lines of *Drosophila* were relatively slow unless their genetic variation was increased either by mutation-inducing radiation (Ayala, 1969) or by crossing with genetically different strains (Ayala, 1965) as shown in Fig. 4.1. (Carson (1964) and Benado, Ayala and Green (1976) failed to obtain results like those of Ayala when using intermittent X-irradiation and mutator genes, respectively, suggesting that a rather delicate balance may exist between mutation and fitness: too little mutation lowers fitness by decreasing evolutionary rates, while too much mutation lowers fitness by increasing load. The position of these thresholds will depend on the intensity of selection and population size.)

Ayala's results have important implication in conservation genetics. First, they demonstrate that the rate of evolution may depend on the amount of available genetic variation. Second, Ayala's results demonstrate the benefits that might accrue from crossing distantly related populations, a subject discussed in chapter 6.

The beneficial effects of induced mutations reported by Ayala contrast, but do not conflict, with the failure of many workers to detect a clearcut heterotic effect of new mutations. For example, Lewontin (1974) reviews several such attempts to demonstrate that the fitness of inbred lines of *Drosophila* could be enhanced by 'instant' overdominance created by X-irradiation. The difference between these results and those of Ayala is

---

* The cornucopian assumption would be a worthy and fascinating study in the history of science. It would appear very likely that Darwinian evolutionary geneticists developed this view as a reaction to the mutationist doctrine that dominated evolutionary thought until the late 1930s and early 1940s. In their attempt to discredit the importance of mutation pressure as a driving force in evolution, the neo-Darwinists needed to show that natural populations contain rich reserves of heterozygosity, and that the 'wild type' ideal of the classical school (i.e. the assumption that virtually all loci are monomorphic for the best allele) was incorrect. In retrospect, it now appears that the neo-Darwinists went too far; the truth, as usual, lies somewhere between the famous extremes.

that Ayala's flies were adapting to novel and stressful laboratory conditions, and there was undoubtedly strong competition among them. The increase in fitness that Ayala demonstrated was the result of selection for new gene combinations, probably additive. In all likelihood, heterosis had little or nothing to contribute.

Obviously, more experimentation along these lines is needed. In the meantime, however, we will assume the validity of Ayala's results, (1) because population genetic theory predicts them (section 4.2.4), and (2) because a deficiency of genetic variation in small populations is supported by the rough but significant correlation of population size and variation discussed in the following section.

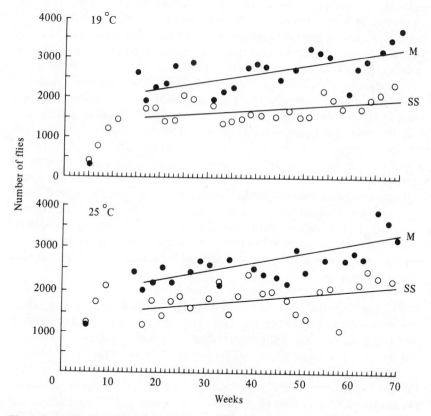

Fig. 4.1 Change in population size due to natural selection in experimental populations of *Drosophila serrata* at two temperatures. (SS, single-strain population; M, Mixed population.) The average rate of increase in population size is about twice as large in the mixed populations (which have more genetic variation) than in the single-strain populations. After Ayala (1965).

4.2.2 Genetic variation in natural populations

A less direct approach to the question of the possible limiting role of genetic variation is to survey variability in natural populations, and explicitly to assume that the estimate of variation so obtained provides information on the 'useful' (for evolution) genetic variation. The recent plethora of surveys of biochemical (enzymes and other proteins) variants in natural populations allow us to make statements about 'raw' allelic diversity, but we are on less firm ground when assuming a correlation between observed variability at the gene product level and the genetic variability (V_G) that underlies the selectable fraction of quantitative variation for morphological, physiological and behavioural traits. We return to this problem in section 4.2.4.

In the preceding chapter (section 3.2.2) we briefly summarized the available data on genetic variation in natural populations for various taxa. One generalization from such summaries that has stood the test of time is that higher vertebrates, especially those that are terrestrial breeders (this excludes fish and many amphibians), are relatively depauperate for electrophoretic variation (see Table 3.5). Even granting that our present techniques underestimate the actual amount of allelic variation (Coyne, 1976; Singh, Lewontin and Felton, 1976; Johnson, 1976), it is still apparent that many reptiles, birds and mammals have less than half the heterozygosity of invertebrates and less than one-quarter that of abundant and widely distributed molluscs, echinoderms and many small insects.

These conclusions should, at the very least, give us some pause in adopting the orthodox or cornucopian view of variation as a cornerstone for conservation genetics. If variation can be a rate-limiting factor for laboratory populations of *Drosophila*, then it must follow that remnant populations of large organisms are often severely handicapped. Section 4.2.3 explores this possibility in some detail.

The cause and effect relationships involving the different amounts of heterozygosity in different populations are very controversial. Many authors prefer to account for the heterogeneity in heterozygosity values in terms of ecological determinism. Differences between taxa or populations have been attributed to (1) the width of the 'ecological niche' or degree of specialization (Nevo, 1978), (2) the ecological amplitude of the population (e.g. Levins, 1968; Selander and Kaufman, 1973b), or (3) the trophic resources (Valentine, 1976). Others (Soulé, 1976; Siebenaller, 1978) are impressed by the correlation between population size and heterozygosity (Fig. 4.2), arguing that the ecological correlations are largely spurious and that the quantity of observable variation is more directly dependent on certain rate processes, especially genetic drift. Section 3.2.2 presents these ideas in greater detail. Luckily, conservationists need not wait for a resolution of this

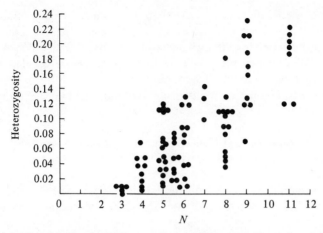

Fig. 4.2 The relationship between mean heterozygosity and rough estimates of population size for many species of animals. Soulé (1976).

controversy. What really matters is the amount of useful variation, not how it came to be there.

4.2.3 *Phyletic optimism and the cornucopian model of variation*

As shown in Table 4.1, bountiful genetic variation is a necessary assumption of a general attitude or thesis we call 'phyletic optimism' – the belief that natural populations have large reserves of genetic variation, and given the need for rapid adaptation to a changing environment, these populations can adapt to all but the most catastrophic challenges by continuous change, that is, by phyletic evolution. Below we review the evidence in favour of phyletic

TABLE 4.1 *Two theories and their implications for conservation genetics (see text)*

heory	Model of variation	Is variation always sufficient?	Major mechanism producing evolutionary novelty	Implications for conservation biology
hyletic optimism	Cornucopian orthodoxy	Yes	Phyletic evolution	Large populations unnecessary for phyletic evolution
hyletic empiricism	Depends on population structure, history	No	Speciation	Large populations necessary for phyletic evolution

optimism, and then attempt to show that the optimist's position, while certainly more applicable to some kinds of organisms than to others, is definitely inappropriate for large plants and animals.

R. J. Berry (1971) presented the case for phyletic optimism. The main points of his argument are easily summarized:

(1) mutation and recombination provide abundant genetic variation and supply the needed genetic tools to cope with environmental challenges;

(2) recent electrophoretic studies of protein variation in natural populations document the existence of abundant genic variation in most species;

(3) natural selection is a powerful force and can override the effects of random genetic changes (genetic drift) and of immigration (gene flow);

(4) adaptation to environmental change is rapid and precise.

The evidence for point (1) is the evolution of resistance, e.g. the marvellous capacity of bacteria to evolve resistance to antibiotics, of insects to pesticides, and of rodents to warfarin. While it is technically incorrect to say that such evolutionary responses are typically dependent on the origin of new mutations or recombinants (where investigated (e.g. McKenzie and Parsons, 1974) the successful genotypes have been found to exist in the wild population all along, albeit at low frequencies), there is certainly no questioning the rapid evolutionary responses that are possible in these organisms when challenged by a single toxic agent.

The evidence for point (2) was reviewed in section 4.2.2. The phyletic optimist sees the results of electrophoretic surveys as supporting the assumption of abundant variation. That is, the observed heterozygosity estimates are prima-facie evidence for cornucopian variation in all organisms.

The evidence for point (3), the power of selection, was already alluded to in the discussion of resistance. Many examples of natural selection could be added, including the rapid evolution of house sparrow (*Passer domesticus*) races in North America (Johnson and Selander, 1964), the subjugation of migration by selection in island populations of water snakes (Camin and Ehrlich, 1958) and the tolerance to heavy metals in plants growing on mine dumps (Antonovics, Bradshaw and Turner, 1971). Point (4) merely summarizes and underlines the first three points.

If phyletic optimism were a universal description of evolution it could warrant complacency about the survival of the earth's large animals and plants, assuming the continued existence of their gross habitats. Unfortunately, this perception of the adaptive process in evolution does not appear to fit all species. One way of explaining this is to consider the 'demographic continuum', the spectrum of intrinsic growth rates, r, of populations.

At the 'high r' end of the continuum are species that can increase in numbers with extraordinary rapidity under favourable conditions. Short generation time and high fecundity are the defining characteristics of these 'bang or bust' species. Defining the opposite end of the demographic spectrum are the so-called 'low r' species; they have the opposite qualities: long generation time, low fecundity, relatively less energy channelled into reproduction and dispersal.

The most obvious correlate of a species' position on the demographic continuum is body size (Pianka, 1970), the larger the size, the lower the population growth rate (Fig. 4.3). The task of conservation would be simple if all target species were small bodied, high r forms such as bacteria, *Drosophila* and house mice. Alas, they are not. To illustrate the problem, let us contrast the response of a high r species and a low r species to a particular evolutionary challenge.

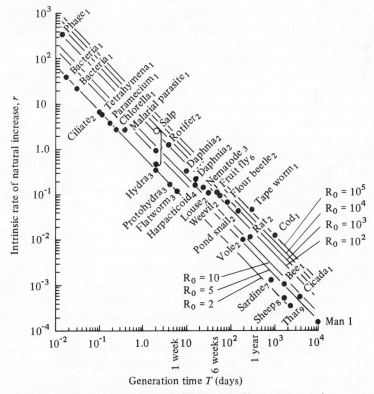

Fig. 4.3 The relationship between the intrinsic rate of increase r, per day, generation time, and reproduction per generation R_0, for a variety of organisms. The parallel lines in the figure represent values of R_0 from 2 to 10^5. Note that absolute body size increases from the upper left to the lower right. From Heron (1972) after Smith (1954) and various sources.

In this scenario we invoke a terrible disaster for the species in question. The application of DDT is such an event for mosquitoes. From the mosquito's viewpoint DDT is a radical and inescapable form of habitat alteration. The result is the dramatic reduction or annihilation of most populations. Between sprayings the populations may recover somewhat (due to the high fecundity and short generation time of mosquitoes) but periodic massacres may repeatedly eliminate all but a tiny fraction of the individuals.

More often than not, however, the mosquitoes survive, at least in small numbers, for the simple reason that there existed (among the original millions or billions of individuals) a small number of variants resistant to DDT. (In a population of 10 million there will be at any given time on the average from 10 to 100 individuals bearing every possible kind of single gene mutation, assuming typical mutation rates.) Even if local populations are wiped out, the newly vacant habitat will be quickly recolonized by surviving resistant mosquitoes from neighbouring regions. Eventually the gene combinations conferring resistance to DDT will come to dominate the mosquito population and the holocaust will have passed.

The reader will have already appreciated the inappropriateness of this scenario if we substitute humpback whale, *Sequoia*, condor, or rhinoceros for the word mosquito. Large, slowly reproducing organisms could not survive even a single knockdown of 99.99%. If some managed fortuitously to escape the first massacre, (1) they might be unable to find a mate, (2) they couldn't survive a second knockdown because the population would not have rebounded in the interval, (3) they couldn't colonize remote unoccupied sites and (4) evolutionary adaptation would be unlikely because there were not enough individuals in the population to begin with to expect the existence of rare mutants conferring resistance in the pre-catastrophe population.

In other words, the four points discussed on p. 84 justifying phyletic optimism are not applicable to large organisms. Point (1) assumes constant availability of genetic variants pre-adapted to many unprecedented environmental stresses. Certainly, populations of microorganisms contain such mutation warehouses; their populations number in the billions and relatively simple metabolic and regulatory mutations (Wilson, 1976) adapt bacteria to an incredible array of habitats. For higher organisms, however, the adaptation process is much more complex, often involving selection for modifier genes, as well as difficult compromises forced by pleiotropic effects, linkage, etc.

Even though small organisms have tremendous advantages in evolution, it is worth emphasizing that local populations of invertebrates often do go extinct, apparently because of a failure to adapt to rapidly changing conditions. Kettlewell (1957) notes the extinction of a local population of a moth during the early stages of the industrial revolution. Ehrlich (1980) cites further examples in the butterfly *Euphydryas editha*. Hence, even insects cannot always respond appropriately to environmental stress, in part because few species and fewer local populations number in the millions.

Point (2) concerns the rich variety of genotypes found in local populations. Indeed, if it could be shown that all species have large stores of accessible genetic variation, and that this variation could be dealt out to offspring in limitless combinations, then species would maintain high levels of fitness under almost any imaginable conditions. In fact, evolutionary and genetic flexibility may not be as great as is often suggested. The assumption of virtually limitless variability may be mistaken, particularly as it applies to allelic diversity. As has already been shown, it is misleading to extrapolate the levels of polymorphism and heterozygosity of insects to large organisms (Table 3.5). The condition of relatively low variability is further aggravated in rare or endangered species; variation in them is likely to be even more seriously depleted owing to the very small numbers in the remnant populations.

Evidence for such depletion in small isolated populations of vertebrates comes from electrophoretic studies of lizards on continental islands. Soulé (1980) showed that many such populations have little or no detectable allelic variation, a condition likely to befall any chronically small population.

Some classes of genetic variability may be antithetical to evolutionary flexibility. For example, many *Drosophila* species are richly polymorphic for inverted sections of their chromosomes. This has led some writers to conclude that all species are provided with copious variability. Such an extrapolation is illogical. Inversions are protected from meiotic recombination. These gene complexes have adaptive significance (Dobzhansky, 1970), yet rather paradoxically, populations with many inversions may have less evolutionary flexibility than those with few or none. This is because the inversions prevent the genes from freely recombining and thus limit the potential number of genotypes available for selection. Carson (1955) stated this long ago, though empirical confirmation is still lacking.

The third point of the phyletic optimist's thesis is quite controversial. It is the hypothesis that natural selection virtually always can overcome the randomizing or counter-adaptive effects of gene flow (migration) and genetic drift. As evidence for the former, ecotypic adaptation in the face of considerable gene flow is cited (Ehrlich and Raven, 1969; Endler, 1977). Others hold to the opposite position that gene flow is a cohesive force (Mayr, 1970, p. 300) preventing local populations from closely adapting to differences in the environment and from fragmenting into distinct local races. Perhaps the best evidence for this position is the correlation between the number of recognized races or subspecies, on the one hand, and the dispersal ability of species on the other (White, 1959; Berndt and Sternberg, 1969; Scheltema, 1971). For example, fossorial mammals with low vagility can be split into dozens of races (Mayr, 1963, p. 396) whereas rodents with excellent dispersal abilities such as the house mouse have much less geographic variation. On the other hand, there are examples of species which are morphologically undifferentiated, even though they occur in different habitats and have poor dispersal abilities, evidence that morphological conservatism may be an intrinsic property of populations (Ehrlich and Raven, 1969).

Whereas the debate continues about the significance of natural gene flow (Jackson and Pounds, 1979), there are cases of unnatural gene flow which demonstrate that selection is not an unbeatable superpower among the evolutionary forces. For example, bobwhite quail *Colinus virginianus* derived from southern races were introduced in the northern states to provide more birds for sportsmen. Instead of helping the quail population grow, this genetic injection actually caused a decline because the hybrid quail were less successful in surviving the northern winters (Clarke, 1954).

Extreme examples aside, the migration or gene flow issue is one of the most important unresolved problems in population biology. Its solution will go a long way in settling another major problem – the explanation for the apparent inertia of gene pools. The observation that gene pools tend to resist change has been made by many systematists. In the words of Ehrlich and Raven (1969), 'The most basic forces involved in the differentiation of populations may be antagonistic selective strategies, one for close "tracking" of the environment and one for maintaining "coadapted" genetic combinations – combinations which have high average fitness in environments which are inevitably variable through time.' Mayr (1970, p. 300) expresses the same belief in the conservatism of genetic systems: 'It is a limited number of highly successful epigenetic systems and homeostatic devices which is responsible [in addition to gene flow] for the severe restraints on genetic and phenotypic change displayed by every species.'

In conservation practice the whole issue of gene flow versus selection may be moot. We don't wish to belittle the significance of the debate over gene flow, but gene flow is in most cases irrelevant when dealing with the capacity of an endangered species to respond to novel selection pressures. The reason is that endangered species are usually isolated remnants of heretofore widely distributed forms. Isolates, by definition, receive no immigration. The exception, an option open to managers of remnant stocks, is that of mixing gene pools from different races or remote areas. Further discussion of this tactic is presented in chapters 5 and 6.

The other issue, the relative strengths of selection and drift, is far from moot, however. As pointed out in chapter 3, p. 36, the impact of genetic drift is proportional to $2N_e$ per generation. For example, consider an isolated troop of twenty-five apes including fifteen breeding adults, in which the gene frequencies at a locus are $p = q = 0.5$ for alleles A and a, respectively. The expected standard deviation due to genetic drift for allele a in one generation is

$$\sigma_{\Delta q} = \sqrt{\frac{pq}{2N_e}} = \sqrt{\frac{0.5 \times 0.5}{30}} = 0.091$$

Thus, we can expect that the value of q will increase to 0.591 or higher or decrease to 0.409 or lower in 32 out of 100 cases in a single generation.

Now, the expected change in the frequency of a due to selection against the homozygote aa is

$$\Delta q = -sq^2 (1 - q)$$

Assuming as above that $p = q = 0.5$, and that the strength of selection (s) against aa is 0.10 (i.e. aa is 90% as fit as the other genotypes), it can be seen that $\Delta q = -0.012$, or about an order of magnitude *less* than the change likely as a consequence of genetic drift. Thus, selection has to be very strong indeed to have a noticeable impact in small populations.

When $2N_e$ is a large number, say > 500 or 1000, the effect of drift will be negligible compared to selection coefficients of measureable intensity. When $2N_e$ is small, say < 100, the randomization of gene frequency changes between generations will not only fix many loci, it will also counteract all but the strongest deterministic forces, particularly directional selection. At a sufficiently small size, genetic drift will begin to extract an appreciable tax on variation, possibly sucking the population into a vicious cycle in which the gradual loss of fitness from (1) decreasing heterosis, (2) fixation of deleterious genes, or (3) the fixation of inferior gene combinations reduces the population fitness and size, which in turn causes more genes to be fixed, reducing fitness further, and so on. The conservation biologist is perforce operating with relatively small populations; he courts disaster when pretending that such populations have the same evolutionary plasticity and potential as do larger ones.

In this section we have pointed out that the arguments used to support phyletic optimism – the position that adaptation of species to environmental change is rapid and precise – are not applicable to many if not most large organisms. The reasons are that:

- large organisms don't have huge populations;
- large organisms cannot withstand repeated catastrophic reductions in population size because they lack the ability to repopulate an area quickly (low r);
- compared to insects and marine invertebrates, most large terrestrial organisms lack efficient dispersal mechanisms that are necessary to occupy empty habitat patches;

- extinction of local populations, even in species with high population densities and growth rates, may be more frequent than conventionally believed;
- genetic variation is less in large animals than in small ones;
- genetic variation is severely depleted in chronically small populations due to genetic drift and inbreeding;
- for higher organisms the capacity to respond to directional selection may be opposed by intrinsic integration of the gene pool and genome, as well as by gene flow between populations.

4.2.4 Phyletic empiricism and the structural model of variation

The purpose of this section is to provide a recommendation for the minimal number of individuals necessary to maintain an evolutionarily viable population, that is, the criterion for long-term fitness. As shown in Table 4.1 the assumption of the 'empirical' approach adopted in this section is that the amount of variation depends on the genetic *structure* of the particular population. In other words, the factors that determine the level of useful genetic variation include the breeding system (e.g. whether outbreeding or predominantly inbreeding), the population size, rate of immigration and the history of the population, especially the frequency and severity of bottlenecks.

In practice it usually will be impossible to know with any certainty the values of these variables for a particular population or species. This leaves us with two options. Either we use some rule of thumb (analogous to the basic rule in chapter 3) or we monitor the population for its genetic variability and somehow base our criteria on the idiosyncratic estimates of genetic variation in each of the species being managed.

The first approach has been attempted by Franklin (1980). He sidesteps theory entirely, arguing, as we do, that the theory of natural selection and genetic variation in finite populations is highly complex and insufficiently mature to serve as a basis for critical application. In order to employ these theories and models it would be necessary to make assumptions about mutation rates, selection coefficients at individual loci, and to decide whether selection is multiplicative or is best approximated by some other model such as the total number of heterozygous loci per individual. It would also be necessary to assign a mode of selection, choosing among such possibilities as dominance, overdominance, frequency-dependence, or even no selective difference among the alleles at a given locus. The last mode, neutralism, is attractive due to its simplicity, but employing it traps one in a logical paradox: the conservation of neutral 'variation' makes absolutely no sense, since only manifest variation produces selectable phenotypic heterogeneity.

Franklin argues that a minimum effective size of 500 is needed to preserve useful genetic variation. His main points follow:

1. The relevant phenotypic traits in conservation biology are quantitative (polygenic). For such traits, the average effect of a gene is small and most of the genetic variation is additive;
2. Weak directional or stabilizing selection does not erode additive genetic variation at a significant rate;
3. The significant evolutionary forces, therefore, are mutation and genetic drift. That is, if a population is below some threshold size, it loses variation by drift at a faster rate than it gains variation by mutation.

Franklin derives his number from the work of Lande (1976) on bristle number variation in *Drosophila*. The evidence is meagre, but Franklin believes his number (500) is about the right order of magnitude. Simple theory also yields this number: in the previous section $N_e = 500$ was mentioned as the lower threshold when considering the joint effects of moderate selection and genetic drift.

In chapter 5 we discuss the application of Franklin's number. Here, we merely caution that the employment of this number is subject to all the qualifications for the use of the short-term number (sections 3.1 and 3.4), namely that an effective size of 500 translates into a much larger number of breeding adults when dealing with real, not ideal, populations.

The alternative to the 'rule of thumb approach' in evolutionary conservation biology is continually to monitor the genetic variation in target species, and adjust the effective size in accordance with the observed rate of change in the level of variation. For most species the impracticality of this suggestion is obvious, at least for the immediate future (i.e. until electrophoretic screening is completely automated, and until tissue samples can be taken efficiently and non-injuriously). Nevertheless, such a management scheme is in the realm of possibility, and it might be useful to begin experimenting immediately using a species with a short generation time.

Such a scheme, however, tacitly assumes a principle we call 'genetic uniformitarianism' (J. W. Senner and Soulé, unpublished) – the idea that genetic variation of all kinds (electrophoretic, quantitative, load of deleterious mutants) is subject to the same constraints based on population structure and history. If this assumption is more or less correct, the relative levels of variation in different sets of genes should be correlated, and every population should have a parametric level of variation (Soulé, 1972) which could be estimated by any reasonably large sample of loci, regardless of kind. Such a theory is supported if variation in quantitative morphological traits is correlated with average variation (e.g. heterozygosity) in a set of Mendelian traits. The problem is the choice of morphological characters, because traits for which the environmental component of variation is large are, *a priori*, inappropriate for estimating the characteristic (parametric) level of variation for the population. Elsewhere (Soulé, 1981) it is argued that the best traits would be those with low coefficients of variation, because they would be expected to have high heritabilities. In addition, for animals one would prefer those traits the phenotypes of which are determined before birth, thereby minimizing post-natal environmental effects. It appears that scale counts in reptiles, dermatoglyphics in primates, and certain bristle counts in insects are satisfactory. In any case, evidence for such correlations exists.

Marshall and Allard (1970) have shown a correlation in variation between two sets of Mendelian characters in grasses of the genus *Avena*: one set was electrophoretic, the other morphological. Soulé, Yang, Weiler and Gorman, (1973) demonstrated a correlation between mean heterozygosity (electrophoretic) and morphological variation in scale characteristics in lizards. Patton *et al.* (1975) found a similar trend in variation among populations of rats; they used electrophoretic and morphometric characters. Morris and Kerr (1974) reported a rather trivial variation correlation in macaques between electrophoretic and dermatoglyphic traits. So far this 'genetic-phenetic variation correlation' (Soulé *et al.*, 1973) has survived several tests, although it does not hold up across species with quite different breeding systems (Jain and Marshall, 1967) nor across distantly related genera (Valentine and Ayala, 1976).

If the 'monitoring' approach to evolutionary conservation genetics is valid and useful, the following scenario suggests how it might be implemented. Assume that a programme exists to track changes in genetic variation in a particular species. Baseline studies gave the following electrophoretic estimates: percentage polymorphism, $P = 0.30$; percentage heterozygosity, $H = 0.05$. Ten generations later the population was again assayed and the results were $P = 0.25$; $H = 0.04$, a decrease in heterozygosity of 20%. Alerted to the possibility of a rapid decline in genetic variability, the managers take immediate restorative steps, possibly including an increase in the effective size of the population (by manipulating area, population structure or the distribution of offspring among families) or the introduction of individuals from other populations, if they exist.

In summary, we have argued that the minimum effective number of individuals necessary for gradual evolutionary change is of the order of 500. Below this number, useful genetic variation will leak away, and as genetic variation is lost, so goes evolutionary potential, the best insurance for long-term survival. In addition, we proposed that conservation biologists should begin monitoring the genetic variation in some captive or protected populations. An assumption of this approach is that estimates of genetic variation based on electrophoretic or similar biochemical methods will be significantly correlated with the kind of genetic variation necessary for gradual evolutionary change in quantitative (most physiological and morphological) traits.

4.3 Environment, genetic variation and extinction potential

Do certain kinds of environments dispose a species toward loss of variation and thence to extinction? If such a phenomenon exists, conservation biologists ought to be aware of it. Many workers have argued that organisms that live in stable and homogeneous habitats should be relatively impoverished for genetic variation, and this led to the prediction that the species on the ocean floor below the edge of the continental shelf should be particularly depauperate in heterozygosity (Bretsky and Lorenz, 1970; Manwell and Baker, 1970). Such impoverishment logically would make them particularly vulnerable to extinction. As stated by Bretsky and Lorenz (1970, p. 2449):

A prolonged period of stability could significantly reduce a species' potential for survival in a more rigorous environment, and hence might be expected to precede a time of widespread or mass extinction. A homozygotic genome would be selected for (homoselection) by species subjected to long-term environmental stability; it is this selection that could ultimately prove disastrous for survival in a later, only slightly less stable regime.

As it turned out, the relationship is either non-existent or in the opposite direction to that predicted: deep-sea benthic invertebrates have relatively high heterozygosities (Schopf and Gooch, 1972; Ayala and Valentine, 1974). Clearly, a uniform habitat does not necessarily produce homozygosity, and experiments which have been interpreted according to this model (Powell, 1971; McDonald and Ayala, 1974) can be interpreted in other ways (Soulé, 1973).

The falsification of the above hypothesis still leaves open the possibility that other kinds of environments tend to diminish variability and produce the predicted vulnerability to extinction, but such a hypothesis would be gratuitous, especially in the light of the fact that this idea was posited originally to explain mass extinction in the fossil record, and mass extinctions have been characteristic of groups such as articulate brachiopods, corals, echinoids and bryozoans living in stable and uniform environments.

Above, it was stated that we needed to qualify the conclusion that there are no environments that dispose a species towards the loss of its genetic variation. While uniform environments apparently do not produce such an effect, it is generally accepted that loss of variation will occur in an environment which exerts strong and continuous directional selection i.e. in very stressful environments where only extreme phenotypes survive and reproduce. Such a population is, by definition, poorly adapted, and conservationists are well advised to avoid rescue operations under conditions that are obviously unfavourable.

4.4 Biogeographic and population size criteria for speciation

It appears doubtful that existing nature reserves are large enough to support the process of speciation within them, at least for large vertebrates (Soulé, 1980a). The basis for this conclusion was a survey of vertebrate taxa on islands, looking for the minimum sized island on which there is evidence for autochthonous (*in situ*) speciation for a particular group. The results suggest that small mammals such as rodents have enough room for speciation on Cuba and Luzon (*ca* 110 000 km²), whereas birds or large mammals as large or larger than jackals or vervet monkeys, require an area the size of Madagascar (the Malagasy Republic) (nearly 600 000 km²). Higher plants appear to fall somewhere in between these extremes. Archipelagos and events of multiple invasion were not considered because the barrier to gene flow in such cases is *between* rather than within land masses. It is important to note that all the islands mentioned above are much larger than existing or contemplated national parks.

If speciation of higher vertebrates is unlikely within reserves, there still remains the possibility that it could occur where a form exists as isolated populations in two or more reserves, assuming that gene flow between the isolates is low enough to prevent swamping, and that extinction rates of reserve populations are negligible. Unfortunately, extinction rates appear to be too high (by one or two orders of magnitude) in habitat patches (islands) the size of nature reserves (chapter 5). That is, by the time that isolates have diverged enough to be incipient species, they will have a negligible probability of remaining in existence. Here is the dilemma – the best means now available to prevent extinction of endangered isolates is the introduction of individuals from other isolates (chapter 5), but this is the very thing that will forestall genetic differentiation between isolates.

These conclusions bode ill for the future natural production of evolutionary novelty, assuming that the majority of significant evolutionary innovations are the products of speciation rather than phyletic evolution (Gould and Eldredge, 1977). For large organisms, natural speciation requires too much space, and probably too much time (Soulé, 1980; cf. White, 1978). For the foreseeable future, the best we can hope for is that the currently existing species will continue to survive and adapt to their changing conditions. Thus, by the turn of the century, man must begin to consider how and when to assume the role of species-maker (we will not speculate on the means), because by then the conditions that have promoted speciation in large terrestrial organisms will no longer exist in most of the planet's terrestrial environments.

In spite of our emphasis on the significance of speciation, the role of continuous evolutionary change should not be denigrated. Even if the effects of gradual changes in gene frequencies are relatively trivial in the overall pattern and course of organic evolution, it would be folly for conservation biologists to fail to insist on the minimum numbers of individuals required to insure the continuation of phyletic evolution. Soon it will be 'the only game in town'.

As stewards of the planet's biota, the task of conservation biologists, for the time being, is to prevent extinction rather than to create new life forms. Perhaps in the distant, more rational future, there will again be enough space for both human and non-human life, and for the continuation of natural speciation in most higher organisms. Until then, it is challenge enough to provide the conditions necessary for the maintenance of fitness in a changing environment.

4.5 Summary

1. Because evolutionary response to environmental change is a condition of long-term survival, species must have the capacity for continuous adaptation. Programmes lacking this potential should not be cloaked in the banner of 'conservation', although they might qualify for the 'preservation' label, in the sense that they satisfy the minimal criteria for short-term fitness and survival.
2. Complacency about the survival of large organisms would be justified if genetic variation is rarely a limiting factor in evolutionary adaptation, and if evolutionary rates under continuous directional selection were indefinitely sustainable, regardless of population size – the position of the 'phyletic optimist'. Unfortunately, the latter thesis cannot be universally applied because large organisms have smaller populations, less variability and, perhaps, less responsiveness to selection than do microorganisms, and relatively small eukaryotes such as *Drosophila*.
3. The sufficiency of genetic variation is the major issue in evolutionary conservation genetics. Do small populations adapt more slowly than large populations as a consequence of less variation in the former; that is, is the 'cornucopian' assumption (unlimited variation) of the neo-Darwinists universally applicable? *Drosophila* experiments suggest that the rate of evolution is related to the amount of genetic variation, and that the latter may be depleted in small populations.
4. Data from hundreds of surveys of electrophoretic variation of enzymes and other proteins in natural populations show that species of reptiles, birds and mammals have less than half the genetic variation found in invertebrate species, and less than one-fourth of that found in many widespread marine invertebrates and insects. There are very few data on the genetic variation in natural populations of outcrossing plants.
5. The role of gene flow in the affairs of local populations is controversial.

Some believe that it is of little significance compared to natural selection; others that it inhibits adaptational specificity. In most situations the debate is moot; the reason is that conservation biologists will be dealing with isolated remnants of species, for which natural gene flow is a virtual impossibility.

6. Compared to genetic drift, the force of selection in small populations is trivial. Until populations reach quite large sizes (> 500), single locus selection coefficients of the order of 1% or less are ineffective.

7. Population genetic theory is not applicable yet to the issue of long-term fitness (the minimum effective population size that provides for continuing evolutionary change in quantitative characters) because of the number of assumptions that must be made before deciding among models and even in the application of any given model.

8. Based on the empirical work of Lande (1976), Franklin (1980) argues that a genetically effective size of 500 is a satisfactory first approximation of the minimum size for the accommodation of continuing evolution. At this size, the loss of genetic variation should be approximately balanced by the gain from mutation. The actual number of individuals that satisfy this criterion may be several times greater than 500 for reasons discussed in sections 3.1 and 3.4.

9. An alternative to the above 'rule of thumb' criterion is the monitoring of genetic variation in the target species, for example, by employing electrophoretic techniques. The assumption underlying this approach is 'genetic uniformitarianism' – the correlation of all measures of genetic variation.

10. Existing and planned nature reserves are too small to provide for speciation of higher vertebrates and plants. Even systems of reserves that contain many populations of the same species (analogous to archipelagos) are unlikely to promote speciation because the rates of extinction of populations of large organisms in reserves are too high, and the movement of individuals among reserves in order to prevent extinction will counteract any tendency in the direction of genetic divergence. Man will have to engineer future speciation in large organisms, at least for the next few centuries.

5

Nature reserves

One of the penalties of an
ecological education is that
one lives in a world of wounds.

Aldo Leopold

5.1 Introduction

Perhaps there is no matter so weighty as the preservation of the most complex systems in the known universe – biological communities. Conservationists have charged themselves with the responsibility for the survival of these systems. Their success depends, in no small measure, on their sagacity and humility. In this respect conservation biology should operate differently from normal (adversary) science. Progress in the latter is usually via the vehicle of competition: somebody proposes a theory; somebody else challenges it and proposes an alternative; battle lines are drawn and schools (armies) gather around the protagonists, many of whom cling to their pet theories even as they die. Thus human science evolves by the natural selection of hypotheses.

Conservationists cannot afford the luxury and excitement of adversary science. The weakness of this parochial style of intellectual progress is that years or decades may pass before a clear resolution is reached and before timid technocrats or politicians decide that action will not bring a storm of criticism. By the turn of the century most options will be closed regarding the design, size and organization of nature reserves. It behoves us, therefore, to give serious and humble attention to all points of view and to be willing to compromise. For this reason and to be realistic, we freely admit that genetics is only one among several perspectives on nature reserves. Accordingly, we give attention to other factors including disease, island biogeography, community ecology and management considerations in this chapter. We will take up these disciplines one at a time, and conclude with an attempted synthesis.

5.2 Design criteria for nature reserves

At this point a definition of 'nature reserve' is appropriate and might prevent misunderstandings. A nature reserve is a region set aside for the protection of the aggregate of species contained therein, as well as the supporting physical environment. Putting it another way, the purpose of a nature reserve is to maintain, hopefully for perpetuity, a highly complex set of ecological, genetic, behavioural, evolutionary and physical processes and the coevolved, compatible populations which participate in these processes.

It is implicit in this goal that a nature reserve must be sufficiently large and self-regulating so as to obviate routine husbandry or feeding of the individual organisms. By this definition, for example, the Gir Forest (1300 km²) which protects the only remaining population of the Asiatic lion, is on the borderline between a nature reserve and a safari park; according to Berwick (1976), when the tribal pastoralists and their domestic cattle are removed, it will be necessary to introduce substitute prey, such as feral zebu cattle, as a food source for the lions.

5.2.1 Disease and reserve design

In the not too distant future it may be common for all the individuals of a given species to be confined to a single reserve. The consequences of an epidemic for a species in a small area are obvious. Paraphrasing a report on the Indian rhinoceros given to the Survival Service Commission of the IUCN in 1978, there are about 1100 individuals, nearly all of which are in two reserves: the Royal Chitawan National Park and the Kaziranga National Park. These parks are at capacity and the establishment of others is problematical. There is concern that disease could wipe out the species before viable populations are established elsewhere.

How serious can epidemics be? Do they ever decimate or exterminate all of the individuals of a species in a region the size of a nature reserve? Indeed they do. Perhaps the best known epidemic in wildlife was the famous rinderpest plague that swept Africa at the end of the last century and has continued to produce local outbreaks at frequent intervals (Scott, 1970). Rinderpest is a viral disease of artiodactyls, but rodents, canids, marsupials and even fowl are not immune. In Africa, entire herds were eliminated, and the East African plains were littered with decaying carcasses. Cattle were usually affected first, followed by buffalo, eland and warthogs. Succumbing later were giraffe, kudu, roan antelope, wildebeest, oryx, waterbuck and duikar, to name the prominent species. In some regions virtually all of the buffalo and giraffe were eliminated; kob were exterminated in parts of Uganda in a later outbreak (Pitman, 1942). In order to stop the southward advance of the disease during the 1937–41 epidemic, all the game was killed in a belt 50 miles wide by 167 miles long between Lake Tanzania and Lake Malawi (Lowe, 1942). Fortunately, the disease never got a foothold in the New World.

Among other diseases of mammals that have potential to cause epidemics are myxomatosis, anthrax, hoof-and-mouth disease, yellow fever (among primates), epidemic haemorrhagic disease and brucellosis. Myxomatosis greatly lowered the population density of rabbits in parts of Europe. Until recently, myxomatosis was an endemic disease of South American rabbits (*Sylvilagus*) in which it was not a serious mortality factor. There may well exist many such localized diseases to which the host is relatively immune, but to which related wildlife species on other continents have little or no resistance. This could explain an episode in Australia. A mysterious, but poorly documented disease apparently caused the decimation or virtual extinction of many dasyurid marsupials in eastern Australia about 1900 (Recher, 1972). One of the species that apparently succumbed in New South Wales was the native cat (*Dasyurus quoll*).

Anthrax has caused epidemics in bighorn sheep (*Ovis canadensis*), and it was suggested that it caused the extermination of these sheep in the Bear Paw Mountains of Montana (Grinnell, 1928). Anthrax has recently been accused of causing 54% of the mortality of ten species of large mammals in the Etoshe National Park in Namibia (South-West Africa) (Anon., 1978). Epidemic haemorrhagic disease decimated white-tail deer herds in North America, wiping out over 90% in some management areas (Hayes and Prestwood, 1969). This disease (and the closely related disease, bluetongue) is also a threat to bighorn sheep, elk and other species.

Brucellosis, or spontaneous abortion, is thought to be a serious threat to the marsh deer (*Blastocerus dichotomus*), the largest deer in South America (Schaller and Vasconcelos, 1978); the source of the disease is probably cattle. Schaller and Vasconcelos attribute the near absence of reproduction over the last several years in the state of Mato Grosso, Brazil, to brucellosis and other diseases, a situation aggravated by serious flooding.

Host-specific parasites can also have devastating effect. For example, bighorn sheep are susceptible to lungworm-pneumonia complex (Forrester, 1971), and up to 95% loss has occurred in some herds in British Columbia. A 50% loss is common during an outbreak.

Epidemics occur in vertebrates other than mammals, although with the exception of birds, they are less frequently observed. Two of the most devastating avian diseases are botulism and avian cholera. Each has caused massive die-offs. Botulism is particularly serious in very wet years because flooding encourages the growth of the pathogen, *Clostridium botulinum*. During the rainy winter and spring of 1952, between four and five million ducks died of the disease in the western US (Rosen, 1971a). Water birds are also most likely to suffer from cholera. For example, over 60 000 waterfowl perished during the winter of 1956–7 at the Muleshoe Refuge in Texas (Rosen, 1971b).

For plants, too, there is no room for complacency. In North America, native tree species (chestnuts and elms) have been exterminated locally by introduced fungus diseases from Eurasia (chapter 2, pp. 18–19). It is likely that such events will become more frequent, particularly in the tropics, as natural populations become more completely fragmented and reduced in area.

The potential for disaster is aggravated by the susceptibility of wild plants and animals to most or all of the diseases that can plague domestic species. Farms and villages with tree crops, rabbits, sheep, cattle, swine and fowl will encircle most nature reserves and these stocks will be a perennial source of contagion for the protected species. Where domestic stock are allowed to range within reserves, the danger is greatly enhanced, as the brucellosis and rinderpest epidemics exemplify.

In recent decades, many if not all of the catastrophic animal and plant die-offs have been directly or indirectly a result of human activities, such as the movement of hoofed stock or the importation of the diseased plants (chapter 2, p. 18). Even without man, however, it is highly probable that these epidemics would eventually be triggered by natural events. In any case, man is here to stay, so there is no room for complacency.

Even if a disease fails to exterminate a rare or protected population, it can so devastate its numbers that the resulting population bottleneck will have serious genetic consequences. In chapter 3 we pointed out that episodic bottlenecks have a disproportionate impact on effective population size and thus on genetic variation. Depending on the rate at which the population springs back to pre-catastrophe size, there will be an interval of varying length during which genetic drift will further deplete variation and hobble the capacity of natural selection in maintaining optimal genotypes.

It is folly, then, to keep all the individuals of a species in a single reserve. Unless there are major barriers to gene flow between populations of the largest species in a reserve (as well as those of disease vectors), the possibility of extinction or severe depletion is immanent, given that our time scale is evolutionary or geological. Furthermore, disease control in reserves presents many problems. In undeveloped habitat, quarantine and other control tactics are costly at best and impossible at worst; a *cordon sanitaire* is more easily established between reserves than within them.

We do not mean that the prime directive of conservation should be to maximize the number and isolation of reserves, but if the choice is between a single reserve containing the last vestiges of one or more significant species, versus several (perhaps smaller) isolated reserves, each with representative populations of the desirable forms and appropriate habitat, then common sense and history give the nod to the latter. The future absence of disease is an assumption we can ill afford to make.

5.2.2 Island biogeography and reserve design

The most seminal branch of ecological theory in recent years has been the theory of island biogeography, first developed by Preston (1962) and MacArthur and Wilson (1963, 1967). This theory predicts that island biota tend to approach dynamic equilibria in the number of species. The forces that bring about the equilibria are immigration and extinction. For a particular island, the precise number of species at equilibrium depends mostly on island size and the distance from the source of immigrants, the 'area effect' and the 'distance effect', respectively. According to the theory, once an island is at its equilibrium for a particular taxon (such as birds or lizards), the rate at which species are lost equals the rate at which unrepresented species colonize the island, by definition.

An example of the area effect is shown in Fig. 5.1. The area effect is a firmly established geographic rule (Preston, 1962) and is usually expressed as $S = CA^z$, where S is the number of species, A is area, and C and z are parameters, the values of which depend on the group of organisms. These parameters are usually estimated by plotting sample observations of S for areas of known size on logarithmic axes. Thus z is the slope and C is the logarithmic base raised to the value obtained at the y-intercept. (See Wilcox, 1980, for a more detailed discussion.)

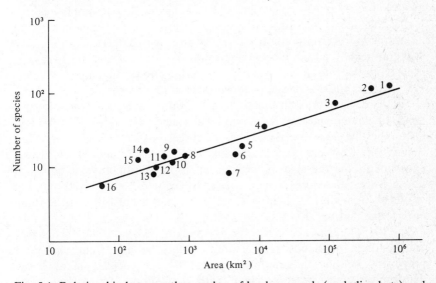

Fig. 5.1 Relationship between the number of land mammals (excluding bats) and island area for the Sunda Islands. (1) Borneo, (2) Sumatra, (3) Java, (4) Banka, (5) Bali, (6) Billiton, (7) Siberut, (8) S. Pagi, (9) Sipora, (10) Singapore, (11) Tanabala, (12) Tanamasa, (13) Pini, (14) Penang, (15) Tuangku, (16) Bangkaru. (After Wilcox, 1980; data from Medway and Wells, 1971 and Chasen, 1940.)

It is an obvious deduction from the species–area relationship that any decrease in the area of an island will lead to a loss of species and the eventual attainment of a new, lower equilibrium. This much is elementary and has been repeatedly documented where a once extensive habitat has been changed at the hand of man, leaving only isolated patches of the original habitat type. Moore and Hooper (1975), Whitcomb, Lynch, Opler and Robbins (1976), Galli, Leck and Forman (1976) and Willis (1980) all censused bird species in isolated woodlands or forest plots. They all found a regular pattern of increasing species diversity with increasing patch size. An interesting artificial experiment to test the equilibrium hypothesis was performed by Simberloff and Wilson (1970). For their islands they used mangrove trees of various sizes. These 'islands' were defaunated by fumigating them under tents (as one would a house) and then observing their recolonization by arthropods. In all cases, the species diversity returned quite rapidly to approximately the number occurring on them before fumigation, and these numbers were close to the predicted values based on area and distance to the source of immigrants.

The crucial issue for conservation, however, is not deducible from theory. This is the question of the *speed* with which biota collapse to lower equilibria once the area is reduced. If the collapse rate is geological in scale, that is, occurring over millions of years, then we need not be too concerned – evolution might produce new species as fast as the old ones are being lost. But if the rate is historical (decades or centuries), then our immediate descendants will live in a noticeably less diverse world.

We are now ready to ask the important questions. There are two:
1. Does the rate of collapse depend on the size of the patch?
2. What is the magnitude of these rates in real time?

These questions might be unanswerable were it not for a fortunate coincidence of climate and the rise of empirical science. Human civilization and technology have developed in the tracks of receding glaciers, and it is one of the aftermaths of this glacial recession that produced hundreds of natural experiments. As the glaciers began to melt about 14 000 years ago, sea-levels around the world began to rise. By around 3000 to 4000 BC, the sea-levels were close to their present heights, and about 140 m higher than before the melt began.

The rate of eustatic sea-level rise was maximal about 10 000 years ago, and hundreds of islands were created around this time from what were previously hilltops and mountain ranges on the coastal plains. These 'landbridge islands' were created with their floras and faunas more or less intact. Examples of some of the large newly isolated pieces of the continental shelves are Britain, Zanzibar, Sri Lanka (Ceylon), Tasmania, Borneo, New Guinea, and Trinidad; there are hundreds of smaller ones.

Theoretically, land-bridge islands should have supersaturated biota upon their creation. This is because a given continental area has more species of a taxon than an oceanic (equilibrated) island of the same size for the reasons stated above: extinctions in a patch of continental habitat are balanced by a high rate of immigration (colonization) from nearby patches, but the immigration rate on an island is so low (compared to what it was before it became an island) that the species diversity collapses to a new, lower equilibrium. The reason that an equilibrium is eventually reached is that as species diversity declines, the rate of extinction also declines (partly because of less competition and partly because of the larger average number of individuals per species) until a point is reached where the severely reduced rate of immigration just balances or compensates for the rate at which species are being lost. By today, of course, most of the 10 000-year-old islands should have lost many species.

The rates of species loss, however, should not all be the same, because larger islands can contain more individuals of any given species, and the per species rate of extinction should decrease as island size increases. Recently, several authors have verified these predictions. Diamond (1972) has shown that land-bridge islands off New Guinea have lost up to 95% of their non-marine, lowland bird species, and that the rate of loss is highly dependent on island size. The largest islands are still somewhat super-saturated compared to oceanic islands of equal size, but even so they have lost about half of their original 325 bird species (Fig. 5.2).

Similar studies in the Caribbean region have produced a parallel description of collapse in avifaunas. Terborgh (1974, 1975) found that many islands have collapsed to an 'oceanic' equilibrium in the last 10 000 years, and he showed (as has Diamond) that the rate of collapse was inversely related to island size.

Land-bridge islands are not the only habitats to have been isolated during the climatic changes following the last glaciation. Brown (1971) studied the small mammal faunas that have been isolated on seventeen mountain ranges rising out of the Great Basin Desert in Utah, Nevada and California. At elevations about 2300 m or so these mountains are cloaked in a pinon-juniper woodland. This habitat supports twelve species of mammals which are unable to disperse between mountain ranges because they are intolerant of the more arid habitats at lower elevations. During the cooler, glacial climate that prevailed prior to 10 000 years ago, the distributions of these twelve species were probably continuous in the desert area, because the pinon-juniper habitat is thought to have extended in a more or less unbroken belt from the Sierra Nevada mountains in the west to the Rocky mountains in the east. As the climate became more arid, the woodlands, including many of their mammal species, began retreating up the mountains.

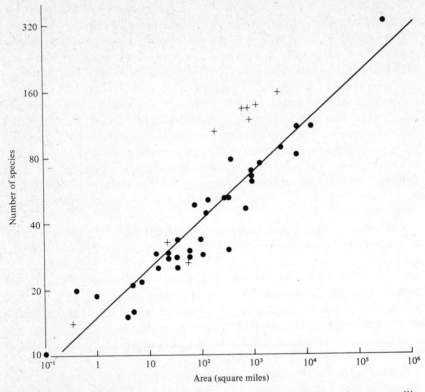

Fig. 5.2 The number of resident, non-marine, lowland bird species on satellite islands of New Guinea, plotted as a function of island area on double logarithmic scales. The ●points are islands which have not had a recent land-connection to New Guinea and whose avifaunas are presumed to be in equilibrium. The + points are islands connected by land-bridges at times of lower sea-level about 10 000 years ago. Up to the time that the land-bridges were severed by rising sea-level, the New Guinea land-bridge islands (+) must have supported nearly the full New Guinea quota of 325 lowland species (point in the upper right-hand corner). (Data from Diamond, 1972.)

Brown analysed the collapse process in these habitat islands; his results are shown in Fig. 5.3. None of the mountain ranges have as many species as do equivalent areas in the Sierra Nevada, but as is the case for avifaunas on land-bridge islands, the number of species that have gone extinct is inversely related to area.

Does the extinction of large mammals on land-bridge islands obey the same rules as the extinction of small mammals and birds? It appears that it does. Soulé *et al.* (1979) analysed the collapse of large mammal faunas on seven land-bridge islands on the Sunda Shelf southeast of the Malay

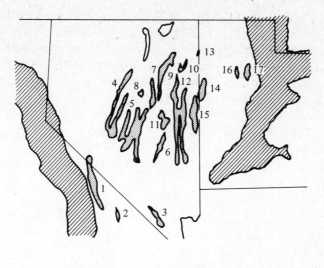

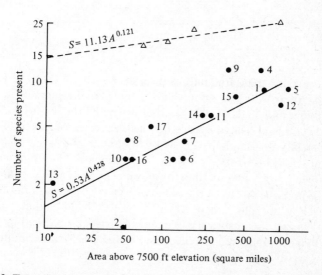

Fig. 5.3 Top, the seventeen Great Basin mountain ranges between the Sierra Nevada (hatched band on left) and the Rocky Mountains (hatched band on right); Bottom, each point (●) represents the number of mammal species and the mountain area above 2300 m (7500 ft) for each of the seventeen mountain ranges. The regression line through these points is plotted. The triangles represent the same kind of data for four sites in the Sierra Nevada, and the dashed line is the regression. In the equations, S is the number of species and A is the area in square miles. From Brown (1971).

peninsula. Table 5.1 shows that island area is indeed correlated with the rate of species loss. Whereas the large islands of Borneo and Sumatra retain about 70% of the estimated original numbers of species, the much smaller islands of Bali and Banka retain only about 10%. It appears, then, that even very large land-bridge islands apparently lose a significant fraction of their large vertebrates in just a few thousand years.

TABLE 5.1   *Estimated number of extinctions of large land mammals on selected Greater Sunda island*

Island	Area (km²)	Initial number of species[a]	Present number of species
Borneo	751 709	51	31
Sumatra	425 485	51	43
Java	126 806	50	21
Bangka	11 952	36	3
Bali	5443	32	4

From Soulé *et al.* (1979).
[a] Estimated from a species-area relation on mainland Malaysia.

Soulé *et al.* predict that large nature reserves ($> 10\ 000$ km²), if free of human intervention, will lose about half of their large mammals in 5000 years or so. Main and Yadav (1971) earlier discussed the loss of wallabies and kangaroos on land-bridge islands off the west coast of Australia; notably, only the largest islands (over 30 km²) contain the large *Macropus robustus* (euro), a close relative of the grey kangaroo (*M. fuliginosus*). According to Main and Yadav, neither the grey kangaroo nor the red kangaroo (*Megaleia rufa*) survive on any of the islands, probably because of insufficient space, and lack of drought refuges or sufficient woodland habitat.

These results are obviously very relevant to conservation, and in recent years many biogeographers (e.g. Wilson and Willis, 1975) have pointed this out. In speaking of reserves, Diamond (1975b) states that

Thus, a small preserve not only will eventually contain few species, but also will initially lose species at a high rate. For preserves of a few square miles, extinction rates of sedentary species are so high as to be easily measurable in a few decades. Within a few thousand years, even a preserve of a thousand square miles will have lost most of the bird and mammal species confined to the preserve habitat.

The term 'collapse' as used in this section and elsewhere (Soulé *et al.*, 1979) is meant to imply a rapid erosion of species diversity, attributable directly or indirectly to the consequences of insularization (Wilcox, 1980). The collapse process will probably proceed at a gradually decreasing rate until all species have been extinguished. Only if there is input into the isolate will an equilibrium be reached. There are two forms of input – immigration and speciation. The former will, for most taxa, depend on human intervention. The latter for large organisms is highly unlikely in areas the size of game reserves (Soulé, 1980).

The collapse process is neither a constant attrition nor a sudden annihilation. It usually will be a punctuated phenomenon. A burst of extinctions might be triggered by the loss of a critical species or a successional habitat. Following this episode of extinctions, the system may settle into an apparent equilibrium during which the extinction rate is low. Sooner or later, however, another critical species disappears, triggering another cascade of extinctions. Thus, collapse is an episodic process. In the early stages, there will probably be a high rate of loss. Later, as species diversity diminishes, the rate of extinction will probably decline since biological interactions, such as competition, will be less significant.

Diamond (1975b) and Terborgh (1975) have proposed some simple rules for the design of reserves. First, reserves should be as large as possible for the compelling reason that the rate of extinction decreases as the size of the region increases. Second, reserves should be as close together as possible. This is because immigration is more likely if reserves are close together, and to the degree that immigration is enhanced, the probability of extinction of a species is lowered. Proximity of reserves also increases the chance that a species will reestablish itself, even if it has gone extinct.

Brown and Kodric-Brown (1977) argue that the rate of immigration affects the rate of extinction for two distinct reasons, one demographic and one genetic. First, a very small population might be extinguished by chance; that is, all the breeding individuals might be the same sex, and only immigration from other reserves could 'rescue' such a pathetic colony. The genetic reason is that immigrants can significantly improve the fitness of a small, inbred population by increasing heterozygosity and 'unfixing' loci fixed for deleterious homozygous alleles (see chapter 3).

It should be emphasized, however, that the 'rule of proximity' is only relevant for species which normally traverse wide stretches of inhospitable habitat. Say, for example, that a nature reserve is established and that after 100 years of deforestation in the surrounding region it is completely isolated from other patches of similar habitat. Some insects, bats and birds might be able to cross the inhospitable terrain surrounding the reserve, but we can pretty well ignore the possibility of immigration for fish, amphibians, reptiles, non-flying mammals and most non-weedy or late successional species of plants. Even for the more vagile groups of organisms, however, it turns out that immigration is not a very important factor, and it can almost be ignored during the early and middle parts of the collapse process (Gilpin and Diamond, 1976).

In general, therefore, the rule of proximity only applies to species with the power of flight or those with efficient wind dispersal mechanisms i.e. birds, bats, some insects and a small minority of plants, most of which are weedy or early successional species (assuming that terrestrial animal species will be unable to safely traverse developed land). Even among winged species, there are many, perhaps the majority, that are extremely sedentary. Many butterflies have intrinsic inhibitions against flying across roads or other sharp habitat boundaries (Ehrlich and Raven, 1969). Many birds, as well, fail to cross even the narrowest barriers. According to Diamond (1975b)

134 of the 325 lowland bird species of New Guinea are absent from all oceanic islands more than a few miles from New Guinea, and are confined to New Guinea plus islands with recent land-bridge connections to New Guinea. Similarly, many neotropical bird families with dozens of species have not even a single representative on a single New World island lacking a recent land-bridge to South or Central America; and not a single member of many large Asian bird families has been able to cross Wallace's Line separating the Sunda Shelf land-bridge islands from the oceanic islands of Indonesia. Such bird species have insuperable psychological barriers to crossing water gaps, and are generally characteristic of stable forest habitats.

Terborgh (1975) emphasizes that water gaps are not the only psychological barriers to forest birds; many forest species will not even cross a few kilometres of unforested habitat, and patches of forest only a short distance from source populations will be bereft of many species.

If one were to rank the relative importance of reserve size and reserve proximity, reserve size clearly comes out on top. Thus, if forced to choose between (a) the establishment of two large reserves at quite a distance from each other, and (b) the establishment of two smaller reserves very close to each other, the nod would go to the former, since only a few highly vagile species would be aided by the latter alternative.

Another argument favouring proximity is that a given vulnerable species is more likely to occur in nearby than in remote reserves because of habitat change with distance. The flaw in this line of argument is that the rank order of extinctions is likely to be similar in nearby reserves (Terborgh and Winter, 1980). In practice, therefore, the arguments for scattering reserves will usually be weightier than those for grouping them into tight archipelagos. Widely dispersed reserves are likely to include more habitats, are likely to infect more people with the germ of nature appreciation and conservation, and are likely to engender less political and economic pressure for their exploitation (Frankel, 1970b). On the other hand, reserves should not be so widely dispersed that redundancy in species lists is precluded.

Notwithstanding the emerging biogeographical consensus in favour of large reserves, arguments supporting the antithesis have been presented. Simberloff and Abele (1976) propose that an ensemble of several smaller reserves might be superior to a single large reserve. They argue that it may be more prudent to establish several smaller reserves because different sets of species will survive in the different reserves, and that overall, the total number of surviving species will be greater in the system of small reserves than in one large one. For example Simberloff and Abele suggest that in a large reserve some species might be more or less equivalent competitively and that one or more species of such a group or guild would be lost through such interactions (competitive exclusion). In a series of smaller reserves, on the other hand, the entire set could persist, although no single reserve would have the complete set due to habitat differences and chance.

There are also subsidiary reasons for preferring many, smaller reserves. These are: (1) fugitive species (those requiring temporary or early succes- sional habitats) might be more easily lost in a single reserve, and as discussed in the next section, (2) catastrophes such as storms or diseases could debili- tate a species if it were restricted to a single reserve, but not in all the reserves in a system.

Simberloff and Abele were roundly condemned for their heterodox views (Diamond, 1976; Terborgh, 1976; Whitcomb *et al.*, 1976). Most of the points raised by the critics are made explicitly in the above discussion about the advantages of large reserves, and it would serve no purpose to reiterate them. It suffices to say that the critics attack the validity of the assumptions made by Simberloff and Abele, while acknowledging that points (1) and (2) in the preceding paragraph are valid (Diamond, 1976).

Another recommendation that has been made by several biogeographers is that, wherever possible, there should be migration corridors between reserves, permitting a continuous exchange of individuals and genes for many species and preventing extinctions for reasons mentioned above. On the surface this is a reasonable suggestion, but in many situations, particularly in the tropics, it is neither realistic nor particularly beneficial. First, corridors must be ample in width and contain a variety of habitats if they are to support and convey any but the most generalized and nomadic species. Granted, corridors should be established wherever possible, but their use should not be permitted to justify the establishment of smaller reserves. Most species will not be helped by them. This is because corridors will usually be unde- velopable topographic features such as rivers along with their flood plains and associated riparian habitats. Such habitats support a unique assemblage of plants and animals but they will not be a conduit for most non-riparian forms. For example, most birds of the climax rain forest will not venture into such habitats. Nevertheless, large predators (because they tend to be habitat generalists) will make use of such corridors, but the longer and thus more useful the corridor, the more likely it is that large animals will be poached or trapped while in transit.

On the other hand, where reserves are very small as well as close together, as they are in parts of suburban California, a network of corridors will be essential. Without them, such species as coyote, bobcat and mule deer will quickly disappear.

Summary and synthesis. To summarize, the evidence from island biogeography permits the following conclusions:

1. In any reserve, the rate of collapse in the number of species increases with the decreasing size of the reserve; nevertheless,
2. The rate of collapse for large animal species in nature reserves is so high that the majority will be extinct in 5000 ± 2000 years.
3. Only a small fraction of plant and vertebrate species, especially in the tropics, will disperse between reserves, even if the reserves are only a few kilometres apart.
4. Before dispersal corridors can justify smaller reserves, ecological and behavioural studies must establish that the corridors will be beneficial to the species for which they are intended, and that changing land-use practices in the future will not disrupt the corridors or decrease their efficacy.

5.2.3 Community ecology and reserve design

The objectives of the ecologically oriented conservationist are the preservation of (1) ecological function (e.g. energy flow and efficient nutrient cycling) and (2) ecological interaction in all its diversity. It is the second objective that distinguishes the conservationist from the sewage treatment engineer, since the latter cares nothing for complexity and diversity and does not shed a tear for the demise of a species so long as his system continues to metabolize waste.

The complexity and interaction (information) content of a community or ecosystem depend on the number of species. Species diversity generally increases towards the tropics where at least two-thirds (Raven, 1976) of all species occur, and where the ecological fabric of the planet is being rent most rapidly (Soulé and Wilcox, 1980, chapter 1). Trophic relationships are the most visible of ecosystem interactions; when A eats B who in turn eats C, we have an elementary food chain. But species interactions are often more subtle than this, such as when males of bees or butterflies collect volatile chemical compounds from certain plants which they use as precursors for courtship pheromones (Pliske, 1975; Schneider *et al.*, 1975; Dodson, 1975). In addition to such chemical mining operations, these particular insects are involved in a complex network of interactions including the pollination of many low-density tropical plants, and the sequestering of distasteful secondary compounds of plants, making them unpalatable models for many insect mimics (Gilbert, 1980).

Two other ecological factors relevant to the stability of nature reserves are the sensitivity of tropical habitats and the structural (microhabitat) specificity of tropical organisms. Tropical ecologists (Terborgh, 1975; Janzen, 1976; Eisenberg, 1980; Gilbert, 1980) often remark on these characteristics of tropical forest ecosystems. The fragility of tropical forest soils, when subject to such disturbances as mechanized logging, is well known (Fosberg, 1973; Sioli, 1975). Gomez-Pompa, Vazquez-Yanes and Guevara (1972) forcibly argue that tropical forests are not renewable (easily regenerated) habitats, especially when compared to temperate coniferous forests.

In theory, there may be no necessary link between diversity and stability (May, 1973), but in practice the disruption of ecological interactions and relationships leads to instability and can result in the local extinction of one or more of the participating species. Further, such local instabilities can start a chain reaction or 'domino effect' leading to the extinction of entire sets of species (Gomez-Pompa *et al.*, 1972; Futuyma, 1973; Gilbert, 1980; Terborgh and Winter; 1980). Table 5.2 is an attempt to systematize some of the phenomena involved in the deterioration of tropical forest ecosystem interactions leading to species extinction.

The causation of deterioration is complex – so complex, in fact, that there is no single correct analytical perspective. For example, one can begin by analysing the impact of eliminating a single species, say a large predator (row 1, Table 5.2). Alternatively, one can approach the problem at the level of habitat, for example by analysing the effects of the loss of particular successional stages. First, however, we shall examine the former perspective – the consequences of the loss of single species.

One effect of predator elimination might be extinction of ground-nesting birds whose nests are often trampled or preyed upon by species grown excessively common in the absence of their predator. This has apparently happened to many of the fifteen to eighteen extinct forest-dwelling birds on Barro Colorado Island (Willis, 1974; Terborgh and Winter, 1980). The island is too small to sustain jaguars, pumas and harpy eagles, and this has permitted such nest robbers as peccary, opossum, monkeys and the coati mundi to become unusually abundant.

Eisenberg (1980) also argues that large carnivores are sensitive indicators of the health of an ecological community, and that they 'define the minimum area necessary to preserve an intact ecosystem'. Paine (1966) and Harper (1969) have demonstrated how the loss of a predator in a marine intertidal ecosystem and a plant community, respectively, can create a drastic collapse of community diversity. The consensus among ecologists is that tropical vertebrates are more specialized than their temperate counterparts, and thus more susceptible to habitat disturbance and destruction (Janzen, 1972; Terborgh, 1975; Eisenberg, 1980). Raven (1976), too, has emphasized the high level of specialization and interdependence of tropical species and

TABLE 5.2 *Some possible cause and effect relationships following the extinction of four kinds of key (low-density, high-impact) species, with particular emphasis on tropical forests*

Ecological category	Cause of local extinction	Effects of extinctions on diversity	Indirect effects
I. Large predators	1. area effects 2. hunting	1. herbivore density increases 2. competitive exclusion among prey species	1. habitat destruction from overbrowsing, compaction, grazing, trampling and predation 2. extinction of ground nesters, plants
II. Large herbivores	1. area effects 2. hunting		1. the *Indirect Effects* in row III

(e.g. bees, butterflies, bats and birds)

2. loss of early successional habitats

reduced pollination and seed dispersal in low density plants lacking specialized mutualists

2. extinction of specialized herbivores

3. extinction of specialized parasitoids and predators

4. destabilization of coevolved food webs

1. *Indirect Effects* in row III

1. starvation and emigration of generalist pollinators

IV. Certain critical plants

1. commercial collecting or selective cutting

2. loss of early successional habitats

predicts that the loss of a plant species can set in motion a cascade of extinctions leading to the demise of from ten to thirty animal species.

An oversimplified description of some of the important ecological processes in such forests is given in rows II, III and IV of Table 5.2. The terms 'mobile links' and 'keystone mutualists' were used by Gilbert (1980) in his recent synthesis of tropical plant–insect relationships. A mobile link is an animal that is a significant factor in the persistence of otherwise distinct plant–herbivore subcommunities. For example, many of the plants that are pollinated by bees, butterflies, hummingbirds, or bats occur as widely separated individuals. The reproductive failure and ensuing disappearance of some of these plants would often lead to the linked extinction of the herbivores dependent on these plants, and thence to the extinction of the parasitoids (usually small wasps) and predators dependent on the herbivores.

Keystone (mutualist) plants are usually sources of nectar, fruit or pollen and provide critical support to mobile link species (Gilbert, 1980). Keystone plants include tree genera such as *Ficus*, epiphytes such as *Heliconia* and early successional plants such as some species of *Solanum* and *Passiflora*. According to Syme (1977) nectar sources can have a beneficial impact on forest stability because a steady supply of nectar will increase the longevity of parasitoid wasps; they, in turn, dampen or prevent the outbreak of defoliating herbivorous insects. It is probable that a necessary condition for the maintenance of maximum species diversity is the persistence of keystone plants, many of which require early successional habitats.

An alternative perspective on the deterioration of community structure is that of the decrease in habitat or patch diversity. It is universally acknowledged that the biosphere is patchy, that is, a mosaic of habitats. Further, it is accepted that the pieces of the habitat mosaic are unstable. A given point on the surface of the earth will be adorned with different plants and animals at different times, depending on the kinds and frequencies of disturbances to which the place is subject. Effective conservation programmes depend on an understanding of patch dynamics (Pickett and Thompson, 1978) in relation to the life histories of the species that depend for their survival on patch heterogeneity.

The more we learn about the life histories of plants and animals (especially in tropical habitats) and about their dispersal patterns, the more apparent it becomes that many species require more than a single habitat, and that individuals must have direct or indirect access to a multiplicity of habitats within their lifetimes. Ecologists now take it for granted that the maintenance of patch heterogeneity (and the survival of species that occupy freshly disturbed habitats, and, indirectly, the many species that depend on them for food or other resources) depends on the dispersal ability of colonizing or early successional species, and that dispersal distances are limited.

Only a small fraction of species will or can cross developed or degraded habitats (p. 109), this being the basis for the conclusion that immigration between nature reserves will be negligible, and that reserves must be large enough so that they always contain habitat reservoirs for all of their species (Pickett and Thompson, 1978). Early successional habitats are critical for the maintenance of diversity in tropical forests.

The most frequent cause for the disappearance of a successional habitat is reduced area. Many reserves are too small to always contain all of the necessary successional stages for the persistence of some herbivore species (Pickett and Thompson, 1978), many of which require the relatively palatable leaves of successional plants (Foster, 1980). Fruit-eating animals are especially dependent on successional species because such plants grow actively most of the year and their fruiting seasons are usually more extended than species of mature forest (Gomez-Pompa and Vazquez-Yanes, 1976).

In addition to reduced area, *per se*, there are other causes of a temporary or permanent loss of early successional habitats. For example, the browsing and grazing activities of herbivores can set back succession and maintain early successional habitats by creating light gaps or bare spots, thus preventing the disappearance of 'open' habitat. A case in point is the extinction of the large blue (*Maculinia arion*), once the most spectacular of British butterflies. The decline in numbers was attributed to the habitat having been overgrown with rank vegetation (Dempster, 1977). The thick vegetation compromises the survival of the ant hosts of the caterpillar. According to Dempster (citing the work of J. A. Thomas), rabbits, which had, until recently, cropped the vegetation enough to permit the existence of the ant colonies, were eliminated by myxomatosis. Here is an example of how an epidemic in a dominant vertebrate indirectly causes the extinction of at least two invertebrates via the process of habitat modification.

Many insects require resources from both mature forest and successional habitats. For example, aroids and orchids are pollinated by specialized species of euglossine bees, but the bees often require early successional plant species for larval resources. Thus aroids and orchids, though they are residents of mature forest, are indirectly dependent on the continuous existence of successional habitats within the cruising ranges of the bees that pollinate them (Gilbert, 1980). Many tropical pollinators have low population densities (Elton, 1975; Gilbert, 1980), and are susceptible to area effects including the stochastic demographic and genetic events which are characteristic of small populations. Thus any reduction in area is likely to result in the extinction of some pollinators and then to all the ensuing after-effects.

The case for small reserves has very few adherents. Of course, it will always be mandatory to protect the nesting sites of colonial birds, the habitats of narrow endemics (Terborgh, 1974; Raven, 1976), bat caves, atolls, and other pin-point features, but any wholesale trend to substitute many small reserves for one or a few large ones makes no sense to most ecologists.

It would be wrong, however, to totally disparage small nature reserves. Although from a purely scientific and conservationist standpoint, small reserves may have few redeeming features, especially in the humid tropics, their sociological significance can justify support from the conservationist constituency. Frankel (1970b) emphasized this, pointing out that 'apart from relieving pressure on larger reserves and giving pleasure and interest to a great many people, the potential educational value of "neighbourhood reserves" is immense. They can contribute materially to the security of major reserves.' Similarly, a certain amount of man-made disturbance can be justified for educational and public relations purposes. Even though the development of roads, for example, is inimical to the survival of certain sensitive species, some reserves, especially those near densely populated regions, will have to be 'developed' in order to educate and enlighten people about the material, psychological and spiritual values of wild flora and fauna.

It is altogether something else, however, to argue that the long-term survival of species and habitats will be much enhanced by the establishment of small reserves. On the surface it may appear reasonable to establish small reserves for the protection of small organisms such as soil and litter arthropods, desert annuals, or even patches of mature forest. The problem is that infrequent natural disturbances such as droughts, cyclones, fires or floods will eventually knock out any given small reserve, and unless a colonization source is very close, many of the species will be permanently lost.

This applies with special force to tropical forests, once thought by most temperate zone biologists virtually to be immune from serious disturbance. In reality, severe droughts (Dale, 1959) and cold snaps (Eidt, 1968) are not infrequent in the tropics. Unseasonal rain can also be disastrous. Following exceptionally heavy dry-season rain in the Panama Canal area in 1970, there was a drastic crash in fruit production, in turn followed by unusual mortality in mammals and the virtual disappearance of parrots and toucans; these birds were seen migrating towards Colombia (Foster, 1980).

For such migratory species it is obviously necessary to provide systems of reserves. These reserves must be close enough to allow migration, but geographically dispersed enough so that no single disturbance will seriously affect all the reserves in the system.

Foster (1980) has systematically surveyed the kinds and frequency of disturbances in tropical forests. He points out that small reserves will need to be intensively managed, both to ameliorate the effects of natural, medium to large-sized disturbances (landslides, hurricanes, volcanic eruptions, droughts and floods) and to create artificial small-sized disturbances to maintain early successional habitats. Intensive management of this kind requires skilled personnel and a large budget. Even the most elementary cost-effectiveness analysis will show that large reserves are more economical than small reserves if one uses reasonable amortization schedules.

Foster also points out that large reserves are safe from all but broad-scale climatic variation and long-term geological changes. Even so, some fruit-eating birds will attempt 'bad-year' migrations outside the park and some large carnivores have dangerously small populations. For example, there are only eight to ten family groups of the giant otter (*Pteroneura brasiliensis*) in Manu Park (15 000 km^2) in Amazonian Peru (Terborgh and Winter, 1980).

Summarizing our conclusions to this point, there may be an episode of rapid erosion of species diversity following (1) extinction of large predators; (2) extinction of herbivores, especially the largest ones; (3) the extinction of certain plants and certain 'linking' species such as pollinators. Second, any or all of these events are likely to be precipitated by a reduction of habitat space because many of the species in the above four categories have low population densities and also because some early successional habitats might be missing entirely from an area of limited size.

We do not mean to imply that the extinction of a key species will create an ecological void leading to the accumulation of organic material or to a breakdown in energy flow. For example, the prey that are normally taken by a newly extinct predator will be eaten by surviving predators whose populations will increase accordingly. Even if such predation compensation is impossible because there are no other predators, there will be detritivores to consume the excess biomass. Thus, we are not discussing gross ecological breakdown. Rather, it is a process of gradual but punctuated attrition of species diversity, marked by the passage of coevolved subcommunities or sets of ecologically linked species. The rate of decay of species diversity, though, is inversely related to area, as previously discussed.

One might argue that we have allowed too much space in this section to the problem of species erosion in small reserves. We would defend the significance of this topic on the grounds that the process of species erosion is relatively rapid. Even the largest reserves may suffer more than 30% loss of large mammals in the next 500 years (Soulé *et al.*, 1979), and the problem is likely to be much more acute in small reserves. The situation is exacerbated by the fact that the vast majority of reserves are very small. Indeed, as shown in Fig. 5.4, 93% are less than 5000 km^2 and 78% are less than 1000 km^2.

Because this situation is unlikely to improve, it is imperative that we understand the dynamics of diversity in these little islands of biotic richness. It is also imperative that we learn how to manage these delicately poised 'vestiges of creation', the subject of section 5.4.

Many biologists weaned and raised on temperate zone forests may find it difficult to accept the emerging picture of tropical forest ecology, particularly the relative importance of biotic interaction in these forests (Robinson, 1979). While it is true that the structure and diversity of temperate coniferous forests might not be drastically altered if wolves, bats or bees were to be eliminated (especially if human hunters control the deer populations), it is probably dangerous to generalize such homeostatic properties to tropical forests. The sensitivity and fragility of the latter ecosystems are thought (Gomez-Pompa *et al.*, 1972) to be of a different order altogether. For example, canopy species of plants in rain forests are rare as well as being animal-pollinated (in contrast to temperate zone forest trees which often form dense mono- or oligotypic stands and are wind-pollinated). Hence the disappearance of a particular species of bee could ultimately result in the disappearance of one or more species of tree, vine or epiphyte such as a bromeliad or orchid (Gilbert, 1980). The time scale of these events, however, is likely to be decades or centuries due to the longevity of the plants.

An even more pressing justification for the preservation of inviolate reserves or core areas in the tropics is that the presence of logging roads left after a 'selected logging' operation encourages human settlement, grazing of domestic animals, hunting and wood cutting by people who are hungry for protein, fuel and land. Thus, while in theory, the multiple use of tropical forests may be ecologically sound, in practice it could well bring about the early demise of that which it is supposed to protect.

This brief overview of community ecology and reserve design leads us to the following conclusions:

1. Ecological homeostasis is generally correlated with size. In order to minimize extinction, it is necessary to maximize area. Small reserves will lose large carnivores, herbivores and possibly certain critical plants. Certain predictable ripple effects will ensue. Whole sets of species will also disappear from small reserves because of stochastic, if ephemeral, eradication of early successional habitats.

2. Some large (and showy) species will not be adequately protected in even the largest reserves. The populations of some large mammals (particularly carnivores) will be subject to stochastic events and will require management (see section 5.2.5). Some tropical migratory birds and bats will emigrate under stressful conditions (such as a failure in the fruit crop) and can only be protected by a system of reserves rather densely scattered over a large area.

3. Reserves, or at least their core areas, should be inviolate in the tropics because of the exacting structural requirements of many tropical species and the vulnerability of tropical forests. The multiple-use concept, developed by and for temperate forests, might be disastrous in the tropics because of the fragility of soils and the exacting ecological requirements of many species. On the other hand, a small amount of

controlled artificial disturbance may be necessary to maintain successional stages, particularly in small reserves.

5.2.4 Genetics and reserve design

Although genetics is the theme of this book and of this section, it is not our purpose to propose that genetic considerations should be paramount in the design of nature reserves. Rather, our goal is to demonstrate that the ignorance of any of the four factors discussed in this chapter (disease, biogeography, community ecology and genetics) can, in the long run, spell doom for reserves and make a burlesque of our conservation programmes, no matter how well intentioned.

Until now, genetics has rarely been considered an important factor in reserve design (but see Frankel, 1970b, 1974). If any scientific arguments are considered, they are usually ecological and biogeographic ones. There are several reasons for this – they include ignorance, prejudice and the mistaken faith in cornucopian genetic variation (chapter 4). An additional reason may have been that, in a moderate to large reserve, genetic problems occur in a small fraction of the species, and such problems are not easily distinguished from reproductive failures stemming from stochastic demographic events or from ecological factors. In spite of this technical handicap, we believe that genetic design considerations are critical to the long-term biological integrity and diversity of protected ecosystems. Even if just a small minority of species suffer a loss of short-term or long-term fitness, the impact could cause ecological havoc, mainly because the genetically debilitated minority will include many of those ecologically pivotal species listed in the first column of Table 5.1. That is, if large predators, large herbivores, mobile link species or keystone plants develop genetic problems, entire coevolved subcommunities (including plants) could be reduced to a much lower level of diversity.

Rephrasing this proposition, conservation genetics is the *genetics of scarcity* – scarcity in numbers. When effective population size is severely reduced, the stage is set for an immediate loss of fitness (chapter 3) and a longer term loss of evolutionary potential and flexibility (chapter 4). Thus, while conservation genetics only applies directly to those species which are rare (although in a small reserve, this may actually be a large fraction of species given the log-normal distribution of abundances), it is highly probable that the welfare and survival of the whole system is inextricably bound to the welfare of the species in the critical categories mentioned in the preceding section.

We address two issues in this section: (1) what are the applicable principles of conservation genetics to reserve design, and (2) how are they to be applied? Two guidelines were suggested in the preceding chapters – a short-term rule in chapter 3 and a long-term rule in chapter 4. The short-term rule (the basic rule) recommends a minimum effective population size (N_e) of 50; the long-term rule (see also Franklin, 1980) recommends a minimum size of 500. The basic rule (p. 72) is not intended to prevent a loss of genetic variation – only a loss of immediate fitness; i.e. vigour, disease resistance, viability and fecundity. It is meant to be used as a guideline for zoos and similar holding operations. The long-term rule was designed to prevent an erosion of selectively useful genetic variation for an indefinite period of time. The latter rule, then, is obviously the criterion of choice for nature reserves, assuming that nature reserves are established to protect their inhabitants for more than a few generations.

The second issue is the application of the long-term rule in the design of nature reserves. Here, several points deserve noting, the most elementary of which is that the rule cannot be applied to ecosystems – only to species. Once a target species is identified, it is in theory a simple matter to determine the area that is occupied by an effective population of 500 or so individuals. Thus, the problem boils down to deciding which species in a reserve should be the targets of genetic conservation. Perhaps it is best to begin by suggesting which species should *not* be directly addressed.

In any reserve there will be a minority of species that are only present sporadically or are marginally adapted to the ecological conditions of the reserve. These forms would normally disappear under stressful conditions, retreating to, or persisting in, regions for which they are better suited. It is unlikely that such transient species will be critical to the well-being of the dominant, keystone species. Nevertheless, for political reasons, it might be tempting to use such transient species in establishing design minima, but this would be scientifically indefensible.

Another controversial category is introduced species. Introduced species are questionable candidates for conservation. Should reserves in Australia and New Zealand, for example, be set up using the population densities of California redwoods or European deer and foxes? Such decisions are arbitrary, but a thorough knowledge of the ecological implications of the alternatives is absolutely essential. For example, in some cases the ecological effects of the extinction of the introduced species might be considered worse than the effects of their survival.

Having eliminated (for most situations) marginal and introduced species, we come to the heart of the matter, namely, which species among the hundreds or thousands in an ecosystem should be considered when establishing guidelines for genetic design and management? What follows are merely suggestions. No two reserves are alike, and the ecological and political exigencies will vary from place to place, from zone to zone, and from continent to continent (or island).

We propose employing the critical ecological categories reviewed in the preceding section. The extinction of species in these roles is likely to cause a drastic reordering and simplification of ecosystem functioning. These roles are (1) large predators (large being a relative term), (2) large herbivores, (3) mobile, generalized pollinators (some birds, some bats, some insects), and (4) fruit- and nectar-producing plants that are important resources for mobile pollinators and generalized fruit and nectar feeders (e.g. insects, primates, birds) and the insectivores and predators they attract. Note that most tropical plants are outbreeders (Bawa, 1974; Raven, 1976); thus, in the tropics, the principles we have adduced in chapters 3 and 4 will probably apply to the fourth category (plants) as well as to the first three.

Other categories are just as important to the health of ecosystems (including decomposers, specialized herbivores, parasitoids and so forth) but such species are usually abundant, and, therefore, not in need of immediate attention. In situations where such a species is both ecologically crucial and in numerical trouble, it will obviously be a candidate for genetic surveillance and management.

The emphasis in the balance of this section will be on large mammals, principally carnivores. There are several reasons for this. The first is vulnerability; these animals are the least dense of all species, and thus will have the smallest populations in any nature reserve. (The human affection and fascination for large animals is a mixed blessing, being both the source of so much popular concern for nature and the reason why so many are in danger from poaching.) Second, the large mammals and birds in an ecosystem are usually sensitive indicators of ecological integrity (Eisenberg, 1980). Third, the repercussions following the loss of these animals are thought to be long-lasting and pervasive (see Terborgh and Winter, 1980). Fourth, measures taken to protect the largest (or rarest) elements in an ecosystem will often provide an umbrella of security for the more dense species.

Some ecosystems lack large mammals altogether, either because they are impoverished due to hunting (Western Europe), isolation (New Zealand), low productivity (Antarctica), or a combination of these factors. Nevertheless, the same principles apply. That is, there will almost always be one or more vulnerable species belonging to one or more of the critical categories. If predators are absent, then the least dense critical species may be a herbivore or a keystone tree, such as a *Ficus* (Whitmore, 1980) or *Cascaria* (Howe, 1977), that provides fruit or nectar in a season when these resources are in short supply for many birds and mammals. It is indeed difficult to imagine a situation where all of the critical species are abundant.

Returning to the long-term criterion, our basic assumption is that an effective population size of 500 is the minimum number that guarantees long-term survival in nature. The justifications for this number (an order of magnitude approximation) are given in chapter 4 (p. 90). Recall that this number refers to an ideal population, the characteristics of which are rarely encountered among large plants and animals. The characteristics include random mating, equal numbers of breeding males and females, the absence of severe fluctuations in numbers, non-overlapping generations, and a random (Poisson) distribution of offspring among families. Because few species satisfy all of these assumptions, it will usually be necessary to increase the minimum number for long-term survival by a factor of two or three, though even this may be too conservative.

As an example of the application of our principle for reserve design, we use the mountain lion (puma, cougar), *Felis concolor*. In Idaho it has a density of about one mature individual for every 26 km² (Hornocher, 1969). Assuming mountain lions satisfy all of the assumptions of a genetically ideal population (they don't), they require an area of 500 × 26 = 13 000 km². In other words, this application of the long-term rule gives us an estimate of the minimum size of a nature reserve that is designed to conserve mountain lions, and indirectly, the ecological dynamics dependent on the presence of this carnivore.

Wolves (*Canis lupus*) are a species for which social structure obviously affects N_e. Wolves have about the same density as mountain lions (Rutter and Pimlott, 1968), but like many other canids, not all of the adults in a pack breed; many serve as 'helpers at the den', bringing food to the young of the dominant pair (Kleiman and Eisenberg, 1973). This and other 'non-ideal' behaviours may reduce the effective size by a factor of from three to six, say, thus increasing the minimum area of a wolf preserve to between 39 000 and 78 000 km². For comparison, the size of Yellowstone National Park in Wyoming (the largest National Park in the forty-eight adjacent states in the USA) is about 9000 km². It would thus appear that there are few if any reserves in the world (outside of the arctic) that are large enough for the long-term survival of the dominant and least dense top carnivores.

At least in theory the genetic principles of reserve design are simple:
(1) employ the long-term rule for the maintenance of genetic variation and
(2) choose one or more species to which the rule should be applied.
The difficulty obviously lies in the latter decision. We argue that the choice should be based on ecological interactions: the guiding principle should be that the appropriate target species are those which are crucial to the maintenance of stability in a significant fraction of the ecosystem. Among this set of pivotal or keystone species, the least dense should be the front-running candidates.

5.3 Design and management criteria: a shift in emphasis

So far in this chapter we have been preoccupied with issues of space, particularly the amount of space and how it is subdivided. In section 5.2.1 on disease, the issue is the optimum way to quarantine the vestiges of wildlife.

In section 5.2.2 on island biogeography, the overriding issue is the rate of extinction of species in the fragments of undeveloped habitats. There is evidence that these rates are both high and area-dependent, so there is justification for emphasizing the importance of large size for reserves. In section 5.2.3 on community ecology, the issues are the long-term ecological integrity and diversity in reserves. Here too, area turns out to be a critical variable. When we turned to genetics in section 5.2.4, area was also the crucial issue, although indirectly. Genetically, the critical variable is population size, but this is transformed to area requirements by employing density estimates.

Our main conclusion is that reserves are, in general, too small to maintain their present diversity, especially in the tropics, where most species occur. As shown in Fig. 5.4, only 3.5% of national parks throughout the world are larger than 10 000 km^2, but even this size is much too small to maintain viable populations of the largest carnivores, and no existing reserve in the tropics is large enough to prevent the wholesale emigration of fruit-eating birds during 'bad years'. Without intensive management, it is likely that the majority of birds and large (> 1 kg) mammals will be extinct in a few thousand years (e.g. Soulé *et al.*, 1979, and references therein) and that these extinctions will precipitate complex chain reactions leading to many other extinctions in all taxa.

Genetic and ecological arguments can also be marshalled to show in general that reserves are insufficient in number. For example, the total number of individuals in some species of large animals in existing reserves is too low, and more reserves need to be established.

In any case, the historical phase of reserve establishment is all but over in the tropics (Amazonia being the only major exception). By the year 2000 ± 10 years, there will be very little natural habitat left to preserve that is not already protected (Whitmore, 1980). The process of reserve acquisition has virtually ended, but the process of attrition is just beginning. This is not to say that conservationists should be less than zealous about obtaining more and larger reserves. Rather, it is to emphasize that more attention needs to be paid to what we already have.

What about existing reserves? We are not sanguine about their success. The data for tropical animals are not encouraging. For example, the Wilpattu National Park in Sri Lanka (Ceylon) contains about 800 km^2 including recent additions. In it there are about 100 elephants and 20 to 30 leopards (Eisenberg, 1980). The Serengeti (Tanzania) is about 13 000 km^2 in area and holds about 2000 lions (Schaller, 1972), probably sufficient, but only about 30 hunting dogs (*Lycaon pictus*) (Frame and Frame, 1976). Not only do the largest national parks contain too few of many of their large predators and herbivores to satisfy our long-term criterion, but many populations fall below the minimum ($N_e > 50$) even for short-term fitness and survival.

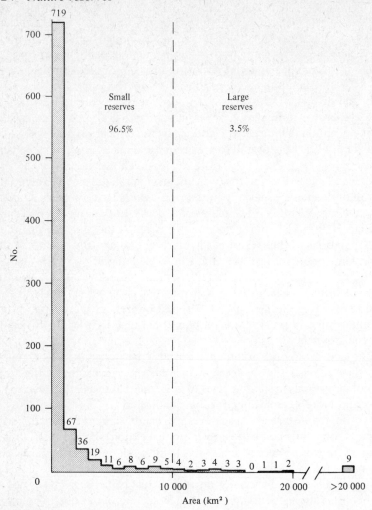

Fig. 5.4 The distribution of areas of nature reserves. The dashed line at 10 000 km²
is an arbitrary division between 'small' and 'large' reserves. (Data from the 1975
United Nations list of National Parks and Equivalent Reserves, IUCN, 1975.)

The biogeographic data on the survival of large predators on land-bridge
islands is consistent with the poor prognosis suggested here (e.g. Terborgh
and Winter, 1980; Soulé *et al.*, 1979). Terborgh and Winter, citing Willis
(1980) conclude that the best predictor of extinction of birds in forest
remnants in Brazil is rarity; i.e. constitutional low population density.
Prominent among these extinction-prone species are the largest forms
(eagles, macaws, large parrots, toucans, tinamous, a wood-quail, a pigeon

and a fruit crow). By itself, the correlation between extinction and density does not prove that genetic problems are a contributing cause, but at least theory and fact seem to be in step.

From genetic considerations alone, then, it would seem that the ecological fabric of most nature reserves is in imminent danger of ripping. Very few reserves have large enough populations of their largest species (particularly predators) to maintain their present ecological structure, assuming that heroic therapies are not instituted in the immediate future.

Fragmentation and insularization are not the only reason for the decay of structure and diversity in reserves. Human activities exacerbate the process, and the trend in reserve quality appears to be negative. Everywhere one reads and hears of serious incursions in nature reserves, including logging, wood cutting, poaching, illegal boundary changes and the killing of game wardens and park guards, particularly in tropical countries. (Bad news can be found in every issue of such conservation-oriented periodicals as *Oryx*, the *IUCN Bulletin, Biological Conservation*, and *Environmental Conservation*.) For these reasons, as well as for related economic and political ones (Myers, 1979; Soulé and Wilcox, 1980), we contend that the issue of reserve design, *per se*, is something of a red herring; design is important, but too much emphasis on design alone is highly myopic.

The issue and problem for the future is *maintenance*. As pointed out repeatedly in this chapter, it is apparent that reserves in the tropics are nowhere self-sustaining systems; they are simply too small. In the absence of colonization from large source areas, a soon to be obsolete possibility (Pickett and Thompson, 1977; Wilcox, 1980), the march of extinctions in nature reserves (assuming they are benignly neglected) will be inexorable. This is not to say that reserve size is irrelevant. The rate of extinction is clearly slower in larger reserves. But the inevitable outcome is just the same.

It is absolutely necessary to be very clear about the real issue in the debate over the size of reserves. What biogeographers have been discussing is whether roughly 50% of the higher vertebrate species will be extinct in 500 years versus their extinction in 5000 years. When seen in this light, it is obvious that size alone is not the fundamental question. Reserves, of course, should be as large as is politically and economically feasible, and reserves should be as numerous and widely dispersed as possible. In addition, biogeographic, ecological and genetic principles dictate the maximization of *both* size and number.

The main issue as we see it, however, is the absolute necessity of careful and continuous scientific management. Unfortunately, the past preoccupation with physical design features may inadvertently promote the false notion that proper size and distribution of reserves will, in itself, guarantee the success of conservation.

Politicians easily gain approval of the international community by drawing some lines on a map in a sparsely populated region, and proclaiming the
establishment of a new national park. This costs virtually nothing, but
without an equal commitment to reserve maintenance, it guarantees the
preservation of virtually nothing. It is more difficult to direct the attention of
politicians, planners and conservationists to the necessity of careful and
continuous management, but it is essential.

5.4 Genetic management of nature reserves

Genetic management is just one aspect of the extraordinarily complex field
of nature reserve management. A comprehensive and authoritative treatment of nature reserve management as a whole is well beyond the scope of
this book, and, we might add, beyond the scope of any book, given our
abysmal ignorance of biological processes in complex ecosystems. To a large
extent, reserve management for many years to come will be a 'seat-of-the-
pants' practice, proceeding by trial and error. This does not justify, however,
ignorance of the state of the art in such fields as insular ecology, biogeography, community ecology, plant ecology, and behaviour. Regarding these
disciplines, we will not add anything to what has already been presented in
sections 5.2.1, 5.2.2 and 5.2.3. Regarding genetic management, however,
there are five recommendations. In one way or another, these proposals all
address the problem of increasing the effective size for species which are too
rare or too fragmented. The proposals are (1) research and monitor genetic
variation in small, natural populations in order to provide baseline and basic
information on the attrition of variation; (2) practise intensive management
on very small populations in imminent danger of extinction; (3) integrate
ecological, behavioural and genetic principles in the culling of large organisms; (4) integrate ecological and genetic principles in the artificial disturbance of mature habitats; (5) initiate artificial migration.

5.4.1 Basic research in genetic attrition

The first recommendation is the immediate establishment of well-designed
genetic monitoring programmes for a carefully selected sample of rare
species. The objective of this endeavour would be to provide data on the
actual time course and correlates of genetic changes in small populations.
The data we have now are skimpy. For example, we know that small
populations of lizards on land-bridge islands lose most of their genetic
variability, the 'small island effect' (Soulé, 1980), and that such depletion is
probably attributable to genetic drift. Yet we know virtually nothing about
the *rate* of loss of heterozygosity and its phenotypic correlates under severe
population size restriction in nature.

The time scale of such a programme would need to be an order of magnitude or more greater than is typical in 'normal' science. For species with an annual life cycle, samples would need to be taken for several decades; a study lasting more than a century would be necessary for long-lived species. This poses unusual but not insurmountable logistical, administrative, institutional and economic problems, the solutions to which must originate in national and international organizations.

Another problem that must be overcome before such a study is undertaken is technological change in analytical techniques used for the description of genetic markers from tissue and blood samples. Because such techniques as electrophoresis (p. 42) will improve, be augmented and finally replaced, it will be necessary to store samples for long intervals for analysis at a future time. The technology for such storage is now available (p. 274).

5.4.2 Intensive genetic management

In many cases it will be possible to increase the effective size of populations within reserves. Several means could be used to accomplish this, assuming that the populations can be observed and manipulated. Among the possibilities are altering the sex ratio, altering the age structure, altering the social structure, altering the offspring distribution among families, and of course, by providing additional or better resources so that the population can grow more rapidly. These therapies belong to the category of management practices for closely managed populations, and are dealt with in more detail in chapter 6.

5.4.3 Rational culling

Culling may have to be practised if the density of a species threatens the integrity of a reserve. Whereas culling can be practised to the demographic advantage of a population (Goodman, 1980), the ignorance of genetic principles could cause irreversible genetic damage. There also may be conflicts between behavioural criteria for culling and genetic ones. For example, in the control of overly dense populations of elephants or other large, social species, it has been considered desirable to eliminate entire social groups rather than shoot out an equal number of selected individuals in many groups. The reason is the social disruption and probably the danger inherent in the former tactic. Unfortunately, such a tactic tends to erode genetic variation faster than is necessary since the genetic relatedness of individuals within social groups will usually be greater than the relatedness of individuals between social groups (Packer, 1978). Hence, in a reserve that contains only a few groups, the elimination of one or more can seriously deplete genetic variation, whereas the culling of selected individuals from all the groups is less likely to eliminate rare alleles or sets of co-adapted alleles. The resolution of such management conflicts for the long-range benefit of the species will require managers who are broadly trained and who have a grasp of basic population genetics.

5.4.4 Rational disturbance and habitat manipulation

Man-made disturbance will be a necessary management practice in small reserves in which there exists the possibility of the temporary loss of early successional species (e.g. Foster, 1980). One danger of this practice is the accidental elimination of too many individuals of certain species. In the selection of sites for disturbance (or trees for cutting), managers must be aware of this possibility, and avoid destroying species with low numbers, especially if they are keystone species.

5.4.5 Artificial migration

One therapy that seems practical at the present time is artificial immigration. If the species occurs in two or more reserves, or captive populations (Conway, 1980), and these populations are part of a management consortium, then routine reciprocal transfers can reduce the rate of genetic erosion and drift. Assuming that a programme of transfers is implemented, it is necessary to know how many individuals should be moved and how often. This is a difficult problem.

Maruyama (see discussion in Kimura and Ohta, 1971, p. 153) has argued that a migration rate of just one or so per generation is enough to establish panmixis for neutral genes in semi-isolated colonies. Lewontin (1974, p. 213) points out that a migration rate as small as one out of a thousand individuals per generation is enough to establish effective panmixis between two populations of moderate (ca 10^4) size. Of course, the smaller the populations, the greater will be the randomization of gene frequencies due to genetic drift. This can be seen from the following examples.

The absolute difference in the frequencies of neutral genes between two populations, each of size N, given that they exchange a proportion m of their genes in each generation will be, on the average

$$d = 2 \sqrt{\left[\frac{\bar{p}\,(1 - \bar{p})}{(1 + 4Nm)} \right]}$$

where p is the frequency of one of the two alternative, unselected alleles (Lewontin, 1974, p. 213). Thus, if N is 1000, and m is 1%, and the average allelic frequency is 0.5 in both populations, then

$$d = 2\sqrt{\left(\frac{0.5 \times 0.5}{41}\right)} = 0.156$$

which is a rather small difference.

The problem is that conservationists will often be managing very small populations. Genetic drift in small populations accelerates the rate of differentiation and the rate of loss of alleles (p. 35ff). Thus, in populations of, say, 50, given the same migration rate as above,

$$d = 2\sqrt{\left(\frac{0.5 \times 0.5}{3}\right)} = 0.577$$

In order to reduce the value of d to its former magnitude (0.156), the value of m would have to increase to 20%. Note that with a given value of d and p, Nm is a constant.

Actually, it turns out that in both of the above cases in which $d = 0.156$, a reciprocal transfer of ten breeding individuals is required every generation. In the former case, however, ten animals represents 1% of the population, whereas in the latter case it represents 20%. The movement of such a large fraction of a colony could have serious consequences in a territorial or social species. The success of this level of forced migration will also depend on the size and tractability of the organisms and the survivorship following introduction.

The above line of reasoning explicitly assumes the neutrality of alleles in contrast to the discussion in chapters 3 and 4 where it was argued that the loss of genetic variation is accompanied by harmful effects. In this case, however, we feel comfortable with a neutralist approximation. The reason is that in small populations the effect of genetic drift is strong enough (p. 88) to neutralize all but the strongest selective forces. In other words, alleles which are strongly selected in large populations will not confer reproductive differences in small populations because the effect of chance on survival and breeding success becomes increasingly significant as population size decreases. Ultimately, however, the loss of genetic variation, such as for alleles conferring resistance to disease or to other periodic environmental challenges, will depress both short-term and long-term fitness.

Beyond what we have said, it would be gratuitous to go much further in speculating on the desirable rate of migration into a population. The minimum is probably a single reproductively successful migrant per generation; a reasonable upper limit is probably five. The sex and age of migrants will depend on the social behaviour and population structure of the species, and this should be jointly determined by the project's scientific advisory team comprised of an ecologist, a behaviourist (for animals) and a geneticist. Populations numbering less than $N_e = 50$ are candidates for immediate artificial gene flow. For those populations larger than $N_e = 50$, the danger of genetic and phenotypic deterioration is less, but erosion of genetic variation must still be controlled, at least as long as the population has an effective size of less than 500. Again, we emphasize that natural populations numbering less than 50 to 100 are in immediate danger and require immediate genetic management.

Artificial migration will produce two effects; the first is beneficial, the second may be beneficial, harmful, or both, depending on the circumstances. The first effect is the increase in N_e and the consequent decrease in the rate of erosion of genetic variation. The second effect, or side effect, of gene flow between populations is an immediate (or slightly delayed) change in fitness. On the one hand, the F_1 hybrids between the two colonies might be heterotic. Such heterosis could be permanent, or transient, depending on the genetic structure of the species and the genetic condition of the stocks before the transplantations were begun. We would expect heterosis if the stocks were already somewhat inbred at the time that the transplant programme was initiated.

On the other hand, artificial migration could produce deleterious genetic changes, either because of chromosomal incompatibilities between the donor and recipient populations, or because of the phenotypic intermediacy of the hybrids. If there is the slightest doubt about the compatibility of the stocks, a programme of genetic, behavioural and ecological testing is mandatory before the transfers commence. The problem is discussed in more detail in the following chapter.

Most nature reserves are much too small to maintain their original diversity for more than a few decades. Even now, many of the largest reserves are in critical need of genetic management. The rate of inbreeding of top carnivores and large herbivores such as the hunting dog, cheetah, giant otter, jaguar, elephant, rhinoceros and large antelopes in many reserves is far above the limits we believe tolerable, and this condition has existed for several years or decades in some species. As part of the management programme of any reserve there should be an up-to-date listing of keystone species with the estimated numbers. For each species with an effective size below 500, a plan should be implemented to eventually elevate the populations to a safe level. Such programmes require expertise and money. Both commodities are in short supply in countries where they are most needed. One survey showed that the tropical countries have, on the average, about one-tenth the personnel and one-tenth the money per unit area of reserved land compared to temperate nations (Soulé and Wilcox, 1980, Chapter 1). The effect is a continuous erosion in the quality of tropical reserves.

5.5 Summary

1. The potential for epidemics should be considered in the design of nature reserves, not only because of the possibility of total annihilation, but also because of the genetic consequences of a severe knockdown in numbers following a major disease outbreak. Both kinds of events are likely in nature reserves, and their probabilities increase as natural populations become fragmented due to human habitat disturbance and development, and as the major reservoir of contagious disease – domestic plants and animals – encircles, ever more snugly, the remaining fragments of undisturbed wildlife.

    Species redundancy, therefore, must be a design criterion in any reserve system based on rational biological principles, assuming that simultaneous epidemics are an uncommon event in two or more reserves harbouring the same species.

2. Several important generalizations relevant to the design of nature reserves have emerged from the studies of island biogeographers. First, extinction rates in habitat patches are area dependent, the smaller the patch (or island), the higher the rates of extinction. Second, even the largest nature reserves, if left alone, will probably suffer major die-offs of species, accounting for a majority of birds and large mammals in a few hundred or a few thousand years. Third, the arguments for proximity of reserves and for dispersal corridors between reserves may have been overstated. The advantages and disadvantages of such aids to dispersal need to be examined on a species by species basis. The existence of corridors should never be used to justify small reserves since (a) few species will disperse along corridors, (b) corridors can transmit diseases as well as genes, and (c) corridors are vulnerable to changes in land use values and practices.

3. The maturing field of community ecology permits the following generalizations regarding design: (a) the stability of ecosystems (particularly in the tropics) may be very sensitive to the presence of a relatively small fraction of the species in an ecosystem, particularly large mammals, certain insects and birds, and certain key plants. The extinction of any of these species (itself a function of the area of the reserve) could precipitate a cascade of extinctions in ecologically linked forms; (b) designers of reserves must consider the size, distribution and longevity of disturbances from tree falls to cyclones and volcanic eruptions, since the disappearance, even if temporary, of a successional habitat from a reserve, will usually cause a sudden and dramatic decrease in species diversity. In most cases, such events will be irreversible due to the limited amount of migration between reserves; (c) the creation of artificial disturbances will be necessary in

small reserves, but the kind and amount of such imposed succession needs to be carefully planned and monitored, especially in the tropics, in light of the sensitivity of tropical species and interactions.

4. The basic genetic criterion for nature reserve design should be the long-term rule of genetic conservation. This rule states that the minimum effective population size for any species is of the order of 500. The choice of which species are to be considered for the application of this rule requires ecological sophistication, but it is evident that few reserves in the world are large enough to protect many of their mammals and birds from immediate decline in fitness (in the worst cases) or a long-term erosion of genetic variation and evolutionary potential.

5. The historical phase of reserve establishment is drawing to an end, it being mainly restricted to the latter half of the twentieth century. Conservationists must now begin to attend more to the problem of maintaining what is already protected – otherwise it will leak away in very short order.

6. We call for basic research in genetic management. In addition, it is imperative that genetic management be applied immediately to many species in many reserves in order to stem the loss of genetic variation and fitness. The following approaches are suggested: (a) intensive genetic management of small populations where manipulations of matings and other demographic variables are possible; (b) culling to minimize loss of genetic variation; (c) the integration of artificial disturbance with rational genetic management; (d) artificial migration between populations in different reserves or between reserves and captive populations.

6

General principles and the genetics of captive propagation of animals

As soon as a species is proved to be on the wane, a captive breeding programme should be set up *automatically*.

Gerald Durrell, 1975

6.1 Objectives and time scale of captive propagation

The purpose of this chapter is to explore the genetic principles of captive breeding within the context of animal breeding in general. No attempt is made to eschew the relevant economic and political issues which lace this activity, since it is our opinion that to discuss the genetic issues in isolation from the economic and political, not to mention the behavioural ones, would be to present a misleadingly simple picture of captive propagation (CP). Thus we have attempted to present a general overview of the subject, preferring to err on the side of over-generalization rather than on the side of scientifically safe but unrealistic certitude.

The breeding of wild animals is practised for many purposes, some of which have nothing to do with conservation. Among the objectives that motivate such programmes are: (1) the production of animals for exhibit and educational purposes; (2) the production of animals for experimentation and research by and for commercial and scientific establishments; (3) the development of new kinds of domesticated animals or the improvement of existing domesticates; (4) breeding of endangered species with the intention of eventually returning their descendents to natural habitats. These four kinds of programmes are elaborated in the following sections.

6.1.1 Exhibit and education

Many zoos are private institutions, depending on gate receipts to a greater or lesser extent for their general operations and running capital. To remain viable, such institutions must compete with other forms of entertainment and recreation such as television, athletic events, carnivals and amusement parks. It is no wonder that the 'gate' is a primary concern to the directors of most zoos, at least in the US, and that expensive, scientific CP of endangered species is a luxury for such institutions if indeed it can even be considered.

Most of the breeding carried out by zoos must have immediate payoffs. Among these are (1) replenishing existing exhibits (a zoo-born animal is usually cheaper than an animal purchased from an animal dealer or another zoo); (2) free advertising in the media that follows the birth of a mammal and is promoted with the ritual photograph of a nubile human female holding an attractive cub or baby; (3) the public relations value if the neonate belongs to a rare or endangered species; and (4) the exploitation of young in 'children's' or 'petting zoos'.

From the standpoint of a conservationist, these practices can all be justified if the zoo is engaged in a serious attempt to educate the public about the plight of animals and habitats. All zoos claim to be 'educational'. Certainly zoos have a tremendous potential impact on public attitudes towards wildlife. The controversy (see Batten, 1976, for an example), however, is not about the potential educational role of zoos, but about the degree to which this potential is realized. Although heroic efforts to educate are made by some zoos, all too often zoo visitors come away with little or no more information about the plight of wildlife in the world than when they entered.

The time scale of breeding for exhibit and educational purposes has been only a few generations. To a large degree, the reason has been the availability of relatively inexpensive replacement stock. As the availability of animals is monthly becoming more difficult, both because of a dwindling supply and an increase in government red tape, it is obvious that zoos will have to go into the business of producing their own specimens.

Even some of the best zoos have yet to face up to this challenge. For example, a survey (Soulé, unpublished) of primates at one internationally recognized zoo showed that of the forty-four forms (thirty-four species, plus ten additional races) only twenty-three (52%) could reproduce at all, and only fourteen (31%) could be expected to be self-sustaining for from three to six generations, given the present size of the group and the potential for increase. Finally, only about four species had the potential to serve as a nucleus of a serious, long-term CP programme. One conclusion of this survey was that a 'business as usual' approach to the breeding of exhibit species, combined with the inability to purchase or trade specimens, would precipitate a decline from forty-four forms in 1977 to about four or five in twenty-five to fifty years. Obviously a self-sustaining collection requires much attention to CP. One need not be a conservationist to appreciate this logic.

6.1.2 Research

CP for the production of subjects for research will become increasingly significant as the supply of wild animals, particularly primates, dwindles due to shrinking population and mushrooming regulation. India has already banned the export of primates, including the rhesus monkey, although publicly funded labs and private entrepreneurs are increasing production. In 1970 alone, 60 000 or so primates were imported into the US; nearly all of these were used in research (Bermant and Lindburg, 1975). Ignoring here the ethical problems that abound in this area, (e.g. Passmore, 1974, Singer, 1975), the supply of wild primates is fast diminishing and it is probable that the research and development establishment in the industrial countries will have to become self-sufficient in primates within ten or twenty years.

A small fraction of important wild species are used for research in sociobiology (e.g. Kleiman, 1980). In most cases these animal colonies were established for other reasons (exhibit, pharmaceutical or physiological research) and are only secondarily used for research not directly or indirectly related to human health.

6.1.3 Domestication

Among contemporary problems, the shortage of protein must rank among the most serious, especially in light of the effect of protein malnutrition on intelligence and learning. Given the serious set-backs met by the 'green revolution' and the dawning awareness of the minor role of 'food from the sea' (Ehrlich *et al.*, 1977), it is apparent that ordinary, labour-intensive agricultural and pastoral practices must provide most of the food for the rapidly growing human population. One would think, therefore, that much money and expertise ought to be devoted to developing new, reliable and efficient sources of animal protein, particularly by domestication of new forms. Actually, no wild ungulates have been completely domesticated as a food source since the time of Christ (except the reindeer in the fifth century) and, in all, only 16 out of 4500 species of mammals have been domesticated to meet man's basic needs (Anon., 1972; Spillett, Bunch and Foote, 1975). This may be changing. Recently some attempts to domesticate new species of African ungulates, particularly the eland and the beisa oryx have met with some success (Coe, 1980), showing that other species of mammal are capable of taming and captive breeding.

A somewhat different approach is championed by R. V. Short (1976) who advocates genetic improvement of existing domestic species, especially the domestic goose, duck and sheep by introducing genes from wild relatives that would increase productivity ('wild genes' from tropical species could promote year-around breeding, and genes from temperate or arctic forms could increase growth rates, body size and the efficiency of food conversion).

Both approaches could substantially reduce the pressure on the remnants of wild game, particularly in tropical countries. Bush meat is a very important protein source, making up more than half of the animal protein consumed in parts of Africa (Coe, 1980). Improvement in the variety, quality and efficiency of domestic animals would lessen this pressure and improve the nutrition of millions of people.

Often, temperate zone species do rather poorly in the tropics. In West Africa, for example, the domestic goat is responsible for the deterioration of much habitat on the outskirts of towns and villages, yet it is a relatively poor producer. Some African rodents might produce much more protein in a given area than goats at much less cost in terms of labour and habitat, but the necessary research and development is lacking (Ajayi, 1975).

The breeding protocols and the genetic criteria for domestication are very different from those for CP and wild species destined for return to nature. Nevertheless, domestication has a potentially important role in conservation. Short (1976) argues that zoos should be engaged in such projects. We agree. Some zoos have the equipment and personnel to deal with the unexpected and the unknown. Zoos and similar institutions should be encouraged to engage in this socially and biologically responsible activity.

6.1.4 Rescue and return

The role of CP in saving endangered species from certain extinction has received much attention. As stated in the first point of the Declaration on Breeding of Endangered Species as an Aid to Their Survival (Jersey, May 1972),

The breeding of endangered species and subspecies of animals in captivity is likely to be crucial to the survival of many forms. It must therefore be used as a method of preventing extinction, alongside the maintenance of the wild stocks in their natural habitat.

Notwithstanding all the arguments that can and have been marshalled against CP as a form of 'extinction insurance', we think that the survival of a species in captivity, even if it becomes 'domesticated' in the process, is preferable to total annihilation. The one qualification we would make is that the resources required for such a project should not cause a critical depletion in resources budgeted for the conservation of natural habitats or the organisms within them.

One ancillary advantage of a CP programme is that it can be used to publicize the plight of a species and arouse public interest and action to save the species and other endangered forms in their natural environment. This actually occurred as a result of a programme to save Swinhoe's pheasant, an endangered species in Taiwan (Martin, 1975). The success of the reintroduction of this bird received so much publicity that the Taiwanese government passed an endangered species act which provided for the protection of thirty other forms.

There appears to be a very general sympathetic response to the imminent extinction of a species among educated people. Many mottoes come to mind: 'Save the whales', 'Save the sea otter', 'Save our redwoods'. Whatever the motivation, CP is the last gasp attempt to save a vanishing and unique expression of biological evolution, and it apparently fulfills an ethical or psychological need shared by many people.

In many places and at some times CP may be the *only* secure means of guaranteeing the survival of a species, particularly large animals. Many North Americans and Europeans will protest that this is a preposterous proposition, believing that sufficiently large reserves and parks have been and will continue to be established to preserve most if not all endangered species and, once established, proper management can virtually assure the survival of the target species. If this were only true!

As discussed in chapter 5, many so-called reserves and national parks in most countries are sanctuaries in name only. Rising expectations and soaring populations daily diminish their security, particularly in the tropics. The villager living near a reserve sees it as a source of cash and food. He sees the land, stretching away for miles, and easily justifies the felling of trees for lumber and firewood. To him, the land would be better used for hunting, farming and grazing. The plight of the gorilla in Ruanda and Zaire (Cousins, 1978), the black rhinoceros in Kenya, and the Arabian oryx, testify to the necessity of CP for certain species.

On the other hand, CP programmes dedicated to the eventual return of species to nature must, in our minds, strictly adhere to certain principles which are discussed in sections 6.2 and 6.3. These principles or *'rules of captive propagation for rescue and return'* are:
1. genetic and phenotypic change should be minimized;
2. inbreeding should be minimized;
3. loss of genetic variation should be minimized;
4. behavioural changes, particularly in the direction of domestication, should be minimized.

The time scale for such CP programmes should be of the order of 100 to 200 years, assuming that it is desirable to minimize genetic deterioration in the stock. Programmes expected to survive for longer times will have to meet much more rigorous standards at the outset, if genetic deterioration of the stock is to be avoided. The details of CP for rescue and return to nature are the subject of most of the rest of the chapter.

6.2 Problems and criticisms of captive propagation

Many conservationists believe that zoos and their recent offspring, wildlife or safari parks and oceanariums, are anathema and that it is a contradiction to speak of such institutions and conservation in the same breath. This feeling is by no means restricted to a radical or militant fringe of conservationists. A past director of the Taronga Park Zoo in Sydney has stated that zoos have virtually no direct role in conservation, although the educational impact of zoos is significant (R. Strahan, personal communication).

In this section, we review some of the criticisms levelled at CP in zoos and the major obstacles to implementing some CP programmes.

6.2.1. Captive propagation and the exploitation of wildlife

'Victory has a hundred fathers and defeat is an orphan.' If zoos had a good track record in CP, most of the critics would be silenced. Successes in CP, however, are still very few. Perry *et al.* (1972) compiled statistics on CP in US zoos. Their survey showed that of the 291 rare or endangered mammal species, 162 are in zoos, 73 have been bred, but only about 30 of these have been bred with sufficient success to justify the hope for their eventual return to nature.

A similar study by Pinder and Barkham (1978) based on data collected by the Zoological Society of London has come to very nearly identical conclusions: out of 274 rare species of mammal, only 61 were considered worthy of study because they had been bred in zoos on more than a few occasions. Out of these, only 26 were considered self-sufficient and out of immediate danger. These species are shown in Table 6.1. Pinder and Barkham also point out that the majority of these species belong to families of domesticated mammals, suggesting that husbandry experience is an extremely important ingredient in the relatively few successful programmes (Senner, 1980).

Conservationists often point out that the decision to breed a rare animal in captivity is a decision to sacrifice a significant number of the few remaining wild specimens to a programme that may fail. There is no guarantee that any particular breeding programme will work, but *if* the specimens are protected in their undisturbed natural surroundings, there is a virtual certainty of breeding success.

Conservationists also argue that once a species has been singled out for captive breeding, policy makers and the public will not be so easily convinced of the need to save the habitat that harbours the species. Other uses for the habitat might now seem more appealing. For example, if the California condor were to be bred in zoos as is proposed (Ricklefs, *et al.*, 1978) then the promoters of mining, logging and recreational facilities would quickly

TABLE 6.1 *List of the twenty-six rare mammal species identified as having self-sustaining populations in zoos*

Vernacular name	Linnean name
MARSUPIALIA	
Parma wallaby	*Macropus parma*
PRIMATES	
Mongoose lemur	*Lemur mongoz mongoz*
Lion-tailed macaque	*Macaca silenus*
CARNIVORA	
Wolf	*Canis lupus*
African wild dog	*Lycaon pictus*
Bengal tiger	*Panthera tigris tigris*
Siberian tiger	*Panthera tigris altaica*
Sumatran tiger	*Panthera tigris sumatrae*
Leopard	*Panthera pardus*
North China leopard	*Panthera pardus japonensis*
Jaguar	*Panthera onca*
PERISSODACTYLA	
Przewalski's horse	*Equus przewalskii*
Onager	*Equus hemionus onager*
Kulan	*Equus hemionus kulan*
ARTIODACTYLA	
Pygmy hippopotamus	*Choeropsis liberiensis*
Swamp deer	*Cervus duvauceli*
Formosan sika	*Cervus nippon taiouanus*
Pere David's deer	*Elaphurus davidianus*
Banteng	*Bos banteng*
European bison	*Bison bonasus*
Lechwe waterbuck	*Kobus leche*
Arabian oryx	*Oryx leucoryx*
Scimitar-horned oryx	*Oryx dammah*
Addax	*Addax nasomaculatus*
Black wildebeest	*Connochaetes gnou*
Arabian gazelle	*Gazella gazella arabica*

From Pinder and Barkham (1978).

point out that the condor sanctuary is no longer needed. The irony in this case and in others is that once there are human incursions into the condor reserve, the species could never be successfully reintroduced. Thus the establishment of a condor breeding programme might, in the long run, endanger the future return to nature. The prevention of such tragedies requires eternal vigilance.

6.2.2 The choice of species for captive propagation

A related argument is that the target species in a captive breeding programme are chosen irrationally. Nearly all of the species selected for CP in
zoos are large, dramatic animals such as rhinos, antelope, apes, cranes and
birds of prey. (Because humans are status-seeking animals that admire
strength or power, they may prize these same qualities in other species.
Curators may themselves prefer a larger v. a smaller animal, just as people
prefer big trees, boxers and television sets.) Still, it is usually true that size
and vulnerability are correlated. As more and smaller species suffer depletion and become endangered, a rational basis for selection will have to be
found.

What other criteria for selection could be substituted for size, assuming
that all candidates are equally threatened with extinction? The list includes
(1) scientific interest, (2) potential economic importance, (3) aesthetics, (4)
prospects for reestablishment in nature, (5) ecological importance (section
5.2.3) and (6) cost of the CP programmes. With regard to the last criterion, it
is obvious that for every large mammal that is captive bred, the equivalent
expenditure of funds could finance dozens of CP programmes for plants and
invertebrates. (This raises an ethical question: does the Indian rhinoceros
have a greater 'right' to preservation than a species of snake or grass?) In
any case, if the concern is safeguarding as much of the biosphere's diversity
as possible, it may be necessary to sacrifice some large forms if it means
saving many smaller ones. We leave open the question of who is to decide
such questions.

6.2.3 Surplus animals

One of the paradoxes of captive breeding is that sooner or later success can
bring more adversity than failure. Surplus animals must be disposed of.
Once the capacity of a facility has been reached, the excess production has to
be sold, traded or 'euthanized'. Another source of excess stock is culling to
maximize the effective population size (section 6.3.3) or to achieve a better
age structure (Goodman, 1980).

One problem is that there are several laws and treaties (section 6.4) that
forbid or strictly regulate the trade in endangered species, thus making it
uneconomic or impossible for zoos to dispose legally of excess stock. In
addition, naive conservationists may not understand the genetic and demographic principles that make such practices as euthanasia necessary. Breeders
will need to work more harmoniously with their fellow conservationists if
they are to be free of legislative and bureaucratic harassment.

6.2.4 Return to nature

Several authors have recently discussed the problem of returning animals to the wild after a sojourn in captivity (e.g. Brambell, 1977; Campbell, 1980; Conway, 1980). Optimists generally emphasize the successful return of the wisent (European bison, *Bison bonasus*) in Poland and the black buck (*Antilope cervicapra*) in Asia, and wolves in the Bavarian National Park. Pessimists can point to cases which have, so far, failed or are yet to be proven successful, such as the peregrine falcon (*Falco peregrinis*) in the eastern USA, the Arabian oryx in Jordan and Israel, and the Hawaiian goose or ne-ne in Hawaii. Assuming, however, that a CP programme for a particular species exists, every effort should be made to insure that the eventual efforts to return the species to its original habitat (or to a new homeland) are successful. On the other hand, the breeders as well as the public should be prepared for occasional failures in the return of a species to nature. This is one of those processes where we may learn from our failures.

Many problems, even subtle ones, can prevent the transplant from 'taking'. Kleiman (1980) discusses some of the non-genetic behavioural changes that are likely to occur in captivity, and that would preclude a successful reintroduction. These include (1) inability to mate, (2) inability to rear young, (3) inability to hunt or forage, (4) inability to escape predators and (5) loss of the fear of man. It should be noted, however, that the likelihood of success is inversely related to the importance of learning and socialization in the species. Reintroduction of invertebrates and lower vertebrates will probably be more successful than reintroduction of mammals, for example, since the latter must often learn how to hunt and how to court.

It goes without saying that every CP programme for a vertebrate needs a consultant behaviourist familiar with the behaviour (social and otherwise) of the species in nature, and who can advise on such matters as (1) which individuals should form the founding nucleus of the colony, (2) the 'furniture' necessary to insure the expression of the entire repertory of behaviour, (3) the kind of enclosure or cage needed to insure the development of normal social structure and normal courtship, breeding and rearing behaviour with virtually no hand-rearing of young, and (4) how to provide for the continuity of hunting and escape behaviours which will be obligatory upon release.

Without such cautions, the release will probably fail, and the CP programme is not worthy of official recognition (section 6.4.3). One suggestion is that any zoo endeavouring to captively breed a species should fund the necessary field work to produce sufficient ecological and behavioural data. With the proper supervision, graduate students could do this work with greater ease and at less expense than could more established professionals.

Many university biology departments would be happy to cooperate in such endeavours.

6.2.5 Limitations of space and resources

Ultimately, the amount of space available for CP sets the upper limit for the total number of such programmes. Available space, of course, depends on funding, and CP must compete with other conservation projects, including the purchase and management of nature reserves, and the latter will continue to attract the biggest piece of a very small pie. Realistically, it is doubtful that the resources for CP will grow much in the next few decades.

Perry and Kibbee (1974) surveyed the capacities of zoos in the US. Table 6.2 shows that there were 32 000 individuals of 885 mammal species distributed among 146 zoos. Considering only the 67 rare species in captivity in 73 zoos, the average total number of specimens per species is only 22, far from the desirable level of 50 to 100. Even worse, the specimens of a given species are widely scattered, there being only 3.2 per zoo on the average. Similar data for European zoos are unavailable.

TABLE 6.2   *The distribution of mammal species and specimens in US zoos*

	All species in 146 US zoos	Rare species in 73 US zoos
Total no. individuals	32 000	1459
Total no. species	885	67
No. zoos exhibiting each species on average	9	7
No. species exhibited by the average zoo	57	6
Total individuals per zoo	220	20
Average no. per species, all zoos	36	22
Average no. per species, per zoo	3.9	3.2

From Perry and Kibbee (1974).

Because most zoos must depend on gate receipts and annual municipal support, political reality prevents these institutions from committing all, let alone half, of their space to expensive CP programmes. The public expects entertainment and a certain amount of taxonomic variety. (Zoos that fail to provide this service will cease to exist, unless they have governmental funding.) After all, what is a zoo without lions, tigers, elephants, bears, camels, kangaroos, giraffes, hippos, crocodiles, pythons, ostriches, parrots, not to mention seal shows, restaurants and other attractions. Following Conway (1980), if half the space now available for mammal exhibition and breeding in the US zoos were to be dedicated to serious captive propagation, only about 150 species at most could be progagated, assuming a minimum of 100 individuals per species (section 6.3.3). Since there are already three times this number of endangered mammals, it is apparent that CP will never completely substitute for protection of animals *in situ*, even making the most optimistic assumptions about space and funds. CP can, however, save many of the largest, and therefore, the most vulnerable forms.

6.3 Genetic problems and management in captive propagation

There are four major genetic problems in CP. These are inbreeding depression, loss of genetic variability, artificial selection and hybridization. The latter problem, hybridization, will arise less frequently than the former three, and in general will be an issue only when the founder individuals are collected in widely separated localities. In this section, we explore the seriousness and pervasiveness of these problems and propose and review some practical mitigating measures.

6.3.1 Evidence for inbreeding effects in zoos

Ulysses S. Seal has written (Seal, 1978) that 'The failure to recognize inbreeding effects, in most cases, can be attributed to the lack of adequate records, not to the overall success of random or inbred breeding programmes.' Until recently, Seal's statement would have been doubted by many curators and zoo administrators.* Whitehead (1978), for example, is sceptical about the deleterious effects of inbreeding, and illustrates his position with Pere David's deer. Among the other species that are often used in an attempt to establish the insignificance of inbreeding are the wisent (*Bison bonasus*), the golden hamster and laboratory rodents such as rats and mice. None of these examples is particularly relevant. With regard to Pere David's deer (*Elaphurus davidianus*), it is, for genetic purposes, a domesticated species, having existed solely in captivity for hundreds of years (Wood-Jones, 1951–52). It is highly probable that most of its deleterious genes have been eliminated as a result of very slow inbreeding and selection. Throughout this long interval, its population size was usually above the threshold for short-term fitness ($N_e > 50$).

The wisent, as already discussed in chapter 3, does indeed show a surprising immunity to inbreeding. In fact, its genetic load is one of the lowest known (Slatis, 1960; Senner, 1980). One explanation for its anomalously

* Traditionally zoo personnel preferred to exhibit the greatest possible diversity of species or, looked at another way, the lowest possible number of individuals per species.

low inbreeding depression is that it, like Pere David's deer, has had a small to moderate size population for a long time, and that its load of deleterious genes has been largely purged by natural selection.

Evidence from laboratory rodents is actually contrary to popular opinion. With regard to inbred mice and rats, the large majority (usually about 90%) of strains died out in the early attempts to produce homozygous lines (Strong, 1978). They died out because of inbreeding depression. The hamster case is irrelevant for another reason – except for the first few generations (a single pregnant female was the founder of the domestic strain), it was not inbred. That is, the fecundity of this rodent permitted its population to rapidly grow into the millions, notwithstanding a bottleneck of $N = 2$ at its inception. Genetically, this 'Adam and Eve' model produces much less inbreeding than does a sustained small population (section 3.1.3).

We do not wish to imply that inbreeding will *always* end in extinction. Not only will some species have very little genetic load to begin with, but chance will, on occasion, bring together as founders some individuals with relatively few deleterious genes, as is shown by the survival of a fraction of lines that are brother–sister mated (chapter 3, p. 69). We would advise, however, that the chances of this happening are so low that prudent breeders would not gamble.

Where sufficient data exist, strong, circumstantial evidence for inbreeding depression usually surfaces in wildlife breeding programmes. Treus and Lobanov (1971) attribute the high level of rickets in the Ashkania Nova zoo herd of eland (*Taurotragus oryx*) to inbreeding. Bouman (1977) and Flesness (1977) agree that the declining fitness of the Przewalski horse (*Equus przewalskii*) herd is due to inbreeding. Most of the present stock is descended from three stallions and seven mares (a bottleneck of $N_e = 8.4$, formula 3.4). Not only is the number of births per mare falling, but the horses are dying younger: before World War II, 41 out of 191 (21%) reached the age of twenty; since the war, only 11 out of 163 (6.7%) reached this age. Flesness showed a significant association between an individual's inbreeding coefficient and both fertility and fecundity.

Apart, however, from anecdotes and isolated cases, we should be looking at data on the frequency and amount of inbreeding depression in an unbiased or random sample of zoo herds. Such information is now beginning to surface. Ralls, Brugger and Ballou (1979) examined records from several zoos, especially the National Zoological Park (Washington, DC) in order to compare juvenile mortality for non-inbred and inbred young of sixteen species of captive ungulates. Overall, juvenile mortality for inbred young was higher in fifteen of the sixteen species ($P = 0.0003$), though the magnitude of the effect varied widely. Ralls and her coauthors were also able to study the effects of inbreeding in selected individuals. For sixteen out of twenty females (belonging to seven species), the percentage mortality of young was greater when the female was bred to a relative than when mated to an unrelated male ($P = 0.01$). Due to the incomplete nature of zoo records, these authors could not correlate the amount of inbreeding with the level of depression in survival or fecundity. Nevertheless, these results should go a long way towards dispelling doubt over the ubiquity of a significant genetic load in wild animals.

6.3.2 Genetic guidelines for maximizing effective size

What is the cost of ignoring the potential of inbreeding depression in CP programmes? We feel there is no room for equivocation – the cost is almost certain failure. Given the evidence from domestic species (chapter 3) and the likelihood that the genetic load is even greater in undomesticated animals, it seems utterly senseless to invest resources in a CP programme unless every reasonable effort is made to manage the group in a genetically rational manner.

Senner (1980), using fairly conservative assumptions about genetic load, has produced a general model of the fate of a captive group maintained at a limit of ten animals ($N_e = 10$), and founded with four animals. Most zoo groups are maintained at far fewer than 10 (Table 6.2) and founded with two or three. Fig. 6.1 shows the general results. The probability of survival remains high until about the twelfth generation, and by the twentieth generation, it is only about 20%. In addition, heterozygosity declines to about 60% of the original level by about the tenth generation.

Senner's model assumes that each female produces five viable offspring. He showed that group survival is very sensitive to this variable – when the offspring number is three, the probability of survival drops to 0.5 in less than seven generations. The smaller the mean offspring number, the more likely it is that genetically handicapped individuals will be parents of the next generation.

Another important variable is the genetic load itself. The known range in the magnitude of genetic loads for viability, so-called B values, is from near zero to as high as 3.74 in the chukar partridge (Abplanalp, 1974; Senner, 1980). Senner used a value of 1.0 in his basic model. Levels of B higher than this result in a significant drop in survivorship of groups. This also applies to female fecundity. Other variables, including the effect of inbreeding depression on the sex ratio, the relatedness (and the level of inbreeding) of founders and the number of founders (so long as it is greater than four or five – see section 3.1.2), are significant but less so than those just mentioned. Senner, like previous authors (Flesness, 1977; Denniston, 1978) emphasizes the overwhelming importance of effective size. Fig. 6.2 illustrates the dramatic effect of increasing the maintenance size of the group.

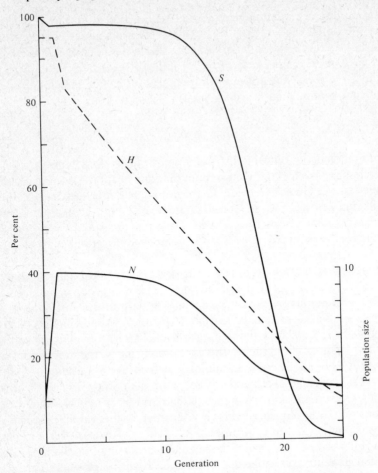

Fig. 6.1 Results from a mathematical simulation of a captively bred population. S is probability of survival. H is mean individual heterozygosity (100% is level in wild-caught specimens). N is population size. (After Senner, 1980.)

These results can be summarized as a list of rules and principles:
1. The inevitable fate of small populations of animals is extinction, although the longevity of such populations responds to or is sensitive to certain variables, all of which are to a greater or lesser extent subject to change and control by the breeder.
2. Among the most important variables is offspring number (fecundity). Breeders should attempt to equalize (see section 3.1.6) and maximize the number of offspring per female; the larger the number, the greater the opportunity for natural selection and for culling (see below), and

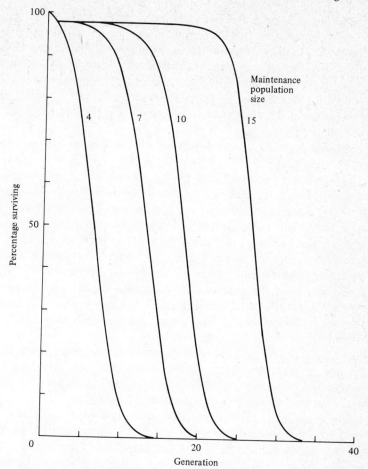

Fig. 6.2 The sensitivity of Senner's model to variation in maintenance size limit (after Senner, 1980).

the lower the chance that stochastic genetic and demographic events will bring about the premature termination of the group.

3. Inbreeding must be minimized because both fecundity and viability, which are critical variables, will be depressed at a rate proportional to the amount of inbreeding.

4. Other variables with significant impacts on group survivorship are sex ratio (it is also affected by inbreeding), the relatedness and inbreeding of the founders, and the number of founders.

5. The most important factor, everything else being equal, is group size. The reason is that the rate of inbreeding is inversely proportional to

population size (section 3.4). Therefore, the principal goals of CP programmes, once established, should be the maximization of both the effective size of the group, and the sizes of families. The latter goal is a matter of skilled and sophisticated husbandry, including behavioural management (see Kleiman, 1980), demographic management (e.g. Goodman, 1980; Foose, 1977, 1978; Miller and Botkin, 1974) and reproductive monitoring and veterinary support (Benirschke, Lasley and Ryder, 1980).

The maximization of N_e also depends on good husbandry. In addition, however, it requires the application of the population genetic principles discussed in detail in sections 3.1 and 3.4. In our opinion, the fundamental guideline should be the basic rule of short-term conservation genetics (section 3.4); that is, a minimum effective size of fifty, not all of which have to be in the same institution. The principal tactics employed to achieve the maximum effective size (whether it is more than or less than fifty) are:

1. Maintaining an equal number of males and females (section 3.1.4).
2. Preventing large fluctuations in the census number (section 3.1.5).
3. Closely managing the reproductive contributions of families so as to approach equality in productivity (section 3.1.6). This may require the promotion of fecundity in the less productive pairs and culling of progeny from the most productive pairs.

To illustrate these principles, we will compare an imaginary 'worst case' with an imaginary 'best case' CP programme. For each case, we begin with thirty-two adult animals, sixteen in one zoo, sixteen in another. Table 6.3 summarizes the management statistics and decisions.

For the 'worst case', the social structure of the two groups is allowed to evolve without interference. The result is that in both zoos, there are two dominant males that manage to sexually monopolize all the females. The effective size, using formula 3.4, is therefore seven in each zoo. Furthermore, there is no movement of animals between zoos for breeding purposes. Assuming a random distribution of progeny among the females, in each zoo the expected rate of increase in F each generation is $1/2N$ or 0.125. This is 12.5 times the maximum acceptable level if we employ the basic rule for short-term maintenance of fitness (section 3.4).

For the 'best case' the animals are handled in such a way that all males mate with the same number of females; that is, there is a sex ratio of 1.0 (instead of 0.25 in the 'worst case'). In addition, in each generation there is a reciprocal transfer of one or more animals between the two institutions, insuring effective panmixis (section 5.4.5). Finally, the progeny are culled so that the families contribute equally to the next generation. Thus, the effective size is approximately $2(N - 1)$ or sixty, and the rate of inbreeding is $1/120 = 0.0083$, or just below the 1% threshold and fifteen times less than the 'worst case'.

TABLE 6.3 *Hypothetical 'worst' and 'best' cases for the genetic management of a captive group of animals*

Variables and results	'Worst case'	'Best case'
Original number of animals	32	32
Number in each zoo	16	16
Number of males breeding per zoo	2 per zoo	8 per zoo
Sex ratio	0.25	1.0
Migration	none	*ca* 1 per generation
Mating system	random	forced pairing
Progeny distribution	random	equal
Change in F per generation	0.125	0.0083
N_e	4 per zoo	32 per zoo

This example shows how the application of the most basic principles of conservation genetics can profoundly affect the rate of inbreeding in captive or closely managed populations. The correct application of each of the three principles (sex ratio, migration, progeny distribution) produces a difference of an order of magnitude or more in the rate of loss of genetic variation and the rate of fixation of deleterious genes. It goes without saying that inattention to these principles seriously jeopardizes the survival of a captive group and is a *de facto* admission that the programme has no scientific merit, and little or no conservation value.

Do the principles always apply? One situation in which it would be tempting to ignore them is where a long-lived organism is to be propagated for a finite time. Say, for example, that a goal of an elephant CP programme is return to nature in 100 years. It might be argued that in such a case a relatively small number of animals would suffice because so little inbreeding would occur due to the few generations (about five) that would pass in the 100-year interval. The flaw in this reasoning is equating the *amount* of inbreeding with the *rate*. Indeed, it is true that an effective population of twenty elephants would only lose about 10% of its heterozygosity in 100 years, and that a species with a one-year generation time would need to have an effective population of 500 if it were to lose no more than the elephant herd does on an annual basis.

Nevertheless 20 elephants are not as 'good' as 500 mice. The reason is that the efficacy of natural selection is a function of $2N_e$ (section 4.2.3). When $2N_e$ is less than 100, natural selection is almost powerless to counteract the fixation of mildly deleterious genes, but when $2N_e$ is greater than 1000, genetic drift is negligible and selection can 'cleanse' a population of such genes.

Thus, the per generation rate of change in F is crucial. For the elephants in this example, it is $(1/2N)$ 0.025. For the mice, it is 0.001. This difference means that some deleterious genes are likely to be fixed in the elephant group, but that many fewer would be fixed in the mice. This is important because the results of attempts to return animals to nature following a sojourn in captivity depend in part on the fitness (e.g. fecundity, viability) of the founders.

6.3.3 Domestication and selection

It can be convincingly argued (Spurway, 1952) that domestication is an insidious and corrupting force in every CP programme and that it is impossible to avoid selective breeding by the humans who manage such programmes. A corollary of this is that the longer the duration of a CP programme, the more 'domesticated' the group has become. Evidence for this proposition comes from a study by Barnett and Stoddart (1969) of the behaviour of male rats (*Rattus norvegicus*) that had been in the laboratory for six to nine generations. The behaviours associated with conflict in these rats were compared to those for wild-caught males. Barnett and Stoddart found that the captive bred rats were significantly less aggressive by several criteria. They argue that the differences are probably genetic, although their evidence did not rule out an environmental effect stemming from differences in the history of the experimental subjects.

It is possible to recognize at least four types of selection that are likely to affect the genetics and phenotypes of captively bred animals:

1. *Selection for increased productivity.* We are aware of at least one incident in which a curator has consciously selected for an increase in the parity of a species that typically has twins. The reason for the selection (for triplets) was economic: The extra offspring could be sold at a large profit. While granting that such artificial selection might enhance the survival potential of the group while it is in captivity, such a change in fecundity would probably be non-adaptive in nature, assuming that natural selection had optimized the states of reproductive characteristics.

2. *Selection for (perfect) type.* Such selection may reduce the variability found in nature (it is similar to normalizing or stabilizing selection), and

could reduce the fitness of the population. Such selection will also change the averages of traits. In whatever direction the departure tends from the pre-captive morphology and behaviour – towards larger or towards smaller size, towards earlier maturation or later, towards showier colours or duller – the change will be deleterious to a newly established wild population. Biogeographic studies provide copious evidence for the subtlety of morphological and behavioural fine-tuning which is the result of natural selection.

A good example of phenotypic change occurring in captivity is the Przewalski horse. Zoos have had considerable success breeding this wild horse which was native to Mongolia but is now thought to be extinct there. At least two changes, both apparently genetic, have been noted in the 5–8 generations of captive breeding (Volf, 1975). The first is early sexual maturity. The second is an increase in the foaling season.* Whereas the original stock foaled only in the spring, the captive descendants now foal all through the year. These changes are neutral or even advantageous in captivity. But in nature, the animals that foal too early or too late would constitute a genetic burden on the population because they would use up resources but would leave no offspring. Foals dropped too soon or too late would starve or die of inclement weather. In nature, early maturing individuals would be too small or too naive to successfully rear young, or the young might be born out of season.

3. *Selection for tractability.* The breeder may select relatively tame, tractable animals to breed, leaving out the skittish or aggressive individuals because of the difficulty he has had in handling them. In nature, of course, the sensitivity and 'nervousness' that is such a hazard in zoos, is just the behaviour that keeps many prey from becoming meals of stealthy predators. An analogous kind of selection will be automatically practised against easily stressed animals.

4. *The practice of 'non-selection'.* Sometimes we forget how few newborns survive to adulthood in the wild and, accordingly, how strong the selective forces can be, A lion pride, for example, suffers a surprising mortality (Schaller, 1972). Starvation, disease and accidents are imminent dangers. The sick usually die. When food is scarce, all the young may perish. Consequently, there is continual and oppressive natural selection acting against the weak, the small, the slow, and those born with the slightest physiological or anatomical defect.

The probability that sick, aberrant or phenotypically abnormal animals will reproduce in zoos (with their medical, nutritional and environmental

* These changes may be partly or fully attributable to genes in the Prague Zoo's herd that originated from a cross between a Przewalski stallion and a domestic mare (Kurt Benirschke, personal communication).

support services) is much greater than is the probability of their survival and reproduction in nature. Thus, the frequency of deleterious genes is bound to increase in captivity. The longer a group in captivity is bred, the more handicapped genetically it will become, on the average, although, as in urban humans, many of these 'handicaps' may enhance fitness in the artificial environment of a modern zoo.

A related danger, and an ever-present temptation, is hand-rearing. It is common zoo practice to hand-rear young. Many of these young are sick or rejected by their parents. Many others are removed to display in nurseries or children's (petting) zoos. Whatever the reason for removal, however, such young will nearly always suffer psychological trauma or social deprivation, and will not be able to function as sexually and socially normal adults (e.g. Goldfoot, 1977). Because of hand-rearing, apes must be coached in maternal behaviour.

6.3.4 Racial purity and the sub-species problem

Races and sub-species are geographic entities. Most species differ from one place to another within their ranges, a phenomenon called geographic variation. Unless populations are isolated, such as on islands or mountain tops, the change in phenotypic characters or gene frequencies is usually gradual (clinal is the technical term); abrupt changes in characteristics are relatively rare except where there are barriers to gene flow. The more information we have about the geographic variation within a species, the more difficult it usually becomes to draw the lines on a map that indicate the borders between races or sub-species. Nevertheless, it often appears to be satisfying or convenient to ignore the continuous nature of geographic variation in many cases, and to formally recognize a geographic section of a species by giving it a latin name and publishing a description. It is widely acknowledged, however, that the borders that define and separate sub-species are usually quite arbitrary.

Mayr (1969) defines sub-species as a 'geographically defined aggregate of local populations which differ taxonomically from other such subdivisions of the species'. The circularity of this definition is obvious. All one needs do to create a sub-species is to define it taxonomically, and this only requires a little motivation and less information. Any diploid, outcrossing species can theoretically be broken down into finer and finer groupings, and this process ends only at the individual, no two of which are exactly alike. Finding statistical differences between populations is no barrier to erecting sub-species, as long as one has large enough samples or uses relatively trivial characters whose variation is controlled by one or just a few loci (such as colour in mammals and some scale counts in reptiles). An example of the unlimited opportunities in this field is the taxonomy of man himself; some authors have recognized up to thirty races (Coon, Garn and Birdsell, 1950; but see Ehrlich and Feldman, 1977). Among zoologists, ornithologists have been more 'successful' at this practice than other taxonomists; there are 3.3 sub-species of birds per species (Mayr, 1970). Aside from the large number of practising ornithologists, the principal reason for the existence of so many sub-species in the birds is that, compared to other vertebrates, birds rely more on visual signalling for recognition purposes. Humans, too, are predominantly visual and this makes it easier for humans to detect subtle differences in the showy plumages of birds.

As mentioned above, one of the reasons for the arbitrariness of sub-species is that the borders of adjacent sub-species are indistinct. Aside from the arbitrary nature of many sub-species, there is another problem. A quarter century ago, Wilson and Brown (1953) argued that where one draws the lines between sub-species depends on which characteristic one happens to be looking at. Wilson and Brown pointed out that concordance in geographic patterns is the exception rather than the rule, and that sub-specific boundaries change with the characteristics used.

Geneticists have long recognized these problems. As Dobzhansky (1970, p. 290) stated, 'It should be made unequivocally clear that the number of races or sub-species which one chooses to recognize . . . is largely, though not completely, arbitrary.' As far as we can see, the only cases for which the bestowal of a latin trinomial is not arbitrary is where a population is completely isolated as on an island. But for island populations, the name of the island suffices to identify the population.

Because of habit and tradition, many taxonomists have chosen to ignore all of these problems. From the point of view of conservation, this would be irrelevant and relatively harmless were it not for an unwritten commandment that is almost always adhered to by the large majority of zoo curators and wildlife managers. The commandment is: *Thou Shalt Not Mix Sub-species* (a suspiciously familiar, if politically moribund idea in the human species; incidentally, there is no evidence for genetic problems resulting from crosses between the most phenotypically different human populations). It is almost as if the conferral of a latin name were a benediction carrying with it the insurance of genetic purity.

Both aesthetic and genetic arguments are used to defend the taboo against crossing sub-species. The aesthetic argument is the argument of taste or values: a pure Bornean orangutan is more attractive, say, than a cross between a Bornean and a Sumatran orangutan. Everyone has his standards of beauty and quality, but it could not be claimed that this argument is relevant to the issues of fitness and adaptation.

The genetic consideration is the very real possibility of partial or complete sterility and inviability of the offspring of genetically distinct parents. In theory intra-specific hybridization should rarely be a problem because sterility and inviability are commonly used as criteria to differentiate between species.* The occurrence of such symptoms would lead many to split the species into two. Nevertheless, entities referred to as sub-species do quite often manifest genetic incompatibility. In crosses between sub-species of *Drosophila equinoxialis*, for example, the males may be sterile (Ayala, Tracey, Barr and Ehrenfeld, 1974), and A. E. Gill (personal communication) has reported male sterility in the offspring of two populations of the meadow vole, *Microtus californicus*, from localities about 600 km apart in California. In general, it seems that the probability of hybrid sterility or inviability in crosses between populations increases as the geographic distance between localities increases. This apparently applies to *D. equinoxialis* and to *D. pseudoobscura* (Prakash, 1972). In amphibians, crosses between remote populations often fail (Fowler, 1964), thus demonstrating that the biological species concept is not applicable to these species.

When problems do occur in intraspecific crosses, they are usually attributable to chromosomal differences in the hybridizing individuals. If the chromosomes are sufficiently different, they may fail to pair properly during meiosis; when such *chromosomal sterility* occurs, few or no viable gametes are formed. Sometimes viable gametes are produced but development fails at some point before or after fertilization; this is sometimes referred to as *genic sterility*. Sometimes hybrid sterility or inviability may not appear in the F_1, showing up in the F_2 or later generations instead. In such cases, it is assumed that recombination between the parental type chromosomes has produced poorly integrated recombinant linkage groups.

It is becoming increasingly apparent that cytogenetic analysis is an indispensable tool for the captive breeder (Benirschke *et al.*, 1980). Many cases are coming to light for which a high frequency of spontaneous abortion or the sterility of a breeding pair of animals is attributable to differences in the karyotypes of the parents. For species in which the rate of chromosomal evolution (inversions, translocations, fusions) is high, breeders will need to be on the lookout for infertility and to take extra pains to karyotype the parents or determine the precise geographic origin of their specimens.

South American primates, for example, appear to have an extraordinary amount of chromosomal polymorphism. Owl monkeys (*Aotus trivirgatus*) vary from $2n = 46$ to $2n = 56$ in chromosome number, and matings between animals with similar karyotypes result in improved reproduction (Cicmanec and Campbell, 1977). Cytogenetic surveys will have to be a routine procedure in the propagation of such chromosomally polymorphic species.

Some might argue that nature should be allowed to take its course, and that breeders should not interfere in the pairing of individuals. This

---

* Here, the controversy over the species concept (Ehrlich, 1961; Sokal and Crovello, 1970) is relevant, but not relevant enough to justify further comment. The question is the genetic differences between the organisms, not their taxonomy.

approach only makes sense if the colony members stem from the same geographic area and, *a priori*, should be karyotypically very similar. All too often, though, the origin of specimens is questionable or unknown.

The detection of chromosomal anomalies during pregnancy by amniocentesis is another use of karyotyping that the breeder can sometimes employ, assuming there is reason to suspect a problem. For example, trisomics (the chromosomal correlate of mongolism or Down's syndrome in humans) have already been detected in all three species of great apes. The early detection of such cases could save precious time and money.

One basis for the taboo against the crossing of sub-species is an implicit assumption that the probability of karyotypic difference increases with taxonomic or geographic distance. In general, this is probably true, but it is not axiomatic. Members of the cat family all have the same number of chromosomes, as do the bears. Modern cytogenetic techniques, including banding studies, can and should be used to screen for karyotypic differences – if none are detected, the prohibition must be maintained on some other basis.

Another potential hazard of mixing sub-species is the production of artificial strains whose adaptations do not precisely fit the habitats of the two or more parental populations. The extent to which intraspecific hybridization actually decreases fitness depends on many factors. Among the most important of these is the behavioural flexibility of the organisms. Invertebrates and lower vertebrates are likely to be more affected than mammals, because the behaviour of the former groups has a higher level of innate determinism, while the behaviour of mammals is more 'open' and plastic. Where there is reason to suspect that inappropriate behaviour might result from a cross, it is obviously advisable to investigate the behavioural ecology of the natural populations.

What are the *advantages* of ignoring racial or sub-specific classifications? There are two – the elimination of inbreeding depression (chapter 3), and the enhancement of evolutionary potential (chapter 4). Whether these advantages outweigh the possible disadvantages will depend on the specifics of the particular situation.

An interesting idea relevant to this topic is the concept of 'optimum outcrossing distance' (Bateson, 1978). That is, if a breeding pair are too closely related (genetically similar), the offspring may be less than optimally fit because of excess homozygosity or homozygosity for deleterious genes. On the other hand, if parents are too distantly related, there can be less than optimal chromosomal and developmental integration in the F_1 and following generations. The former is called inbreeding depression; the latter is outbreeding depression (Darlington and Mather, 1949; Falconer, 1960). For example, Bateson (1978) has shown that male Japanese quail (*Coturnix coturnix japonica*) prefer to court and copulate with unfamiliar but phenotypically similar females, eschewing both females to which they were exposed as young as well as birds with grossly different plumage.

Price and Waser (1979) were able to test the optimum distance hypothesis directly. They found that seed set in the meadow perennial *Delphinium nelsoni* was highest when the flowers were fertilized by pollen from plants growing at a distance of about 10 metres. Seed set dropped off if the parents were growing closer together or further apart than this distance.

If such a phenomenon occurs in animals, the optimum physical distance is probably much greater and will depend on, among other things, the vagility of the species. At this stage in our art, we cannot rule out the possibility that crosses between races could be beneficial in some situations, especially when the last vestiges of a particular sub-species are showing signs of inbreeding depression, and the only available unrelated animals belong to another 'sub-species'.

Those who are categorically repelled by the idea of mixing races might consider the following rhetorical question. Given that the sun was about to go nova and man had the technology to send a single, one-way colonizing rocket to a planet in another solar system, what would be the optimum 'racial' makeup of the colonist crew? Say that we know very little about the climate of man's new home, nor whether the planet is inhabited by an indigenous, intelligent life form. The return to nature of captive animals after one or two hundred years in captivity may be analogous to the scenario. Can we assume that the climate, flora and fauna will be the same as now and that today's sub-species will have the best chances of survival?

6.4 Politics and economics of captive propagation

Captive propagation does not proceed in a social vacuum. Ethics, economics and legislation are at least as important as is the art of husbandry and the application of genetic principles. Thus, we feel obligated to make some comments about the social milieu of CP, and to make a few proposals.

6.4.1 Ethics

Some conservationists (the 'better dead than bred' variety) would applaud the extinction of zoos. We would not. Zoos and similar institutions have a tremendous potential for public education in the area of conservation goals, problems and programmes. With some legislative and fiscal 'encouragement' this potential could be realized in a much deeper public understanding of natural history and conservation. We do not deny that some, probably most, zoos exploit their inmates and do a less than adequate job of education, let alone scientific captive breeding. In the worst examples, the psychological welfare of the animals has a low priority, and even in the best zoos there are cramped, antiquated exhibits which produce neuroses and social deprivation. As Campbell (1978) states, 'Improperly kept under poor conditions, they languish in misery and boredom.' For example, keeping great apes in cages against their will is a practice of doubtful morality, given their hominid behaviours (Gallup, Boren, Gagliardi and Wallnau, 1977) and remarkable linguistic abilities (Premack, 1971). (On the other hand, CP may be the only way to insure their ultimate survival.)

Yet, it is both myopic and unfair to blast away at these institutions with the guns of righteous indignation. There are many examples of zoos which, given their limited resources, are doing excellent basic research in conservation and related fields (e.g. Benirschke *et al.*, 1980; Conway, 1980; Kleiman, 1980) and which are making serious attempts at public education and scientific captive breeding. These institutions need encouragement, not harassment.

The less conscientious institutions will either have to change or be closed. We believe that society at its present stage can neither afford nor long tolerate zoos that are 'in it strictly for the money' and which ignore the welfare of wildlife in general and individual specimens in particular. The exploitation and consumption of wildlife in zoos is only tolerable if there is a compensating effort made to explain the status of wildlife to the customers and awaken in them a feeling of responsibility for their non-human relations.

We do not expect that all zoos will be able to afford to initiate and house scientifically managed CP programmes. We would expect, however, that all institutions that hold and display wildlife will at the very least cooperate with institutions that engage in CP. This cooperation would take several forms, including:

1. breeding loans to enlarge the gene pools of CP programmes;
2. the temporary housing of surplus or reserve stock;
3. the sharing of information and expertise.

6.4.2 Economics

CP is expensive. It costs several hundred to several thousand dollars a year to feed and care for a large animal. Therefore, one of the main obstacles to the establishment of CP programmes (which generate little in the way of 'gate' receipts) is cost. Zoos cannot rely solely on the beneficence of wealthy patrons and conservation organizations to endow such programmes, so it appears that public institutions will have to underwrite the majority of CP programmes.

The California condor (*Gymnogyps californianus*) breeding programme may set the precedent in the US. Following intense efforts to protect North America's largest bird in a small remnant of its original range, the US Fish and Wildlife Service, with the backing of the National Audubon Society and the American Ornithologists' Union have agreed that the population, now numbering about thirty, is continuing to shrink to oblivion, and that CP is a necessary component of a comprehensive recovery programme. The plan now is to remove about nine birds from the wild and to breed them (along with the single captive individual at the Los Angeles Zoo) at the San Diego Zoo's Wild Animal Park and at one other facility. Already $662 000 in federal funds are requested for the programme, but this is only a fraction of the eventual costs. Over the next ten or twenty years, the CP aspects of the recovery programme, including capture, breeding and release, will probably cost several million dollars. The California Department of Fish and Game as well as the Audubon Society are participating in the funding and administration of the programme.

This programme has hardly begun, but already there are some obvious morals and probable consequences. First, it pays to be a big species – such treatment would not be lavished on an ordinary vulture, let alone a species of grass or toad. Second, the number of CP programmes that can be underwritten is probably very small. It appears that such programmes will often require an extraordinary level of administrative and financial cooperation between public and private institutions, and there are low and unpredictable ceilings on the budgets of such 'non-essential' expenditures. Third, only a handful of zoos have the financial margin to commit themselves to CP projects unless there is some form of guaranteed governmental assistance and leadership (Warland, 1975).

6.4.3 Legislation

John Milton said that 'Men of most renowned virtue have sometimes by transgressing most truly kept the law.' If the survival of biotic diversity on this planet is a moral imperative, then breeders are almost forced to civil disobedience. The plethora of laws and regulations controlling every aspect of CP are at best stifling and at worst a paralysing mass of red tape. In the US, for example, there are twenty-seven different offices within the federal government agencies that have some authority over the use of plant and animal species (Anon., 1979), not to mention the countless state and local bureaus. These offices obtain their authority and direction from statutes and treaties including the Endangered Species Act of 1973, the Lacey Act, the Marine Mammal Protection Act, the Migratory Bird Treaty Act, the Black Bass Act, the Animal Welfare Act, and the Convention on the International Trade in Endangered Species. Designed to prevent or limit the commercial exploitation of wildlife, these statutes also effectively cripple legitimate scientific activities that can increase our knowledge and help to protect rare and endangered species. The difficulties are compounded when organisms must be moved across international or state borders, and months or years are required in most cases to obtain the necessary permits and approvals for the collection, shipping and exchanging of specimens for breeding purposes. The time delays and expenses are fast becoming prohibitive. Thus, the very laws that were intended to protect organisms against excessive suffering, exploitation and extinction are themselves barriers to the research and projects which are focused on these very ends.

It has been suggested (Anon., 1979) that a blanket exemption from these regulations be applied to the scientifically legitimate and ecologically sound activities of museums and universities. This proposal has merit but it should also be extended to scientifically designed and managed captive breeding programmes.

Safeguards, of course, are essential. With regard to CP of gazetted rare, threatened or endangered species, the primary qualification for such exemption should be participation in a *Designated Captive Breeding Programme* (DCBP). In other words, we are proposing that CP become essentially a licensed activity and that the licensing agency be either an international body such as the IUCN (or its Survival Service Commission) or national bodies established for this purpose. In order to qualify for DCBP status, an institution or a consortium would need to submit a proposal detailing its plans for acquisition, propagation and release. Granting of DCBP status would be contingent on establishing that:

1. the target species is truly in danger in the wild and that its extinction would have significant cultural or ecological impact;
2. the programme would actually enhance the chances of survival;
3. data and records will be maintained in a manner consistent with the highest scientific standards and be accessible to all participants and to all qualified scientists;
4. a detailed *breeding plan* is agreed to by the participants, and that the plan meets certain genetic, demographic, behavioural, veterinary and husbandry standards, and that there is a guarantee of expert consultation for these five areas;
5. contractual agreements be submitted that bind the cooperating institutions to the above *breeding plan* and to the decisions of the steering committee;
6. there are means to establish the progress and accountability of the programme.

We think that there are several advantages to this proposal, or something like it. First, the so-called CP programmes now in existence are a hodge-podge of projects, some with considerable scientific and conservation merit, but many which are completely devoid of scientific expertise and control. Yet even the concerned professional can hardly distinguish the sound from the shoddy programmes. Virtually all zoos exploit the birth of rare animals in their advertising and public relations, whether or not the birth is related to a scientifically legitimate conservation programme. Further, it is not uncommon for any group of four or more animals to be exhibited as a 'conservation breeding project'. Many signs on exhibits and birth announcements are misleading to the public. Even those few zoos that are engaged in serious and costly CP programmes often exaggerate the significance of their accomplishments and herald any birth as a great victory for conservation. If a DCBP authority existed, and if the communications media were attuned to the distinction between DCBPs and undesignated 'breeding programmes', the deserving institutions would receive their due credit and the undeserving would receive encouragement to participate in the consortia that comprise DCBPs.

A second advantage of this proposal over the current unregulated or anarchical state of affairs might be that the administrators of designated programmes would be able to collect, ship, trade and even euthanize animals without having every transaction scrutinized or delayed by government agencies. Instead of a never-ending series of hurdles, there would be only one – the approval–recognition process. Once a DCBP status was granted, the blanket exemption mentioned above should be automatically applied. The necessary safeguards against abuse should be built into the proposal – principally in the form of a team of consultants (some of whom should be on the Steering Committee) whose reputations would be sufficient guarantee of high scientific and ethical standards.

A third advantage is that the implementation of such a proposal could cause an immediate rise in the status of CP among conservationists and biologists. One of the reasons that CP does not currently enjoy much esteem is that there are no standards of performance or criteria of evaluation. (Someday zoos might proudly display the number of DCBPs in which they participate, just as universities headline the number of Nobel laureates on their faculties.)

A further advantage of DCBPs would be an improvement in the quality of CP programmes. The best institutions and the best programmes would qualify with relative ease for the DCBP status. Informal encouragement as well as rewards could convince other institutions to follow suit. Not only would there be the publicity value for participation in breeding consortia, but, for reasons given in the preceding section, there should also be positive inducements in the way of grants and contracts for CP. This may be happening already. There is legislation (HR 805) pending in the 96th Congress of the US to establish a 'National Zoological Foundation', one of whose functions is intended to be the granting of funds for CP. It is logical to link such funding to some accreditation process such as the DCBP status, whether or not the designating body is the 'NZF' or some other national or international group.

6.5 Summary

1. Because wild animals are bred for many reasons, each programme will, to some extent, be guided by a special or unique set of rules. Nevertheless, there are four major categories of captive breeding: (1) stocking and replenishing exhibits; (2) supplying specimens for research and experimentation; (3) for domestication or the improvement of existing domesticates; (4) conservation–rescue of endangered forms and return to nature if possible. With the exception of the second category, captive breeding will be largely confined to zoos, although experimental farms and government research institutions should be more involved, particularly in domestication and conservation.

2. For some animals, especially large ones, captive propagation (CP) will be the only practical means of guaranteeing survival. It is doubtful, however, if more than a few hundred species of birds and mammals will ever be bred successfully in captivity, primarily because of economic constraints.

3. Behavioural management should be a major component of CP programmes. Such an emphasis will often improve productivity in captivity. Equally important, however, attention to the entire range of natural behaviours will usually be a prerequisite for successful reestablishment of species in natural or semi-natural habitats.

4. Genetic management is absolutely essential if the captive group is to retain its fitness, evolutionary potential, and if domestication is to be avoided.

5. There is convincing evidence for the universality and seriousness of inbreeding depression in captive colonies of wild animals. Thus, a prime directive of CP is the maximization of effective population size and the avoidance of inbreeding.

6. One of the most important goals of genetic management should be the increase of fecundity – not only because it leads directly to larger populations, but also because it permits more latitude for natural selection and culling. Thus, sophisticated reproductive management and husbandry are essential elements of genetic management. To be avoided, however, is selection for a genetic change in fecundity, such as selection for early maturity or multiple births; such changes might actually decrease fitness in nature.

7. The principles of population genetics should be applied in order to maximize effective size; we recommend that the minimum be $N_e = 50$, and that this be achieved by managing sex ratio, migration between colonies in different institutions, and by equalizing the reproductive contributions of families. In the best cases, N_e can be twice the actual number of breeding adults.

8. Failure to apply the simple principles outlined in this chapter and elsewhere is, to us, evidence that a programme has little or no conservation value. In fact more harm than good can come of poorly managed programmes, because deleterious genes are more likely to be fixed in such groups and this could ultimately decrease the chances of survival for the species as a whole.

9. Artificial selection and hand-rearing must be carefully monitored and limited wherever possible.

10. The taboo against mixing sub-species should be tempered with karyotypic data and common sense. In most cases sub-species have little biological meaning and are only a rough guide to geographic separation. Sub-species hybridization may be beneficial to inbred lines or lines depleted of variation. Cytogenetic studies are indispensable when a cross between geographic populations is being considered.

11. We propose that captive propagation of endangered animal species be licensed by national or international bodies and that such Designated Captive Breeding Programmes (DCBP) meet certain minimum scientific standards. Zoos should receive recognition and financial support for legitimate conservation efforts, and should be encouraged to join in breeding consortia, the organizational scaffolding of DCBPs.

7

The role of botanical gardens in conservation

7.1 The threatened plant species

In conservation, as distinct from their many other functions and responsibilities, the role of botanical gardens relates foremost to the *threatened or endangered species*. In the maintenance of living collections of such species botanical gardens have many analogies with their zoological counterparts. Yet for the systematist and evolutionist, the botanical guardians of endangered species have available a last resort – the preservation of representative *non-living* (herbarium) collections.* Corresponding collections of zoological material – especially of vertebrates – would be a great deal more costly to obtain and accommodate. It is not a mere accident that such zoological research materials usually are maintained by museums rather than zoological gardens, whereas botanical gardens continue their tradition as centres for research in the taxonomy of plants, in part at least based on the wealth of herbarium specimens which have been accumulated and studied over the years. These and other sources of information enable botanical gardens to serve as intelligence and information centres for endeavours to locate, identify and preserve threatened plant communities or species.

A widespread concern for the preservation of threatened plants is more recent than that for the survival of rare and endangered animal species. This is partly due to lesser awareness, partly to the fact that many animal species have been subject to *selective reduction* by man, whereas plants, and of course many animals, are mainly endangered by *deprivation of habitats* through the destruction of natural ecosystems, rather than by man's impact on particular species. A concern for threatened species was strongly expressed by the United Nations Conference on the Human Encironment held at Stockholm in 1972. In 1974, the International Union for the Conservation of Nature and Natural Resources (IUCN) established the Threatened Plants Committee (TPC) with the objective to enlist worldwide participation of botanists in gathering information on threatened species and their

* The conservation of seeds, discussed in section 7.2.1, is not a 'last resort', since the half-life of seed of many wild species is rather limited.

location, and on means for their preservation. To this end TPC decided to set up three types of organization – (1) regional, on the basis of geographical association, (2) for special plant groups of particularly threatened taxa, and (3) institutional, based on particularly concerned institutions such as botanic gardens, university departments, etc. (Heslop-Harrison, 1974).

A great deal of information on the conservation of threatened plants in living collections and the more general role of botanic gardens in nature conservation was brought together at a conference held at the Royal Botanic Gardens, Kew, in 1975 (Simmons *et al.*, 1976). The conference proceedings have been extensively used in the preparation of this chapter.

How are threatened species defined? The TPC has proposed a modification, for use by botanists, of the categories introduced by IUCN for both animals and plants. The proposed terms are 'endangered', 'rare' and 'vulnerable'. Endangered species are in serious risk of disappearing from the wild state within a few decades if present land use and other causal factors continue to operate. The distinction between rare and vulnerable is neither clear nor permanent. A rare species is one 'that either occurs in widely separated, small subpopulations . . . or is restricted to a single population' (Drury, 1974). Drury points out that species which have developed 'the ability to exploit an extreme habitat but to sacrifice the ability to compete with other organisms in variable habitats, [will have] restricted and discontinuous distributions', i.e. they will be rare and highly specialized. Adaptations to extreme habitats may be of considerable ecological and physiological interest and, in the age of recombinant DNA, one may contemplate the possibility of attempting the transfer of such adaptations to economic plants. This is an exciting new prospect which we cannot afford to overlook in considering the preservation of rare species. Though rare, they may be reasonably safe but become vulnerable or endangered when their habitats are altered or otherwise challenged. This is happening today over large parts of the earth, through all-out destruction or drastic modification of habitats, such as clear-felling of forests and replacement by grassland or wasteland; through overgrazing or fire; through shifting cultivators abandoning the traditional system of short-duration cropping and long forest fallow; through water and air pollution, herbicides and pesticides, mining, highways, hydroelectric or other water storages, airports and urbanization.

But it is not only rare species which under present-day conditions become vulnerable or endangered. Their fate is shared by many species which are – or were – common components of tropical ecosystems, and especially of tropical forests which are being destroyed at a rate of millions of hectares each year. Little is known of the distribution of many of the species, of their potential usefulness, their ecology or their reproductive physiology. One can only make a rough estimate of the number which will be threatened and Raven (1976) suggests that by the end of the century they may amount to something like one-third of the 150 000 plant species of the tropics.

About the plant species of the temperate zone we know more because they have been studied over long periods, and because of the stronger institutional base, including a much larger number of well-equipped botanical gardens. Of the 85 000 temperate species, about 4500 can be considered to be threatened (Raven, 1976). Yet information which could serve as a basis for decisions on specific action is as yet scarce. There is the *Red Data Book* for the flowering plants (Melville, 1970) which collects worldwide information, a list of endangered and threatened plants in USA prepared by Ayensu and De Fillips (1978), and Given (1976) has proposed a register of rare and endangered indigenous plants in New Zealand and listed 314 taxa believed to be in need of protection.

Thus we are faced with two kinds of plants which are, or are likely to become, rare and potentially endangered. The first includes those which are rare because, as we have seen, they fit into restricted ecological niches to which they are peculiarly adapted; we might call them the 'old rare' species. By contrast, the depleted, or 'new rare' species become rare and endangered because of the wholesale alienation of habitats occupied by the ecosystems to which they belong. Thus the former are rare because of their extreme specialization, while the latter are rare in spite of their greater adaptability and wider distribution. The former may have few, if any, companion species which are similarly threatened, the latter may be involved with many hundreds of species in a common fate of extensive habitat destruction. Ecological as well as logistic considerations suggest that the former, if they or their habitat are threatened, as a rule can only be preserved under some form of cultivation, whereas for the latter the only practicable means of conservation would be *in situ*, i.e. in nature reserves which would provide refuges for a whole range of species represented in the threatened communities. Where appropriate reserves are unobtainable or their tenure or maintenance is in doubt, individual species will have to be preserved under cultivation, which would necessitate a choice of species.

Preservation under some form of cultivation will in the first instance depend on initiatives which may come from botanical gardens, university departments, research organizations, international agencies, horticultural firms or associations, community groups or concerned individuals. In any concerted effort, such as that initiated by the Threatened Plants Committee, botanical gardens inevitably have a central place, not mainly because they have capabilities for growing plants (and maintaining seeds), but because of their ability to gather and disseminate relevant information, to coordinate widespread efforts, and to establish international, national and community links which are essential for an effective and sustained effort.

7.2 Botanical gardens and the conservation of threatened species

The role of botanical gardens in conservation comprises two main areas, one relating to the *ex situ* conservation of threatened species, the other to the gathering, recording and dissemination of information. Both are important and interdependent.

7.2.1 Ex situ *conservation of threatened species*

Threatened species are commonly preserved as living plants or, less commonly, as seeds. Let us first consider preservation in the form of seeds. The potentialities, limitations and technologies of *long-term seed storage* are discussed in section 9.4.1. Here we discuss its role in the preservation of threatened species. In annual and short-lived perennial species a supply of stored seed has distinct advantages over frequent cultivation which imposes recurrent expense and exposes the material to biological and other risks (see section 9.4.1).

Has seed conservation a place also in the preservation of longer-lived plants? Compared with preservation in the form of living plants with the attending restraints of space and expense, seed conservation has obvious advantages. Yet seeds have a time limit of vitality; and, anyway, seed storage is not an end in itself, but a means to recover plants at a later date. Hence preservation as seeds – or in the form of tissue culture (see section 9.4.5) – postpones but does not circumvent the need for finding a niche for reestablishment as live plants, presumably under cultivation. (Here there may be a need to emphasize that this discussion relates to the conservation aspect of seed storage, but not to distribution and exchange which is the main purpose of existing seed storages in botanical gardens.)

The conservation of *living plants* in botanical gardens (or equivalent institutions) involves substantial investments in facilities and maintenance, a staff with diverse expertise, sustained interest and concern, and public and administrative support. Hence the scope of the conservation effort has to be limited in accordance with the facilities, space and care that can be devoted to conservation, although some of the material will serve a variety of purposes, e.g. as botanical exhibits, or for educational, horticultural or aesthetic display. The inevitable selection of material to be conserved will be influenced by such interactions, by the logistic realities of demands on space – especially greenhouse space – and on maintenance, by the traditions and responsibilities of the institution, and by the scientific interests of the staff and of collaborators. General criteria likely to influence decisions could include the economic significance, aesthetic value, and last, but one may hope not least, a rating as 'endangered' or 'threatened' by the Threatened Plants Committee or similar organizations, or by the institution itself.

Many of the species which are now threatened had not previously been cultivated. They may present problems which have to be solved. Wolliams (1976) describes the methods developed at the Waimea Arboretum, Hawaii, for the collecting, propagation and establishment of seeds or vegetative material taken from the wild. Many species require individual methods to be developed for the germination of seeds, propagation of cuttings, or for establishment including the provision of mycorrhizal fungi, but help can be derived from a detailed ecological description of the original site. Once worked out, the method of propagation can be made available to assist other public or commercial institutions to establish a threatened species in cultivation. Research into the reproductive physiology and the ecological requirements of a species will have a similar effect in facilitating the establishment and reproduction of threatened species in *ex situ* conditions. Such studies may indicate the most appropriate (and economical) conservation sites, possibly in satellite gardens with specific conditions.

In his comprehensive review of the role of botanical gardens in conservation, Raven (1976) emphasizes that 'the simplest, most effective, and least costly way in which botanical gardens can ensure the perpetuation of particular species is to introduce them successfully into the horticultural trade'. Obviously, this applies only to species of horticultural merit. Raven cites *Ginkgo biloba* L., *Franklinia alatamaha* Marsh., and *Encephalartos woodii* Sander, as successful examples of the salvage of rare and endangered species through their introduction into commercial horticulture. *Swietenia humilis* Zucc., virtually wiped out in its habitat in the lowlands of Panama, is now a common street tree in the Canal zone. The beautiful wildflowers of Western Australia, at one time threatened by the extensive trade in cut flowers, are now being saved by the establishment of commercial plantings of the most popular, and hence most endangered species, combined with protection in the wild.

It is recognized that the introduction into cultivation is likely to narrow the genetic diversity of a species and to change its variation pattern in response to drastically altered selection pressures. Yet it is obviously preferable to keep alive a seriously endangered species as a semi-domesticate, rather than allow it to disappear altogether. Similarly, it seems justified to collect – as far as possible non-destructively – material of extremely rare and endangered species for reestablishment in cultivation. Such action, if conducted by a responsible and authorized institution, is altogether different from the wholly destructive collecting of specimens for sale which is having the effect of seriously depleting, and in some cases endangering some rare and beautiful species of orchids, succulents and many others which are collected for commercial use.

Two questions arise. First, to what extent is the preservation of rare and threatened species in botanical gardens justified, considering the expensive effort involved? This question is discussed in Chapter 1. Second, is it likely to be effective? This will depend in the first instance on the suitability of the environment and on the management – including the absence or control of diseases and pests, availability of information on water and nutrition requirements, etc. It will equally depend on the individual and socio-political circumstances. 'History has shown', says Raven (1976), 'that collections of particular groups rarely survive the retirement or transfer of the individual who had the greatest interest in them' or, one may add, of the director who often is keen to pursue his own policies rather than those of his predecessors. Poppendiek (1976) questions whether 'even the European gardens are able to guarantee long term stability'. He points out that some very valuable collections have been accumulated many times, to be lost with the departure of the gardener or scientist concerned. 'The motivation and sense of purpose is necessary and it is to be hoped that the motivation will be increased if there is some international confirmation that it is really worthwhile.' We conclude that in the short term botanical gardens may serve as refuges for threatened species, but that security is tenuous, subject not only to the policies and interests of those directly involved, but to the social, economic and political attitudes of the community on one hand, to international responsibilities on the other. Both are discussed in the next sections.

7.2.2 Information, documentation and education

Information and its documentation extends in two directions, internal and external, the former relating to the collections of a particular botanical garden, the latter to activities and contributions of its staff within the country and beyond.

Inventories of collections, both of live plants and of seeds of threatened species, should be compiled, and all information relating to their origin, ecology, reproductive physiology, flowering and seeding, diseases and pests, distribution of material to other institutions, entered in computer records, or at least in computer readable form. This will facilitate exchange of information between botanical gardens and with other interested institutions (Simmons *et al.*, 1976, section IV on documentation).

Externally botanical gardens may assume responsibility for ascertaining the status and location of threatened species in their own country. Depending on interests, expertise and available resources such activities may – and do – extend to other parts of the world, and especially the tropics where, as has already been stressed, the needs are greatest because of the unparalleled diversity of species, the acuteness of the threat, and as yet the scarcity of institutions and personnel. The effort may be centred in a geographical area, or on one or more taxa of particular interest or concern. Findings should again be recorded, but botanical gardens may assume a wider responsibility for recording information received from other sources, including university and other institutions, and from independent botanical and other collaborators and observers. Wide community participation, encouraged by appropriate publications, would be valuable to both the scientists and the interested public.

Indeed, extension of interest and involvement in conservation in general, exceeding the preoccupation with threatened plants, should be, and in some instances is, a major concern of botanical gardens as an important part of their programme in education and public information. As in some zoological gardens, exhibits of species that are rare and threatened in their natural habitats draw attention to the need for their protection, and to the conservation of plant species in general. If emphasis is placed on the need for the *prevention* of species becoming endangered, in addition to the preservation of those which have reached this sorry state, the lesson which threatened species hold for the uncertain future of so many wild plants and animals will be the more effective. Community support grows with understanding, and here the botanical gardens, with their large numbers of visitors from schools and the general public, have a singular opportunity and responsibility for communications with young and old.

7.3 Return to nature

In the rehabilitation of endangered or extinct species from material reproduced in botanical gardens, two kinds of questions arise, one relating to the siting of the release, the other to the population genetic structure of the released strain.

The release may be sited in one (or more) of the original habitats of the species, in a site similar to the original one if the latter has been modified or destroyed, or in an altogether different habitat. In the first case, the ecological factors which threatened or destroyed the species are likely to check the replacements, as is the case with the Texan species *Styrax plantanifolia* Engelm. planted on the Edwards Plateau where it is endemic, and where the seedlings are eaten by deer and goats (Raven, 1976). Protection against those predators is unlikely to save the species in the longer term. Raven quotes a number of other species as being multiplied prior to being reintroduced to the original habitats or their proximity, but only one, *Clarkia franciscana* Lewis & Raven, as having been successfully reestablished in its habitat – a single serpentine-covered hillside. Introduction to alien habitats calls for the precautions which are normally applied in the introduction of wild biota to new habitats. From being a rare and threatened plant, a species could turn into an aggressive weed in a different environment, e.g. on agricultural land under cultivation.

Species which have become endangered through large-scale destruction of their habitats may have lost the opportunity for reestablishment in or near their original sites. Reestablishment within a suitable community would provide the best and perhaps the only chance of long-term survival. Indeed, recourse to such man-made or man-influenced nature reserves may be both useful and inevitable, especially in the tropics, where the justification for the preservation of a wide range of species is strengthened by the possibility of future economic use (see chapter 8, p. 211).

What guidelines can be established for the population structure of a release? As already discussed, both sampling procedures in collecting, and propagation procedures for multiplication, are likely to narrow the genetic base. This may have consequences for the ecological adaptability on one hand, the genetic vulnerability to pests and diseases on the other. On both grounds genetic heterogeneity is likely to be an advantage and should be safeguarded by representative collecting and propagation as far as the available material permits. Such propositions are open to experimental test, which should provide useful information on the process of rehabilitation. In such experiments it might be instructive to include some 'old rare' species (see p. 165) to examine the suggestion that they are rare because of a high degree of specialization, i.e. a low level of adaptability.

Considering that ultimate rehabilitation under natural conditions is at least one of the reasons for the *ex situ* preservation of some threatened species, it is perhaps surprising that the record of rehabilitation efforts, and especially of population genetic changes, is not more extensive. Such efforts could provide important information on the genetics of attrition and of reestablishment of a species. Now that electrophoretic surveys of proteins are available as an unambiguous and readily obtainable measure of genetic diversity, it is possible to examine the population genetic structure of species at various stages of attrition, cultivation and reintroduction, as was apparently done in the case of *Clarkia franciscana* briefly mentioned by Raven (1976). If available for a range of species such information would provide useful evidence on the adaptive process involved in the occupation or reoccupation of a site, and possibly on some of the reasons for success or failure in the rehabilitation of threatened species.

7.4 International networks

The acute threat to a rapidly growing number of plant species in many parts of the world, and particularly in the tropics, is a powerful reason for moves to place the existing informal cooperation between botanical gardens on a more generally effective basis. This was the tenor of a discussion on international cooperation which took place at a meeting of the International Association of Botanic Gardens (IABG) in Moscow in June 1975. It called upon 'all the workers of botanic gardens of the world actively to participate in nature conservation, in particular in protection of plant life' (Lapin, 1976). However, it remains to be seen what practical steps will emerge in the face of restricted facilities and finance, preoccupation with many other responsibilities, and in the absence of an official international organization to foster cooperation and integration of activities.

The areas in which international coordination can materially contribute to the conservation effort have already been reviewed. They are briefly summarized in the international context:

(i) Regional planning and coordination. Agreements on shared responsibilities would enable botanic gardens to concentrate on taxa within their climate zones, and jointly to preserve a fuller range of the genetic diversity of widespread taxa than would be possible for any one institution.

(ii) International coordination of information. The American Horticultural Society's Plant Sciences Data Centre has accumulated the accession records of twenty-six botanical gardens and has become a major information centre for threatened species under cultivation (Ayensu, 1976). The magnitude of the task may make it necessary to have more than one recognized data centre. Moreover, there is a need for a computerized records system for the register of threatened plants, with information on habitats, conservation measures taken, cultivation sites, etc. Lists of threatened plants are being compiled by the IUCN Threatened Plants Committee. Such information should be freely exchangeable, hence an agreed system would be a matter of great convenience, especially if individual institutions were also to adopt it. It might be of mutual advantage if the system called EXIR (see chapter 9, p. 242), developed for the documentation of genetic resources of crops and their wild relatives, were adopted also for threatened plant records.

7.5 Conclusions

We can now attempt to assess the direct contributions – as distinct from indirect ones such as documentation, information and education – that botanical gardens are able to make to the preservation of threatened species. Five distinct areas have emerged in which the facilities and personnel of botanical gardens are well equipped to perform important tasks which are relevant to conservation:

1. Research on the reproductive physiology, ecology, propagation and maintenance of selected species.
2. Assembly of material and distribution through commercial channels to encourage the survival of threatened species as horticultural domesticates.
3. Maintenance and propagation for reestablishment in natural habitats.
4. Conservation in seed banks. Its use is limited by difficulties and costs of

regeneration and by the prevalence of recalcitrant seeds in tropical plants.

5. Live conservation in botanic gardens, their associated field stations or collaborating institutions. It is generally recognized that the most effective way to preserve threatened species is within the communities in which they have survived. However, where the ecosystem as a whole is threatened or replaced and nature reserves either cannot be established or are subject to strong social or economic pressures, then *ex situ* conservation in botanic gardens or equivalent institutions may be the last line of defence. Indeed, to provide a refuge for rare and threatened species is often regarded as one of the important scientific responsibilities of botanical gardens.

How do these contributions rate in terms of the preservation of *genetic diversity* on one hand, of *security in time* on the other? In all forms of conservation, choice of site, sampling procedures, or habitat restrictions will inevitably result in a narrowing of the genetic diversity, accompanied by natural and, in the case of domestication, by deliberate selection. Again, as in all forms of conservation, security of tenure depends on decisions which can be revoked at any time, and by any generation. It may be that *ex situ* forms of conservation, requiring continued effort and expenditure, are more exposed to change of support than are large nature reserves. They may also be more exposed to biological risks. Yet in some instances the reverse may be the case. Some species of *Eucalyptus* are cultivated around the globe, on areas far exceeding their Australian habitats; and the few restricted areas of *Pinus radiata* and the relic populations of *Cupressus macrocarpa*, both in California, have given rise to hundreds of thousands of hectares of the former in Australia and New Zealand, and to innumerable hedges and shelter belts of the latter in New Zealand. Such instances are relatively rare, yet commercialization may be one of the most effective means of preservation, albeit at the cost of some of the genetic diversity. As for the prospects of reintroduction to natural conditions, there is too little evidence to justify expectations, but further experiments should be closely followed.

There remains the preservation within botanical gardens themselves, offering a refuge for a greater diversity of species and with somewhat better security than is available through any other means. Yet we have recognized the very real limitations imposed by available space, staff and facilities, by competing responsibilities and demands, and last but not least by shifts in policy due to new staff, or to changes in political control or public concern. In terms of the time scale of concern (see Chapter 1) preservation in botanical gardens might not rate beyond a generation or two. Yet international agreements may succeed in strengthening the security of tenure which individual institutions are able to provide. Mutual responsibilities impose on participants at least moral obligations which, in case of default, could be taken over by others. Ultimately the long-term validity of preservation in this form depends on the interest, perseverance and dedication of individuals, in addition to traditions and obligations. In the long term, all preservation is a social as much as a biological issue.

7.6 Summary

1. The role of botanical gardens in conservation relates to the preservation of species whose survival is endangered, for the most part through loss or drastic modification of their habitats (section 7.1).
2. A rare – and potentially vulnerable – species may occur in a single habitat or in a few separated populations because of its ability to exploit an extreme environment. Other species, though common components of major ecosystems, may yet be exposed to extinction through extensive habitat destruction, affecting also large numbers of companion species. Ecological and logistic considerations suggest that the former can be preserved only in cultivation, the latter, apart from a limited number of selected species, only in nature reserves (section 7.1).
3. *Ex situ* conservation of threatened species can take various forms, including living plants, seeds, or eventually tissue culture. Seed conservation is mainly useful in annuals or short-lived perennials. Seed conservation of longer-living species must ultimately result in the cultivation of living specimens involving substantial space and costs (section 7.2.1).
4. Botanical gardens and equivalent institutions can contribute to *ex situ* cultivation by generating information on reproductive physiology and ecology, and on the techniques of propagation, maintenance and protection (section 7.2.1).
5. The most effective and least costly way to perpetuate a species is to introduce it into the horticultural trade (section 7.2.1).
6. Botanical gardens can play a part in reintroducing threatened species to natural environments. Such efforts have been few, though others are in preparation. Information on the structure of populations prior and subsequent to the release may provide useful information on processes of establishment in natural environments (section 7.3).
7. Conservation of threatened species within botanical gardens may be regarded as the last line of defence for species which are not preserved in nature reserves or adopted by horticulture. Selection on logistic grounds – size, maintenance, protection – and on economic, scientific, aesthetic grounds is inevitable. Competing responsibilities, changes in staff and in policies place serious doubts on the long-term prospects of preservation, though international agreements may succeed in strengthening it, as will an international system of documentation of threatened plant species and of holdings of botanical gardens (sections 7.4 and 7.5).

8. The role of botanical gardens as sources of information on threatened species, their distribution and rating of vulnerability on one hand, information and education for schools and the general public on the other, may possibly make as significant a contribution to conservation as does the cultivation of threatened species (section 7.2.2).

8

The genetic diversity of plants used by man

8.1 Evolution of genetic diversity

8.1.1 The evolutionary continuum

All domesticated species derive, directly or indirectly, from wild species some of which were among the much larger numbers of species used by hunter-gatherers for unknown millennia. The ancestors of many domesticates have been identified, although those of some important crops, among them maize and common wheat, are still subject to research and argument. Wild species, including most of the forest and pasture species and some tropical tree fruits, are used and extensively planted without having undergone the more or less drastic morphological and physiological changes which are associated with the domestication of many crops. Thus there is an evolutionary continuum linking the prehistoric pre-domesticates with the present-day cultivars; and an ecological continuum linking wild and semi-domesticated species with those which have been modified beyond the point of no return to natural conditions.

These relationships are relevant for two reasons. First, wild species continue to be taken into some form of domestication. With the increasing pressures for intensification of land use on one hand, for new industrial raw materials on the other, species exploited in their natural state and others as yet not used at all are taken into cultivation and, ultimately, into full domestication. Second, there is a growing realization that wild progenitors and other relatives of crop species constitute an important and as yet scarcely exploited extension to the gene pools of domesticated species available to plant breeders. Transfers of genetic material between distant relatives have become possible through the application of modern genetic techniques, and these are likely to be increasingly used.

8.1.2 Levels of adaptation

The domestication of plants was not the drastic event which is conveyed by such terms as the neolithic or the agricultural revolution, which are as inappropriate as the term 'green revolution' for the technological developments in recent times. The evolution of neolithic agriculture was spread over a millenium or more, the evolution of modern agriculture started more than a century ago and is still proceeding. Domestication grew out of food gathering which almost imperceptibly led to elements of cultivation. In the lives of gatherers 'the advantages of growing plants on purpose are not conspicuous at the beginning and the differences between intensive gathering and cultivation are minimal' (Harlan, 1975a). But once cultivation became a part of the food gathering process, it set up new evolutionary dynamics which almost inevitably led to a progressive transformation of the plant. In the earliest domesticates, the cereal grasses, the non-shattering of seeds was the most drastic adaptation to cultivation which distinguished and separated the self-seeding plants of gatherers from the domesticated relatives seeded by cultivators. In genetic terms this step was simple, in most instances due to one or two genes, and non-shattering types have been easily selected in wild grasses. Once incorporated in the genotype, an evolutionary chain reaction led to the accumulation of adaptations to the conditions of cultivation and to the needs of the cultivator (Harlan, de Wet and Price, 1973).

We can thus discern two levels of adaptation. First, there are the adaptations of the wild species used by food gatherers and early cultivators, adaptations mediated by natural selection under natural conditions. Second, there are the adaptations acquired under conditions of domestication, also resulting mainly from natural selection, but presumably complemented by some deliberate selection on the part of the cultivator. The difference is essentially ecological and sociological, social patterns of settlement and cultivation setting the pattern for adaptive changes in the plants on which settled community life depended.

8.1.3 Domestication: species-specific adaptations

We have already recognized adaptations resulting from selection under domestication as distinct from those acquired under natural conditions. We can now examine the patterns of genetic diversity which are generated under domestication.

The variation patterns of domesticated plants are the result of two distinct but interacting selection pressures. One resulted from the impact of cultivation itself, i.e. from the ecological factors which distinguish an agricultural from a natural ecosystem; the other from the impact of the diverse environmental factors – physical, biological, cultural – to which a species is or has been exposed. The first kind of adaptations may be common to a species or sub-species; it can be called 'species-specific'. The second will differentiate a taxon into races or varieties, but may result in analogies *between* species exposed to similar ecological selection pressures; it can be called 'ecotypic'.

Harlan *et al.* (1973) list taxon-specific adaptations in the cereals, 'adaptation syndromes resulting from automatic selection due to planting harvested seed'. These syndromes evolved in response to selection pressures associated with harvesting and with seedling competition. They resulted in an increase of seed recovered through non-shattering of grain, an increase in seed production (larger inflorescences, uniform ripening, increase in proportion of seeds set), and an increase in seedling vigour resulting from larger size of grain and from more rapid and synchronous germination through loss or reduction of germination inhibitors. Some crucial adaptations, e.g. non-shattering, are simply inherited, hence more easily and rapidly acquired than others, such as the number and size of grains, which are polygenically determined. Non-shattering, as we have seen in the previous section, is perhaps the most crucial adaptation in the domestication of the cereals. It has been obtained in every grass species in which selection was attempted, usually with great ease, e.g. in *Andropogon hallii* by harvesting seed a month after the usual time (Harlan *et al.*, 1973).

In some of the cereals non-shattering is more complex. The wild wheats have an inflorescence which at maturity breaks up into segments (the spikelets), with grains that are tightly enclosed by bracts (the glumes). Domestication set up selection pressure for a non-fragile inflorescence which remains intact in harvesting, and for glumes which readily release the grain during threshing. In the common wheat, *T. aestivum* ssp. *vulgare*, *both* adaptations are controlled by one, rather complex, supergene called *Q* which evolved under domestication, and is carried by all varieties of common wheat. Hence a non-breaking inflorescence and free-threshing grains are 'taxon-specific' characteristics. However, the *degree* of free threshing is modified by other genes which are subject to selection pressure, hence resistance to 'grain shedding' is a characteristic of cultivars which evolved under dry and windy conditions such as those prevailing during wheat ripening in New Zealand. This is an example of the effect of 'ecotypic' selection pressure on a 'species-specific' characteristic.

A shift from sexual to vegetative reproduction is a rapid and effective way for establishing a domesticate. In this way important characteristics, such as a distinctive size of the fruit, can become fixed and adopted as the standard and hallmark of domesticated races. This, according to Zohary and Spiegel-Roy (1975), was the case in the domestication of the ancient fruit species of Near Eastern origin, olive, grape, date and fig. All these species are cross-pollinated and extremely heterozygous, hence sexual reproduction yields a multitude of undesirable types. Vegetative reproduction is simple in all of them, and outstanding individuals were readily selected to initiate clones which may have been maintained for centuries. The authors suggest that in their five or six millenia these species may have undergone very few sexual cycles. Such instant domestication occurred in many more fruits as well as in crop species which can be vegetatively propagated such as yams, manioc and many others (see Harlan, 1975a, p. 143).

8.1.4 Species differentiation: ecotypic adaptations

The area, or areas of origin of many crops are known though not the precise sites. Some originated in one of the centres of diversity identified by Vavilov (1949–50), some others in no centre at all but within a wide region.* Archaeological evidence shows that many domesticated species spread fairly rapidly throughout the known world. Wherever they went they were modified by the environment and by the cultural methods adopted by different civilizations. Dispersal played a major role in diversifying the gene pools of crops. Potentialities for recombination within primary gene pools (Harlan and de Wet, 1971) were given diverse opportunities for realization, as were mutations conferring resistance to parasites encountered and/or evolving in various environments. Introgression from wild and weedy relatives – the secondary gene pool – enriched the gene pools of crops and broadened the scope for the selection of adaptations. Such introgression could still be observed in recent times (Harlan, 1951; Zohary and Feldman, 1962). Thus there emerged distinctive local races, adapted to the many variants and interactions of natural and cultural environments to which crop species were gradually exposed. We have come to call them 'land races' or 'primitive cultivars' in distinction from the 'advanced cultivars', the products of scientific plant breeding in the last 100 years, initially by selection from land races.

There is a marked difference in the nature of the adaptation of land races and of advanced cultivars. Land races had evolved, largely by natural selection, under conditions of traditional, i.e. low-input cultivation, and adapted to the particular environment in which they were grown. Such localized adaptation resulted in diversification between land races which is still recognizable in local names for ancient cultivars of crops or fruits (e.g. grape vines). In contrast, advanced cultivars are bred for high performance under intensive, high-input cultivation which to a considerable degree tends to even out environmental differences of site and season. Environmental levelling is paralleled by genetic levelling: modern varieties are deliberately selected for performance over a *range* of environments, thus reducing the need for specific local adaptation. Genetic levelling *within* cultivars is even more drastic: the heterogeneity of land race populations is replaced by the relative homogeneity of near-pure lines, hybrids or blends. Clearly, advanced cultivars, like land races, are ecospecifically adapted, though on a

---

* For a brief outline of Vavilov's centres of diversity see Harlan (1975a, pp. 52–7).

different scale. Over vast areas, even different continents, a few modern cultivars can take the place of the plethora of land races which had been in use until a few decades ago. In some species of livestock, as seen in Chapter 10, this process has gone even further. In consequence genetic resources of both crops and livestock have been placed at serious risk.

8.1.5 The dynamics of genetic diversity

From the plants gathered by pre-agricultural hunter-gatherers to those by which we live now, the levels of genetic diversity have undergone dynamic changes at specific and infraspecific levels. The trend has been for a decrease in the number of species used, accompanied by an increase in infraspecific diversity, until, as we saw in the preceding section, the impact of intensive agriculture and scientific breeding drastically restricted also the infraspecific diversity of crop species. The general pattern which is apparent derives from a progressive transition of a generalized natural ecosystem to a specialized artificial ecosystem, involving a drastic reduction in diversity (Harris, 1969). Fig. 8.1 illustrates schematically the changes in genetic diversity, from the level of the wild species to that of the advanced cultivar.

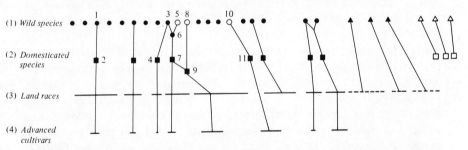

Fig. 8.1 Schematic description of relationships between wild and domesticated species and of changes in the genetic diversity of domesticated species (indicated by length of horizontal lines). Wild species now used ▲ and potential domesticates (△ ☐) on right of diagram. *Wild species:* ●used by gatherers; ○not used by gatherers; ▲ used, but not domesticated; △ potential domesticate. *Domesticated species:* ■ domesticated species; ☐ potential domesticate. *Land races*, extensive genetic diversity. *Advanced cultivars*, genetic diversity reduced.
Examples: 1, *Hordeum spontaneum*; 2, *H. vulgare*; 3, *Triticum boeoticum*; 4, *T. monococcum*; 5, *Aegilops sp.* ?; 6, *T. dicoccoides*; 7, *T. dicoccum*; 8, *Ae. squarrosa*; 9, *T. aestivum*; 10–11, *Trifolium subterraneum*. From Frankel (1977).

During the transition period from hunting–fishing–food gathering to domestication of plants and animals, a great variety of plant and animal species were used, with the distinction between 'gathering' and 'cultivation' gradually emerging. At Tepe Ali Kosh, in south-west Iran, Helbaek found in excavations dating from the early beginnings of domestication (7500–5600

BC) seeds of forty-odd plant species and bones of about thirty-five animal species; two of the former and two of the latter belonged to domesticated species (Fig. 8.1, line 1). The total number of species greatly exceeds that used today in the same region. During the following millennium the inhabitants of Tepe Sabz, also in south-west Iran, used at least seven domesticated plant species (including nine distinct types) and three domesticated animal species. As the number of domesticated species increased, the number of wild species used for food decreased (Flannery, 1969) (Fig. 8.1, line 2).

With the spread of agriculture in Afro-Asia and in Meso-America, plants, animals and man, as Darlington (1969) observes, evolved in close interaction. Selection produced crops from wild plants or from weeds. The latter, as colonizers of disturbed environments, were already pre-adapted for cultivation (Hawkes, 1969). Others, like our hexaploid wheats, originated from the combination of wild and domesticated species. The number of cultivated species increased with the rise of civilization and the differentiation of human needs, and with the spreading of settled agriculture into regions which made their own contributions, such as the numerous tropical fruit and vegetable species originating in South-East Asia and Meso- and South America. Quoting Mangelsdorf (1966), 'during his history, man has used at least three thousand species of plants for food and has cultivated at least one hundred and fifty of these to the extent that they have entered into the world's commerce'. But, Mangelsdorf continues, there has been a tendency to reduce the number of crops to those which are the most efficient; and 'today the world's people are actually fed by about fifteen species of plants'. It should be noted, however, that this process applies in the main to staple food crops. A number of wild species have been taken into cultivation even in the last hundred years – as pasture plants or for soil conservation, as industrial or medicinal plants, or as ornamentals; and, as Brücher (1968a) points out, there is as yet a reservoir of potentially useful indigenous plants in South America and this is also the case in tropical regions of Asia (see section 8.2.8).

As the number of major food plants declined, variation *within* species increased (Fig. 8.1, line 3). This has been the case especially with the ancient staple food crops which spread early in their history from the region in which they originated into others where they acquired further diversity. In this way the gene pools of our major crop plants were greatly enriched. When wheat reached China 3000 years after its domestication in the Near East (Darlington, 1969) many types evolved which are peculiar to that region. Some, very likely evolved under irrigation, have ears with unusually large numbers of spikelets and spikelets with unusually large numbers of grains. Wheat, like so many other good things, was carried across the sea of Japan and one may guess that it was this character combination which possibly centuries later reappeared in a modern Japanese wheat, Norin 10, to find its way into

Vogel's record-yielding hybrids (Vogel, Allan and Peterson, 1963) and ultimately into the high-yielding Mexican varieties which helped to introduce the 'green revolution' in countries of Asia and Latin America. The process of infraspecific diversification continued throughout recorded history. The momentum was maintained by social and technological changes and by migrations, the most momentous being the post-Columban exchanges across the Pacific and Atlantic oceans. With the advent of scientific agriculture, plant introduction greatly extended the transfer of genetic materials on a worldwide scale. The extension of Old World gene pools had marked effects in the 'new' continents, North America and Australia.

In recent times scientific plant breeding had the effect of restricting genetic diversity everywhere. Purebred cultivars began to replace land race populations from the middle of last century, and some fifty years later few if any of the land races of major crops remained in what we now call the developed countries. Indeed, this progression was a major part of 'development', and is so now in the developing countries. Introductions, selections from indigenous varieties, hybrids between them, and recently varietal composites are rapidly replacing the land race populations which remain. The new cultivars have one feature in common – a high degree of genetic homogeneity (Fig. 8.1, line 4). The drastic upgrading of productivity which resulted from the replacement of traditional forms of agriculture was as essential as it was inevitable, but it placed in jeopardy the reservoirs of genetic diversity on which the continuing evolution of cultivated plants to a large degree depends. This dilemma will be discussed in later sections.

8.2 Genetic resources today

The evolutionary processes outlined in section 8.1 have resulted in the great diversity of genetic resources which are now recognized, ranging at the evolutionary level from wild ancestors to advanced cultivars, at the ecological level from components of primeval ecosystems to those of high-input agriculture and horticulture. A functional classification was first introduced by the International Biological Programme (IBP, 1966; Frankel and Bennett, 1970b):

Land races
Advanced cultivars
Wild relatives of domesticated plants
Wild (i.e. non-domesticated) species used by man

This classification is used throughout this chapter and the following one.

8.2.1 Land races

'Land races', in the words of J. R. Harlan (1975b), 'have a certain genetic integrity. They are recognizable morphologically; farmers have names for them and different land races are understood to differ in adaptation to soil type, time of seeding, date of maturity, height, nutritive value, use and other properties. Most important, they are genetically diverse. [They are] balanced populations – variable, in equilibrium with both environment and pathogens, and genetically dynamic . . .' The genetic diversity of land races has thus two dimensions: *between* sites and populations, and *within* sites and populations, the former in the main generated by heterogeneity *in space*, the latter in addition by heterogeneity *in time*, i.e. by short-term variation between seasons, and by longer-term climatic, biological and socio-economic changes.

In the main, land races evolved at low levels of cultivation, fertilization and plant protection, subject to selection pressures for hardiness and dependability rather than for productivity. That genetic diversity within land races provided some protection against climatic extremes and epidemics (see quotation from Harlan, 1975b in the preceding paragraph) is plausible, though we lack experimental or observational evidence; and epidemics did occur throughout recorded history. Yet in less extreme circumstances land races were capable of producing at a level which would keep the cultivator alive until the next crop. This can be seen as a symbiotic relationship between crop and man resulting from interactions between biological and cultural evolution, in juxtaposition to the strict control over crop evolution exerted by the present-day plant breeder.

It does not, however, follow that land races were solely selected for survival of crop and man. The extent and the effect of *deliberate* selection is unknown, except in horticultural crops where value characteristics are readily discernible (cf. section 8.1.3). However, the fact that the first advanced cultivars, whether selected from European land races in Europe and North America, or in Asia from indigenous land races, produced greatly improved yields goes to show that land races contained high-producing components.

Today land races have outlived their usefulness in agricultural production, and wherever selected varieties attuned to local conditions are available land races should give way to them. Their role now is to serve as *sources of genetic materials for plant improvement*. Such materials can be either direct selections of individuals for the establishment of a cultivar; or they can be specific genes (or characteristics) to be incorporated in existing genotypes or cultivars. The former has been and is the first step in the genetic improvement of any crop. Today the major agricultural crops are past this stage wherever plant improvement has taken root. And even where this has not been the case, as in some parts of Asia or Africa, there is now a tendency to replace indigenous land races with introduced 'high-yielding varieties' rather than with locally selected ones. However, even now a selection from an ancient land race of a major crop may be introduced for direct cultivation, such as the barley yellow dwarf virus (BYDV) resistant barley variety, Benton, which was derived from a collection made in Ethiopia (Foote, 1966). In the minor field crops, and in horticultural crops, especially in the tropics, there is still considerable scope for direct selection from land race populations, and these remain a precious heritage.

For the most part, however, land races are used, and are likely to be used in future, as *donors of genetic components* which in some way enhance the biological or economic adaptedness of a crop plant or the content or value of a plant product. Such contributions may be well defined, simply inherited and easily transferred, such as many resistances to diseases and pests, or the contents of desirable or undesirable substances; they may be adaptations to specific – often to extreme – environments; or they may be co-adaptations inherited as supergenes, or held together by linkage, by the breeding system, and/or by selection pressures.

It should be recognized, however, that ever since Vavilov, land races had been regarded not only as genetic resources, but also as *sources of information* on the evolution and genetic differentiation of crop species. Vavilov's interest in patterns of variation, which led him to the 'law of homologous series', has recently been revived in studies of variation in crop species, using germplasm collections as sources of information (e.g. Knowles, 1969 and Ashri, 1973 on safflower; Jain, Qualset, Bhatt and Wu, 1975 on durum wheat; Holcomb, Tolbert and Jain, 1977 on rice).* A common interest in these studies is the identification of geographical centres of genetic diversity, because of their historical role in crop evolution and their present-day significance as sources of germplasm for plant breeding and genetic conservation.

Before reviewing the information on land races as represented in germplasm collections, we must consider evidence relating to the population structure of land races as they occur in the field. The fact that variability is extensive is unquestioned, but the genetic structure of land races requires definition in terms of breeding system, selection pressure, polymorphism, seasonal variation, etc. As we shall see in the section which follows, sampling and maintenance of land races for inclusion in germplasm collections have not as a rule been adequate to facilitate such studies, and, regrettably, research specifically designed for this purpose does not appear to be on record. However, a good deal can be learned from closely related wild species, and we take recourse to two studies of crop relatives, *Hordeum spontaneum*, the wild progenitor of barley, and a wild species closely related to the domesticated tomato, *Lycopersicon pimpinellifolium*. Both studies used isozyme and morphological markers, and both were conducted in well-defined areas with numerous, accurately pinpointed sampling sites.

---

* Germplasm collections are assemblies of genotypes or populations representative of cultivars, genetic stocks, wild species, etc., which are maintained in the form of plants, seeds, tissue cultures, etc. Individual samples are called 'entries' or 'accessions'.

In the study of wild barley, twenty-eight populations were used. Of the twenty-eight isozyme loci tested, twenty-five were polymorphic, thirteen of them with strong geographic differentiation. Genetic differentiation of populations showed clinal, regional and local patterns, with environmental correlations which make ecological predictions possible, mainly on the grounds of combinations of temperature and humidity, but also soil type and vegetation. The evidence favours a hypothesis of natural selection acting directly on the marker loci and linked sequences. The extremely rich genetic diversity in wild barley is thus at least partly adaptive, and the authors believe that this wild species warrants extensive utilization in plant breeding especially for extreme environments (Nevo, Zohary, Brown and Haber, 1979).

In *Lycopersicon pimpinellifolium*, Rick *et al.* (1977) studied eleven isozymic and two morphological loci in forty-three populations. Variation is even richer than in wild barley. Each locus was variable, with from two to seventeen alleles per locus. Whereas in predominantly self-fertilized barley heterozygosity is at a minimum, in *L. pimpinellifolium* it substantially exceeds the level to be expected from the rate of outcrossing, possibly as the result of annual fluctuations in outcrossing. The rate of outcrossing varies extraordinarily between populations. Regional differentiation of allele distribution included a clinal displacement of one allele by another along a north–south axis, and single and double peaked clines. Only two of the many alleles exhibited random distribution. In spite of environmental differences in the region, there is no clear trend seen in ecological relationships which could explain the striking genetic differentiation, hence an adaptive role for the observed variation has not been established.

Brown (1978) considered ways in which conclusions from population genetic analyses of wild species can assist in the formulation of strategies for the exploration and utilization of genetic resources. On the basis of an examination of allelic distributions in populations of twelve species he concludes that alleles with a localized distribution, yet high frequencies in particular locations, represent a substantial fraction of the variation. This justifies a strategy of collecting moderately sized samples in a large number of sites within the target area as proposed by Marshall and Brown (1975) (see p. 219). These conclusions need qualification for being derived from observations on wild species (wild relatives of crops, or wild species used by man). However, as Brown (1978) suggests, the information on population structures of wild species may provide leads which could result in reducing the vulnerability to parasites of related domesticated species

8.2.2 Germplasm collections as sources of information

Following the example of Vavilov's Institute of Plant Industry (now the N.I.

Vavilov Institute of Plant Industry) in Leningrad, germplasm collections have been established and maintained in many countries. Primarily intended as genetic resources for plant breeding and associated sciences, many, like Vavilov's collections, have been used as research material, to elucidate evolutionary processes and taxonomic relationships, identify centres of origin or of diversity, discover distribution patterns of general variability or of particular characteristics or genes, and to relate these to the environments in which they occur. Studies of this kind range from multi-character analyses of world collections to detailed studies of more limited proportions of geography or subject. As a result of the recently intensified interest in genetic resources the interest in the study of germplasm collections has also intensified, hence a discussion of strengths and weaknesses of existing collections is in place, partly also to draw attention to weaknesses which might be avoided in those now being established. The preoccupation with material in collections is both natural and useful: natural, because the material is readily available as against material in the field which is less accessible and, more likely than not, is there no longer; and useful, because such studies can add to the information on processes of diversification and sources of adaptive variation.

The use of germplasm collections as research material is based on the premise that accessions representing a country or locality had evolved being exposed to the selection pressures prevalent in the particular environment. As a rule the material will consist of land races or selections drawn from them. In many collections this is only partially the case, hence discrimination is needed between material that in this discussion is termed indigenous, authentic, or valid, and material which is derived from introduction, hybridization with introductions, or induced mutation. Discrimination is not always applied, or even possible, and this and other limitations of collections call for serious consideration.

First, in some major collections, especially of the temperate cereals, *land races are poorly and unevenly represented*, Vavilov's own collections being notable exceptions. The biological status of accessions, whether land race, selection, hybrid variety, introduction, etc. often is not adequately documented. Second, *sampling procedures during collecting and maintenance* are varied and often poorly documented. They may range from a random sample, or a random plus a biased sample (representative of 'distinctive' types), to a biased sample, often without a record of actual procedures. But even if a population was adequately sampled in the field, as a rule its diversity is reduced by selection of one or a few 'representative' individuals, the remainder being discarded. Genetic integrity may be further infringed by natural hybridization, natural selection, or genetic drift occurring in the course of *reproduction for maintenance*, which until the advent of cold-storage facilities was carried out at least every three to five years (see

Chapter 9, p. 232). Ashri (1973), for example, suggests that a divergence between observations made on the same collection of safflower in subsequent years may be the result of natural hybridization, as might be, one might add, some of the 'polymorphisms' within accessions which are noted by some authors. Last, but not least, not many collections have precise *information on the geographical origin* of all their accessions. This restricts evolutionary and ecological interpretation of variation patterns and the planning of repeat collecting of further material.

To what extent a germplasm collection represents the indigenous genetic variation of a region depends to a large degree on the impact that plant introduction and breeding had made *at the time of collecting*. Fifty years ago, at the time of Vavilov's collecting expeditions, land races prevailed throughout the geographical centres of diversity, but this was no longer the case when many other collections were assembled, at least not for those crops which were subject to early breeding efforts, such as the temperate cereals. There is a 'vintage' factor in the composition and status of collections: older collections should provide more valid information on evolutionary patterns, recent ones on areas where genetic diversity may still persist. This must be watched especially when different collecting 'vintages' are combined in one collection.

To this day species differ in the extent to which they have been modified by plant breeding, and this is reflected in germplasm collections. In this regard crop species range from those only in the initial stage of intensive plant improvement, like safflower, chickpea or pigeonpea, to species like wheat or barley which have been transformed by introduction and intensive breeding on a near-global scale. In general, and subject to exceptions already indicated, germplasm collections of the former species are likely to have greater authenticity as research material than those of the latter.

From these observations it should be evident that *germplasm collections which are used as research material must be subjected to critical examination.* Few entire collections will come up to the high standard set by Holcomb *et al.* (1977) that 'only an ideally representative and well-documented collection would be useful for evolutionary and ecological studies of variation', but many contain substantial numbers of accessions which can be identified on grounds of validity and appropriateness of individual accessions. Material of uncertain or inappropriate status or origin should not be included in surveys or analyses. So-called world collections often have defects of the kinds mentioned, because as a rule they derive from a multiplicity of sources. They contain well-authenticated and appropriate accessions alongside others entered with little discrimination for origin or status. We shall discuss elsewhere the question of size of germplasm collections as *genetic resources* (see section 9.6), but authenticity rather than numbers must be the main criterion for *research material.* In the absence of such critical examination one may have more reason for concern at seemingly excessive than clearly inadequate numbers. In the durum wheat collection used by Jain *et al.* (1975), for example, the small number (7) from Sicily, on which the authors comment, is more likely to be genuinely Sicilian than the 836 from Turkey are likely to represent genuinely Turkish cultivars. Similarly, there never have been 24 cultivars of safflower cultivated in Australia, and one may even question the number of 135 for USA, both derived from the USDA world collection of safflower (*Carthamus tinctorius* L.) (Ashri, 1973). The inclusion of accessions in country or regional statistics should rest on the assurance that they originated, or at the very least are, or have been, cultivated there, i.e. have demonstrated adaptedness to the environment.

In an exceptionally authentic collection such as that of safflower (see below) such minor blemishes are of little importance provided numbers are reasonably large and the *questions asked are of a rather general nature.* Indeed, the need for authenticity of the material increases with the *specificity of the objectives* as defined in geographical, ecological, physiological or evolutionary terms. Authenticity is not as essential in multi-character surveys of divergence between regions or countries than, for example, in studies of character or gene frequencies in different ecological zones within a country, or of distribution patterns of parasite races and host resistance.

A collection which is highly representative of the germplasm of a crop is that of safflower referred to above, thanks to 'the dedication and foresight of Professor P. F. Knowles and the USDA', whose timely collecting effort only just preceded the main impact of the introduction of improved varieties (Ashri, 1973). The collection was intensively studied for evolutionary divergence between countries (and regions), for variability in different countries, and for geographical sources of agronomically desirable traits. Ashri (1973, 1975) used twenty morphological characters to assess variability and divergence in different parts of the world. Some traits were uniform throughout, others showed variability throughout, others again had localized variability. On the basis of these and earlier observations Knowles (1969) and Ashri (1973, 1975) concluded that in cultivated safflower fairly distinct types can be recognized which are found in what Knowles (1969) calls 'centres of similarity'. Ashri comments that the variability within centres is at least in part due to the data being assembled for political rather than agro-ecological regions. But this may not be the only relevant factor. Indeed, variation *within* countries may be more significant (as indicated in the analysis by Wu and Jain reviewed below) and of greater scientific interest. Knowles (1969) reported on different regional types in India, and Ashri comments that 'it would be worthwhile to record the detailed collection, locations and analyze the data for divergence'. We wholeheartedly agree. Analyses based on detailed information on collection sites and their ecological conditions could be more revealing of adaptive evolutionary processes, and more effective in identifying sites of valuable genetic resources, than large-scale analyses of world collections.

Wu and Jain (1977) used part of Ashri's data in an attempt to verify the centres of diversity proposed by Vavilov, Knowles and Ashri respectively, using a diversity index (the Shannon information index). They found significant differences between countries *within* regions, but not *between* regions (roughly corresponding to the proposed centres). The data provide only partial agreement with the Vavilovian centres, partly, they believe, due to the choice of the eighteen characters used. They favour the use of the 'widest possible array of descriptors of variation in multivariate surveys of available genetic resources', including protein variation which should facilitate information of 'allelic diversity for a large number of loci and its role in adaptational change'.

Diversity in characters of ecological significance can be more revealing of adaptive processes than are multivariate analyses. In the safflower collection, divergence was apparent for both plant height and period to flowering. Divergent germplasm sources were identified, the results being fairly consistent in the four locations where the collection was evaluated. Further, correlations between these two characters and yield and its components were calculated. For the most part correlations were absent or very low, indicating that selection pressures for both seed yield and earliness are likely to be successful (Ashri, 1973, 1975).

Two further diversity analyses of world collections, of rice (Holcomb *et al.*, 1977) and of durum wheat (Jain *et al.*, 1975), may be mentioned. In rice, accessions from nine countries in the germplasm collection of the International Rice Research Institute were collected at random, excluding irradiated, cross-bred or introduced materials. It was not possible to ascertain whether all entries were of local origin, but in the case of rice these are likely to constitute the overwhelming majority. The rice collection had been assembled in the course of the preceding fifteen years, although some parts had come from other (and presumably older) collections. The durum wheat collection, obtained from the US Department of Agriculture consisted of 3000 entries from all parts of the world, presumably assembled over a period of years from a variety of sources; details of origin and biological status were not specified.

In rice, a multivariate analysis of thirty-nine qualitative and quantitative characteristics confirmed the previously recognized variation patterns in *indica* and *japonica* rice, and indicated divergence between countries, but with a good deal of overlap. The corresponding analysis of the durum wheat collection, using six qualitative characters, indicated divergence between and within regions, and centres of diversity, in Ethiopia, the Mediterranean region and also in India, the last presumably an artefact resulting from the uneven nature of the collection. The lower level of diversity in Near and

Middle Eastern countries, without a doubt ancient centres of diversity, is, as the authors suggest, due to small numbers but possibly also to the time of collecting. In this case the shortfall can scarcely be corrected – as the authors suggest in this and other analogous cases – by additional collection, since there is little left to collect (see e.g. Frankel, 1973), but by recourse to other germplasm collections.

These studies were of an exploratory nature, to test the potential of multivariate analysis for the evaluation of geographical patterns of diversity in existing germplasm collections. The results obtained reflect the nature and the limitations of the material – in the case of rice the complexity of variation patterns, in that of durum wheat the uneven representation and uncertain biological status of the available accessions. However, they indicate the usefulness of a measurement of divergence for identifying areas of diversity and areas of concentration for particular characteristics. These in turn can be submitted to a more detailed analysis, provided material and information are adequate. Such an analysis was conducted by Witcombe and Rao (1976) on farmers' wheats indigenous to Eastern Nepal and collected by Witcombe. Thirty-nine characters were used in a multivariate analysis which identified five genocological regions associated with, but not solely determined by altitude. Seed exchange within the region influenced the variation pattern.

Regarding the identification of targets for exploration and conservation, it must be concluded that the analyses of world collections referred to did not succeed in refining information on the geographical distribution of centres of diversity. However, surveys of the distribution of specific characteristics can provide useful leads where local land races have not been displaced. For a discussion of exploration targets see p. 213.

A somewhat different approach was followed by Nakagahra, Akihima and Hayashi (1975) who used esterase isozymes as genetic markers for tracing geographical patterns of diversity in rice. They used a representative collection of native varieties derived from eight regions of rice cultivation in Asia, from Sri Lanka in the south to Northern China and Japan. Of the nine easily distinguishable bands two occurred throughout the area, the others in a characteristic geographical pattern, with simple zymograms in the extreme south and north, and the most complex, and variable, patterns in the centre, i.e. in the Himalayan region, Burma, Vietnam and Yunnan. This coincides with the area designated by many authors (see Chang, 1976) as the centre of origin of *Oryza sativa*. It would now be of great interest to trace zymogram patterns *within* the areas of diversity and to relate them to environmental, ethnographic and cultural patterns. A great deal of material has been collected in tribal areas of the region in recent years which could be useful for such studies (see Vairavan, Siddiq, Arunachalam and Swaminathan, 1973; Singh, 1977a).

Another study of isozyme variation patterns in a domesticated species produced similarly revealing results. Examining allozymes at fifteen loci of four enzyme systems, Rick and Fobes (1975) found that in the cultivated tomato, *Lycopersicon esculentum*, there was next to no variation in cultivars derived from major areas of cultivation, including South America, with the exception of Peru and Ecuador, where introgression from sympatric *L. pimpinellifolium* is the likely source of the atypical diversity. The high degree of uniformity in the cultigen is in sharp contrast to the diversity in wild species of the genus, suggesting that only a small fraction of the available variation was included in the evolution of the domesticate.

Only preliminary information is available on an extensive worldwide survey of isozyme polymorphisms in barley (Allard, Kahler and Weir, 1971). It indicates extensive allelic variability in the species, with a clinal distribution from East to West across the Eurasian land mass. That selection plays a part is supported by the evidence of allelic frequencies of the same loci in a composite barley cross (see p. 199).

The usefulness of the isozyme method for detecting patterns of variation is illustrated by these examples. As Brown (1978) remarks, 'it is the most convenient [method] available for detecting genetic differences at a number of loci, and it allows the examination of differences close to the DNA level'. One may also emphasize that two of the studies mentioned had limited objectives for which authentic and reasonably representative material was available. Earlier in this section we suggested that analyses of specific target areas based on well-authenticated and documented material may be more productive than large-scale surveys of global dimensions.* Similarly, the distribution of specific characters or genes – such as the isozymes discussed above – has an important part in unravelling evolutionary relationships, and in many instances in providing information of great practical importance. The information perhaps most generally sought from a collection relates to the geographical distribution of genes conferring resistance to pathogens of economic plants. It has certainly received the most intensive attention in the evaluation of collections.

It is natural that one should turn to the centres of genetic diversity (see section 8.1.4) for sources of host resistance. Although, as Leppik (1970) emphasizes, 'the origin and evolution of [crop] parasites are still unexplored', centres of genetic diversity of pathogens, and their geographical relationships to those of crops, have been identified for a number of species. In many instances, centres of diversity of host and pathogen coincide, as is the case with the rusts of Old World cereals, whereas the cultivated potato, *Solanum tuberosum*, domesticated in the Andes, is susceptible to the blight fungus, *Phytophthora infestans*, which evolved in Mexico where the indigen-

---

* One is reminded of the famous phrase, coined by the late Dr E. F. Schumacher, 'small is beautiful' (E. F. Schumacher (1974). *Small is Beautiful*. Abacus, London).

ous wild species of *Solanum* are resistant to blight (see Leppik (1970) for a review of host and pathogen evolution). It needs, however, scarcely to be stated that many pathogens found new and extensive evolutionary opportunities in the vast regions to which the major – and some of the minor – crops expanded in the last 100 years, necessitating ever diversifying genetic defences for the host. This is most prominent – and most extensively studied – in the rusts of the cereals, where the quest for more extensive protection has increasingly led to the search for, and use of, resistance derived from wild relatives which had been exposed to the broad genetic dispersion of the parasite in its own centre of diversity.

However, the need to resort to wild species as resistance sources does not apply to all crop species, nor to all pathogens of any one species. Indeed, land races continue to be important sources of resistance, and in some instances the only ones. A survey for barley yellow dwarf virus (BYDV) resistance among 6689 entries in the USDA world collection of barley showed that of the 117 resistant entries 113 came from Ethiopia, 3 had an Ethiopian parent, and 1 came from China (Schaller, Rasmusson and Qualset, 1963). In a subsequent survey, which also included 5 entries from Sudan, 2 of these, presumably of Ethiopian ancestry, were resistant (Qualset and Schaller, 1969). These results show that resistance is confined to Ethiopia where it is likely to have originated. So far only one gene for resistance has been identified (Qualset, 1975).

An extensive survey for resistance to three major insect pests of sorghum, shoot fly (*Atherigona varia soccata*), stem borer (*Chilo zonellus*) and sorghum midge (*Contarinia sorghicola*) was conducted over a number of years on the world collection of sorghum in India. A number of resistant lines were identified, and promising breeding work aiming to incorporate resistance in well-adapted varieties is under way (Pradham, 1971).

8.2.3 Land races – specific contributions

So far the discussion has been concerned with information which has been or can be derived from germplasm collections of land races. We must now consider their *contributions to plant improvement* as donors of genetic components, to which brief reference has already been made. We shall first deal with specific characteristics which can be precisely defined and which, more often than not, are simply inherited and readily transferable. The material is usually obtained from germplasm collections which, as we have seen, do not always discriminate between land races and selections or hybrids derived from them. At any rate, collections made in centres of diversity as a rule contain a large proportion of indigenous germplasm.

This certainly is the case in the collection of rice assembled by IRRI, the International Rice Research Institute, from all rice-growing areas in the world. A most comprehensive search for resistance sources to diseases and insect pests of rice was conducted over a number of years by IRRI scientists (Chang, Ou *et al.*, 1975). The screened material included many thousands of accessions in IRRI's collection, ranging from wild and weedy species to primitive cultivars and modern varieties. Since in rice there has not been extensive breeding for race-specific resistance, there is still scope for identification and application of non-specific types of resistance (see pp. 205–6). Extensive research on host-pathogen interactions had preceded or accompanied the survey. Cooperation with scientists in countries where pathogens are most prevalent assisted in the search. Some sources of resistance have been identified for all rice pathogens, and in several instances a number of different sources, but their genetic relationships are mostly unknown so that there could be some duplication.

The world collection of safflower, *Carthamus tinctorius* – referred to earlier in this chapter – was screened in Israel for a range of diseases, including safflower rust, *Puccinia carthami*, the leaf spot diseases *Ramularia carthami* and *Cercospora carthami*, and phyllody which is a gradual reversion of florets to vegetative organs, induced by a mycoplasma disease transmitted by a leaf hopper (Ashri, 1971a). As far as the infection levels permitted – which were high to adequate in most instances – resistant lines were identified for all diseases. Resistance to rust was prevalent among lines from the Middle East, while lines from India–Pakistan and East Africa were more susceptible, which may be due to racial diversity in the pathogen. In *Ramularia* the situation was reversed, resistance being centred in parts of India where the disease is common, as against the Middle East where it is absent. A number of lines had resistance to two or more of the diseases. There was good correspondence in the geographical distribution of some disease resistances and of some morphological characters with which they are associated.

This collection was also screened for resistance to the safflower fly, *Acantophilus helianthi*, a severe insect pest in Old World safflower areas, and endemic in Israel where the tests were made. It attacks many other plants, in addition to *Carthamus* spp. No reliable sources of resistance were found in the domesticated species of safflower, but three lines had relatively low levels of infestation. In the nine wild species tested several species contained fly-free lines, including two species, *C. palaestinus* and *C. flavescens*, which are cross-compatible with cultivated safflower (Ashri, 1971b).

Two leaf spot diseases of groundnut, *Cercospora arachidicola* and *Cercosporidium personatum*, cause substantial damage world wide. No adequate resistance has been found in the cultivated species. *Arachis hypogaea*, but is present in cross-compatible wild species (Abdou, Gregory and Cooper, 1974). Several sources of resistance to the groundnut rust, *Puccinia arachidis*, have been identified in land race material from Peru

(Hammons, 1977), including some made by W. C. Gregory and evaluated at the International Crops Research Institute for the Semi-Arid Tropics (R. W. Gibbons, personal communication). Resistance to the peanut rosette virus, which is widespread in Africa south of the Sahara, was found only in a group of cultivars from the Ivory Coast, Upper Volta (Gibbons and Mercer, 1972).

Germplasm collections have been screened for a wide range of characteristics, some of which were discovered at very low frequencies. A few examples must suffice. A survey of a large section of the barley world collection, conducted in Sweden, led to the discovery of two sister lines, collected in Ethiopia in 1923, both with a high protein content, and one of them also with a high lysine content. The latter, named Hiproly, had, under Swedish conditions, shrivelled grain and low yield, but its nutritional quality was transferred to high-yielding varieties (Munck, 1972). A corresponding search for high lysine in the world sorghum collection yielded two high-lysine lines (Singh and Axtell, 1973). Stølen (1965) tested a collection of 630 varieties of barley for tolerance of low pH in the soil. A small proportion were moderately tolerant, and one did significantly better at low than at high pH.

There is no country with a greater range of environments than India, or a greater diversity of farming systems and crops, with a multiplicity of diseases and pests which in their long association with their host species evolved a diversity of host–pathogen relations. Even prior to the advent of high-yielding varieties of wheat and rice, cultivars of Indian or hybrid origin reduced the genetic diversity of the principal crops. But to this day India retains extensive reservoirs of ancient diversity in farmers' fields in many parts of the sub-continent, but especially in mountainous and in tribal areas which physical, ecological or social barriers have excluded from the advances of modern technology. Recent exploration in the Himalayan foothills of north-east India yielded large numbers of primitive rice cultivars with resistance to major diseases and pests, including blast, bacterial blight, tungro virus, gall midge and stem borer, and some with combined resistances (IARI, 1972; Seetharaman, Sharma and Shastry, 1972). Similarly, exploration in maize growing areas in the north-eastern Himalayan region yielded cultivars resistant to leaf blight, downy mildew and corn borer (Singh, Bhag, 1974). The author stresses the immense variability in the maize populations found in this region. As many as fifteen individual races of maize have been identified and fitted into the race system developed by Mangelsdorf. They included highly promising material: some races equalled or even exceeded high-yielding hybrids or composites used as checks (Singh, Bhag, 1977b).

Until recently the grain legume (or pulse) crops had received little attention from the Indian plant breeder, hence local cultivars have remained relatively undisturbed. Now that the reduced area under pulses, due to the expansion of high-yielding cereals under irrigation, is causing serious concern, a concerted breeding effort for superior varieties of pulses has led to a search for extensive germplasm resources, and especially for sources of resistance to the many diseases and pests. A systematic search for resistance to yellow-mosaic virus of *Phaseolus* species has produced resistant stocks of mung bean (*Ph. aureus*) from the Punjab (Gill *et al.*, 1975), and of urd bean (*Ph. mungo*) from Gujerat (Singh, H. B. *et al.*, 1974). In chickpeas (*Cicer arietinum*) five Indian cultivars with resistance to *Fusarium oxysporum* f.sp. *Cicer* and eighteen with *Ascochyta* blight resistance are available. One cultivar from Afghanistan is wilt resistant (ICRISAT, 1976). All grain legume breeders in India emphasize the urgent need for more comprehensive germplasm collections to be made in the main areas of genetic diversity of the various crops.

All Indian cultivars contain a large proportion of indigenous germplasm, not excluding the recent varieties of wheat and rice which have absorbed more exotic germplasm than the current cultivars of most other crops. In many instances the Indian germplasm contributes not only general environmental adaptation, but specific characteristics including grain quality adapted to Indian food preferences, and resistance to many diseases and pests. The multi-resistance lines developed by the Central Rice Research Institute derive the resistance to most of the major diseases and pests of rice – gall midge, brown and plant hopper, tungro virus and grassy stunt virus – from Indian land race material. Similarly, land races provided the resistance sources of important diseases of wheat such as *Alternaria* leaf blight.

The chickpea (*Cicer arietinum*) is one of the most ancient crops that originated in the Middle East. It spread from there throughout the Mediterranean, to the East African highlands, to Central Asia, and to India where today 75% of the world's chickpea area is located, with another 10% in Pakistan. Primitive varieties are still widespread in many parts of the extensive distribution, though being replaced by selected cultivars (Maesen, 1972, 1973). It would, however appear that this process is accelerating, since collecting and conservation of chickpea have recently received a high priority by the International Board for Plant Genetic Resources (IBPGR, 1976a). Such rapid changes emphasize the need for constant vigilance on one hand, preparedness for action on the other.

In contrast with the chickpea, the lupin is a newcomer as a major crop, although two species have a long history of human use. *Lupinus albus* was cultivated by the Greeks and Romans, *L. mutabilis* by mountain Indians in Peru and Bolivia for unknown periods of time. Unlike the wild lupin species, both have characteristics common to the domesticated pulses – non-shattering, non-hairy pods, and soft, permeable seed coats. But both have high alkaloid contents, hence are toxic to man and beast. The mountain Indians remove the alkaloid by leaving sacks of seed for weeks in running water; occasionally they use alkaloid-containing water as insecticides

(Brücher, 1968b). In Europe several other species were grown sporadically, or locally, presumably in less favoured environments where they were used for human consumption, and even for making flour. They were also widely used as ornamentals. It appears likely that in this way larger seeded types gradually evolved in both *L. luteus* and *L. angustifolius* (Gladstones, 1970). In the middle of the eighteenth century both species began to be used as green manure crops for sandy soils, but their use as forages was restricted by toxicity.

In the 1920s, Sengbusch (1934) succeeded in selecting alkaloid-free individuals of *L. luteus* and *L. angustifolius* from unselected land race material. He also succeeded in selecting individuals with the other essential or desirable characteristics, non-shattering pods, soft seed coat, and white seed colour (Sengbusch, 1953), each determined by one or two recessive genes. There are a number of recessive alleles for low alkaloid content in these two species and in *L. albus* (Gladstones, 1970). Low-alkaloid individuals were found at frequencies between 1 in 10 000 in *L. luteus* and 1 in 1 million in *L. albus*. As von Sengbusch (1953) points out, the susceptibility to pests and diseases, together with the high rate of cross-fertilization in these species would cause selection pressure against the retention of low-alkaloid recessive mutants. It is therefore highly probable that the low-alkaloid selections were the result of recent mutational events rather than regular components of the population retained by frequency-dependent selection (Harding, Allard and Smeltzer, 1966). However, agronomic characteristics such as non-shattering seed pods and soft seeds, were found in land race populations of *L. luteus* with little difficulty, suggesting that alleles controlling these characteristics were regular components of the genetic structure of the land race populations.

We are indebted to Dr J. R. Harlan for information on an unusual genetic contribution derived from a rare land race of sorghum which he collected near the Sudan–Ethiopian border. It is grown for medicinal purposes as a cure for whooping cough. The strain exhibits twin-seededness which is not uncommon (0.15% in the world collection of 14 000) and is usually controlled by a single gene which does not confer increases in yield. However, Dr F. Miller, of Texas A & M University, who directly received the strain, records yield increases of between 20% and 30% derived from the 'whooping cough gene'. Dr Harlan records that this gene is the only one of its kind in the world collection and was found in only one small population grown for medicine and not for food, but it may be grown more widely among cattle keeping tribes in the region.

8.2.4 Land races – adaptations and co-adaptations

From such well-defined genetic elements which affect a particular structure or function we turn to genetic contributions which have more generalized effects on the adaptedness and efficiency of genotypes. Most prominent among these is resistance to, or tolerance of climatic stresses such as drought, heat or cold, or edaphic stresses such as salinity, alkalinity, or mineral toxicity.

Reactions to climatic stresses are highly complex phenomena involving various organs and processes, with effects differing with the time of incidence and the intensity of the stress. Screening for resistance may therefore necessitate a number of complementary approaches, including field tests in stress environments, laboratory or glasshouse tests designed to reproduce the effects of natural stresses under controlled conditions, or laboratory tests developed from physiological or histological research into the effects of the stress. All these were applied by Russian workers in screening many of the collections of the Vavilov Institute of Plant Industry for frost and drought resistance. Among 19 000 accessions of winter and spring wheat, some 200 were highly resistant to frost and 150 to drought. As might have been expected, highest frost resistance is not found in the centres of genetic diversity (see section 8.1.4) for this crop, but in wheats from USSR, Canada and USA. The drought resistant types came from Central Asia, India, Syria, Chile, Mexico and, somewhat surprisingly, from Canada (Dorofeev, 1975).

The germplasm collection of the International Rice Research Institute, to which reference has already been made, was extensively screened for drought resistance in mass field screenings. The largest proportion of resistant types was found in upland rices from Africa and South America, in hill rices from Laos, and among early ripening varieties from Bangladesh. Field tests were supported by greenhouse techniques which separated the effects of early (vegetative) and late (reproductive) stresses. They also made it possible to discover associations between plant characteristics and drought resistance. Drought resistant varieties have more and thicker deep roots than susceptible ones, and this is now being used in screening techniques (IRRI, 1975).

IRRI (1975) is also concerned to find sources of tolerance to adverse soil conditions which limit yields or prevent the cultivation of rice. Large numbers of lines have been tested for tolerance of salinity and alkalinity which are the most extensive soil problems in Pakistan, India and other Asian countries. A small number of resistant lines were identified. In many areas flood waters cannot be controlled and periodically rice crops are exposed to submergence. Survival and productivity depend upon the rate of elongation, and of many hundreds of accessions some elongated by more than 40 cm during 7 days, starting from complete submergence. IRRI workers also discovered varietal differences in tolerance of iron toxicity and zinc deficiency. Heavy-metal tolerance in populations of the pasture species browntop (*Agrostis tenuis*) has become well known through the studies of A. D.

Bradshaw and his colleagues (see Gregory and Bradshaw, 1965, for a review).

Another kind of adaptation, the control of maturation by photoperiodic response, has evolved in land races of sorghum. Bunting and Curtis (1968) reported that in comparative yield tests conducted in different regions of Northern Nigeria, it was always the local varieties which yielded better than those from elsewhere, except varieties flowering about the same time as the local ones which flower at the end of the rainy season. Earlier flowering varieties suffer from grain being damaged by moulds or insects, late flowering ones fail to mature. The authors suggest that flowering dates are not controlled by photoperiod varying with latitude, but by the number of successively shorter days. Photoperiod control of locally adapted land races, as Andrews (1970) points out, is an important character which needs to be incorporated in high-producing dwarf sorghums adapted to Nigerian conditions.

Like day length, temperature during a critical developmental phase can have drastic effects on productivity, but this can be remedied by the introduction of a genetic component which confers tolerance of the inhibitory temperature level. Experiments in the Pasadena phytotron revealed that the yields of tomato varieties grown in California were substantially increased by lowering the night temperature six weeks prior to harvest to levels well below those of California (see Went, 1957). Varieties from areas with high night temperatures were screened, and an otherwise worthless strain from the Philippines was discovered to have the required tolerance, and this was incorporated in commercial varieties.

In the central Indian State of Madhya Pradesh wheat has been grown for several thousand years under some of the driest conditions in the world, with hardly any rain falling during the growing season. Wheat is clearly marginal and should have been replaced by a summer crop making use of monsoon rains, but being part of a subsistence farming system, wheat growing has continued and acquired adaptations to extreme drought conditions. One of these appears to be the shedding of the lower leaves to form a mulch which reduces evaporation (H. K. Jain, personal communication).

Of particular interest for an understanding of the evolution of land races, and for their contribution to future crop evolution, are the correlated adaptations, or *co-adaptations*. Their importance has been emphasized by several authors. For example, Frankel and Bennett (1970a) suggest that 'the enormous diversity of gene complexes determining adaptation and productivity, assembled and incorporated over centuries of cultivation in different environments . . . must be recognized as the outstanding and unrivalled characteristic of primitive varietal populations'. This seems highly plausible; and Darwin, in *The Origin of Species*, emphasized the common occurrence of co-adaptation in domesticated animals and cultivated plants as strong evidence of selection. Correlations between characteristics were listed by the early writers on scientific plant breeding, especially any correlations between yield and morphological characters which could be more easily scored than yield itself.

Recently, Qualset (1975) estimated the frequencies of combinations between eight characters in the Ethiopian barley collections noted previously. the characters used were heading time, seed and ear characters, and BYDV resistance (Fig. 8.2). For only five of the twenty-eight combinations the probability of independent distribution exceeds 0.05. But, as Qualset remarks there is no explanation for the selective advantage of the characters less still of their combinations, with the exceptions of BYDV resistance and heading time. Since BYDV resistance is higher at higher elevations, as would be late flowering, the association of these characters can be understood as due to natural selection.

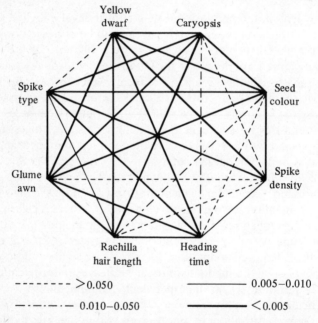

Fig. 8.2 Association polygon for eight characters in a collection of 650 Ethiopian barleys. The type of line connecting two characters indicates the probability of independent association. From Qualset (1975).

No doubt such co-adaption would be more readily understood if the inheritance of the characters were known. In fact relatively little is known of the genetics of co-adapted characters. Genes can be held together as supergenes (including inversions), by linkage, by the breeding system, by selection, or by combinations of any of these. Some of the best genetic evidence comes from the work of Dobzhansky and others in *Drosophila* (summarized by Dobzhansky, 1970). In populations polymorphic for co-adapted gene sequences, the latter are maintained intact by the high frequency, i.e. the fitness, of the inversion heterozygote. Chromosomes from ecologically distant localities may not be co-adapted, hence crossing-over in heterozygotes may break up the complexes. In the wild oat species, *Avena barbata*, Allard, Babbel, Clegg and Kahler (1972) discovered strong evidence for co-adaptation of a number of allozyme loci, with adaptive responses of particular allelic combinations to macro- and micro-environmental differences. They found gametic-phase disequilibrium for five loci, three linked and two independent, assorted in two balanced five-locus gametic types, one prevailing in drier, the other in wetter micro-niches and macro-environments. Selection for these favoured assortments is assisted by restrictions on recombination due to linkage and/or the breeding system.

In natural populations of wild barley (*Hordeum spontaneum*) in Israel, Brown, Nevo and Zohary (1977) found widespread polymorphism and marked linkage disequilibrium for the four esterase loci studied in composite barley crosses (see next paragraph; also cf. p. 199). Co-adapted allelic complexes differed between sites, but no consistent ecological pattern could be discerned in these preliminary data.

In cultivated plants there are few genetic studies of co-adapted complexes on record. Genetic evidence of correlated genes – four esterase genes, three closely linked and the fourth on another chromosome – was obtained in two barley composite crosses with different sets of parents (Allard, Kahler and Weir, 1972; Weir, Allard and Kahler, 1972). From an initial random distribution of the three (effective) alleles at each locus, the three most favoured combinations gradually increased, to contribute 40% in generation 26 in one cross, and nearly 80% in generation 41 in the other. The two most favoured gametes were the same in both crosses. It is shown that the interaction between natural selection and restriction of recombination associated with predominant selfing, was the primary cause for the formation of coordinated units. It remains uncertain whether it was the allozymes themselves, or loci associated with them, which were the object of selection, with the allozymes serving as convenient markers.

Adaptive complexes in wheat are suggested by the work of Law and Worland (1973). By using single chromosome substitutions they found associations between genes for vernalization and cold resistance, for resistance to eye-spot (*Cercosporella herpotrichoides*) and freezing (presumably in three chromosomes), and for grain weight and height. In all three cases there is evidence that the associations are at least partly due to linkage. It is clear that the mapping of genes on chromosomes, and, by the use of telocentrics, on chromosome arms (Sears, 1966) as is now possible in wheat, is a powerful tool for uncovering the genetics of adaptations and of co-adaptative complexes.

Genetic associations between morphological characters and yield itself are not only of interest to the plant breeder, but they may contribute to knowledge of the physiology of yield (see the next paragraph). It had been shown in wheat that in the absence of lodging and of dwarfing genes there is in general a genetic correlation between plant height and yield. Gale and Law (1977), pre-quoting Law, Snape and Worland (1978), state this correlation to be due to tight linkages and/or pleiotropy. However, Gale and Law were able to show that the dwarfing genes 'are associated with yield rather than the reverse', due to linkage or pleiotropy. Since the semi-dwarfing genes, *Rht 1*, *Rht 2* and *Rht 3* are associated with insensitivity to gibberellin, it is possible to identify the *Rht* status of any plant in crosses with tall standard varieties, and thus to establish the relationships between plant height and yield as influenced by dwarfing genes on one hand, other genes affecting height on the other. In this way it was shown that the dwarfing genes themselves are associated with yield increase, but that in their presence other genes for plant height further increase yield. The authors conclude that in the presence of dwarfing genes – which can be ascertained by gibberellin insensitivity – selection should be for increased height, and that a dwarf ideotype could be counterproductive.

Of particular interest is the suggestion that the insensitivity to gibberellic acid (GA) associated with the semi-dwarfing genes is due to a block in the utilization of endogenous GA, with a GA antagonist acting on the 'active sites' of GA action. As Gale and Law (1977) point out, the evidence on which these suggestions are based indicates that the GA-insensitive/dwarfing genes 'affect a basic change in the metabolism of varieties that carry them'. This expresses itself in other processes such as an abnormal tillering response to applied GA.

This discussion of the action of the *Rht* genes takes us finally to a consideration of adaptations acquired by land races under *favourable* conditions of cultivation, rather than under stress conditions with which we started this discussion. Here the syndrome of characters which contribute to the productivity of the 'high yielding' varieties of wheat with Norin 10 ancestry comes to mind. In addition to short height, discussed in the preceding paragraphs, there are increases in grain weight and in the number of grains per area. All of these can be seen as adaptations to conditions of ample water supply and high fertility. The syndrome was derived from a Japanese variety which, in spite of its somewhat mixed ancestry, seems to point to China as a likely source of the syndrome. Wheat had been grown in China under irrigation for thousands of years, and Chinese land races embodying features of the syndrome were not uncommon some forty years ago, as was evident from a collection of Chinese wheats assembled in New Zealand.

There can be little doubt that an adaptive response to favourable conditions has been of widespread occurrence. As already pointed out in section 8.2.1 high-producing selections from land races in western and north-western Europe contributed the *genetic* component of the original 'green revolution', which started a hundred years ago.

It must be recognized that on the whole plant breeders do not appear to have extensively used their germplasm collections as sources of new or improved adaptations, perhaps assuming that they had already been incorporated in well-adapted existing cultivars. Yet Vavilov emphasized that the combination of parents adapted to widely different environments may provide opportunities for major advances – presumably through the combination of different adaptive complexes. This is a precept which has been practised by some of the most successful plant breeders – William Farrer in late nineteenth-century Australia, and Norman Borlaug in Mexico in our time.

There is, however, yet another consideration which may lead to a broadening of the genetic base. Recent developments have drawn attention to the dangers – in juxtaposition to the well-recognized advantages – of genetic homogeneity in crops (see Marshall, 1977). The growing interest in the resistance potential of multilines and varietal composites is one result; and the discovery that cytoplasmic DNA in addition to nuclear DNA controls processes determining resistance to pathotoxins has added a new dimension to the needed defences of plants, including heterogeneity in crop cytoplasm. Cytoplasmic inheritance of resistance to *Helminthosporium maydis*, the corn leaf blight, is unquestioned, as is the differential effect of the pathotoxin on mitochondria of resistant and susceptible genotypes (Petersen, Flavell and Barratt, 1975), although it is as yet uncertain which particular cytoplasmic genome controls the reaction. It is less certain whether the recently discovered differences between species of *Nicotiana* and *Triticum* in the polypeptide composition of Fraction I chloroplast proteins (Gray, Kung and Wildman, 1975; Chen, Gray and Wildman, 1975) are indicative of intraspecific differentiation. No differences have so far been found within species of *Nicotiana*.

8.2.5 Land races – conclusions

(i)   Genetic diversity *between* land races is well established from observations in the field and in germplasm collections.

(ii) Genetic diversity *within* land race populations is also well established, from direct observation in field and experiment station, and from the results of selection. Yet there is little knowledge of genetic structure and population dynamics in land races, other than observations on regular introgression from wild relatives in some circumstances ('microcentres', Harlan, 1951; Zohary and Feldman, 1962). Harlan's description of land races as balanced populations in equilibrium with the environment (see section 8.2.1) is plausible though so far unsupported by direct evidence. Detailed studies of land race populations in their natural and social setting would be relevant for crop evolution and ethnobotany and for the exploration and utilization of crop genetic resources.

(iii)   The enormous genetic diversity which is contained in land races is unquestioned. It is evident in the existing germplasm collections which, in spite of limitations, are our main source of information on variation in crop species. The main limitations derive from the way collections have been assembled and maintained, including uneven geographical representation, inadequate documentation of origin and biological status, frequent reproduction resulting in genetic erosion. Analyses restricted to well-authenticated material, e.g. from specific regions, may be more productive than less discriminating surveys of world collections.

(iv)   Today the remaining land races still in cultivation, together with land race accessions in collections, constitute the reservoir of natural genetic variation. They complement and diversify the gene pool of advanced cultivars in the continuing quest for improved efficiency, productivity and adaptation to changing environments and human needs. This reservoir, as we have seen, continues to contribute selected genetic materials: selected individuals, at the early stage of genetic improvement; specific genes or characters important for survival, productivity, quality, etc; adaptations, or co-adaptations which are beginning to be understood in genetic and physiological terms. With our growing understanding of vital processes such as the relations of pathogen and host plant, the utilization of genetic contributions from land race material is likely to increase.

(v)   In many parts of the world where land races persist – mostly in isolated or otherwise disadvantaged localities – they are under threat of displacement. For most crops the continued existence of the genetic reservoirs will devolve on the material which can be brought into germplasm collections, and on the manner in which these collections are maintained. These matters are presented in Chapter 9.

(vi)   This discussion will have made it abundantly clear that the interest in land races is based on their current and future usefulness in plant improvement. The usefulness of germplasm collections depends on the extent to which material they hold is described and evaluated, and the information is made available to users through an effective information and documentation system. These matters will be discussed in section 9.7.

(vii) It is highly probable that the interest in land races as crop genetic resources will greatly increase in the course of the next decade. For this there are two reasons: first, the growing information on the physiology and genetics of adaptations; and second, the growing pressure for discovering multiple resistances to parasites, whether for incorporation in single genotypes or in multilines, whether in the form of horizontal or vertical resistance (see p. 205). New sources are, and will be at a premium, with the largely unexplored land races as the obvious reservoirs of resistance genes accumulated in evolutionary history.

8.2.6 Advanced cultivars

There is a curious contradiction in the relative prominence given to advanced cultivars, the products of scientific plant breeding and the mainstay of modern agriculture. On the one hand they have been, and continue to be the principally used genetic resources of plant breeders; on the other they have received little attention in the discussions and publications on genetic resources during the last ten or fifteen years. For this there are several reasons which were recognized early in the discussions on genetic conservation (e.g. IBP, 1966). Advanced cultivars are well represented in many existing germplasm collections, whereas primitive cultivars are for the most part poorly represented in relatively few collections. Many advanced cultivars are closely related, differing only in minor respects, in contrast with the genetic diversity of primitive cultivars. Advanced cultivars can as a rule be obtained from their breeders or from national collections, primitive cultivars are not readily available either from collections or from the areas where they are still grown – even if these are known. Above all, primitive varieties are vanishing rapidly. Hence the collection and preservation of primitive material receives, and should continue to receive, foremost consideration.

Yet the advanced cultivars must not be neglected. One may assume – though perhaps with some doubt – that the currently used cultivars incorporate most of the co-adapted complexes which had been assembled in their predecessors in the earlier stages of plant selection. This may be partly justified, but it is not unlikely that the early advanced cultivars, which had been selected prior to the widespread use of fertilizers, intensive cultivation, or irrigation, may have possessed adaptive characteristics which later became redundant, yet may be needed again for less favoured environments or low-input agricultural systems. Similarly, resistance to currently rare biotypes may again be required should they build up in the future. Another reason for retaining representative collections of obsolete cultivars is that they are – or should be – evaluated and documented for many characteristics associated with performance. The question arises which to preserve, in view of their enormous number. Judicious culling is inevitable, though along what principles?

These remarks apply also to 'genetic stocks' of various kinds – natural or induced mutants, breeding lines with specific characteristics, stocks with specific resistances including resistant components of multilines, tester stocks for parasite biotypes, and, of course, to the large array of cytologically and/or genetically characterized stocks which have become tools of trade in current cytogenetic, evolutionary and breeding research. The maintenance and distribution of the latter – an essential service to current research – has become a burden on individual institutions and calls for cooperative action. These – and perhaps other genetic stocks – should be looked after by appropriate institutions or by cooperatives of scientists. Initiatives might come from international organizations like the International Genetics Federation, Eucarpia, SABRAO etc.

8.2.7 Wild and weedy relatives of domesticated plants

The principal interest in the ancestors and other relatives of crop species is their wealth of resistance genes to parasites, acquired over long periods of host–pathogen coevolution. In many crops, including most of the cereals and their rusts, centres of diversity of the ancestral species and of important parasites coincide. A lack of coincidence, as Watson (1970a) points out, may be due to a number of causes, such as the absence of an alternate host required for sexual reproduction. The potato, as we have already seen (p. 190), was domesticated apart from the blight fungus, and resistance evolved in other *Solanum* species which are endemic in the centre of diversity of the parasite. Similarly, of the fifty-five species of *Nicotiana* tested for resistance to the blue mould fungus, *Peronospora tabacina*, all but one of the nineteen Australian endemic species have a considerable degree of resistance, whereas only one of the thirty-five American species, *N. knightiana*, had moderate resistance, all others, including the putative parental species of *N. tabacum*, being susceptible (Wark, 1963). When the fungus spread throughout the tobacco growing areas of Australia and the United States, and recently of Europe and the Near East, resistance was obtained from Australian species, *N. debneyi* and *N. goodspeedii*, and, in spite of strong sterility barriers, transferred to *N. tabacum* (Wark, 1970).

Work by Harlan, de Wet and their associates, extending over many years, has demonstrated how the use of an extensive gene pool of related species and a close study of hybrid derivatives can throw some light on evolutionary relationships on the one hand, yield valuable results of introgression on the other. Hybrid derivatives from maize × *Tripsacum* crosses support the hypothesis of a close evolutionary relationship between maize and teosinte, with tripsacoid characteristics of maize having been derived from teosinte rather than *Tripsacum* (de Wet, Harlan, Stalker and Randrianasolo, 1978). Selection for resistance to a number of important diseases resulted in the discovery of resistance transfers from *Tripsacum* to some of the maize types recovered from complex recombination pathways (Harlan and de Wet, 1977).

Most of the resistances obtained from wild species are in the form of major genes, especially in some of the best-studied diseases such as the cereal rusts. However, in reviewing the contributions of wild species to resistance breeding, Watson (1970a) suggests that even in species where major genes conferring specific or 'vertical' resistance – i.e. resistance to specific biotypes of a parasite – have been extensively used, there is likely to remain a genetic system of minor genes as a source of non-specific or 'horizontal' resistance. As evidence he quotes *Solanum demissum* as a source of both specific and non-specific resistance to *Phytophthora infestans*. *Avena sterilis*, in addition to specific resistance genes to crown rust, *Puccinia coronata*, which have been extensively studied (Murphy *et al.*, 1967; Dinoor, 1975), appears to have a non-specific type of resistance effective against all races in Israel.

Are specific resistance genes derived from wild species more effective than those from domesticates? Watson (1970a) makes two points. First, genes from wild species 'are often effective against all strains of the organism in the area where the work has been done', and, one may presume, where the wild host species has evolved. Second, there is no evidence that the host–pathogen relationship is more stable when resistance genes are transferred from wild species. He quotes as examples the specific resistance to *P. recondita* in *Aegilops umbellulata* transferred to wheat by Sears (1956), and resistance to *P. striiformis* derived from *Ae. comosa* (Riley, Chapman and Johnson, 1968) both of which can be easily overcome when in a background of hexaploid wheat. He concludes that the most effective and economical application of specific resistance genes is in combination with each other, i.e. in the form of multiple resistance (see Watson and Singh, 1952).

Whatever the potential pool of resistance sources, both known and as yet unknown, it is inevitably finite, hence the rate at which resistances are being 'used up' under the impact of pathogen evolution is of greatest importance for the future of agricultural productivity. In past experience, both long-term stability as well as fairly rapid breakdown have occurred with any type of resistance system, although there are substantial differences in the relative frequencies. The degree of stability depends on gene action and organization in the host, and, as Pryor (1977) points out, on the selection pressure that a resistance system imposes on pathogen survival. Without a doubt, specific single-gene resistance in homogeneous cultivars constitutes the most vulnerable of the resistance systems, although there are examples of stability (see Watson, 1970b), and BYDV resistance, found in Ethiopian land races, may be one of them (see p. 183). At the other end of the scale, non-specific, horizontal, polygenic resistance (Van der Plank, 1975) tends to be a great deal less vulnerable, but there are examples of fairly rapid, as well as of slow and gradual breakdown. In fact, there appear to be many links between the two, with the genetic background affecting the effect and stability of single genes on one hand, of components with gene-for-gene resistance in horizontal resistance on the other (Ellingboe, 1975). It is likely, though an over-simplification, that horizontal resistance, being generally more stable, is the more desirable system. But it must be recognized that single-gene resistances, singly or in intra- or intergenotypic combinations, are the principal defences against many pathogens which attack most of the crops which feed the world. A number of approaches have been introduced which are designed to increase the effectiveness and the stability of resistance systems based on specific resistance. Multilines are composite varieties consisting of lines carrying different resistance genes, each of which is either resistant to all biotypes ('clean approach') or to single, but different biotypes ('dirty approach') (see Marshall, 1977). Both approaches make heavy demands on resistance sources, hence economy of use, i.e. minimum exposure to selection pressure by the pathogen, is claimed to be one of the objectives. Experience is not sufficient to permit conclusions, but an examination of models based on the 'dirty approach' indicates that stabilizing selection of the pathogen will occur only in limited and rather rare circumstances (Marshall and Pryor, 1978). So far there is no general solution in sight, hence the need for intensifying the search for resistance sources.

The importance of wild relatives as resistance sources is enhanced rather than diminished by such considerations. In a growing number of crops the resistance genes stored in related wild species are the main, in some instances the only resources available when known sources are 'used up', as had been the case in resistance to some biotypes of crown rust until resistance was obtained from *Avena sterilis* (Murphy *et al.*, 1967). In other instances resistance so far has only been found in wild species. A case in point are two diseases of the sunflower, *Helianthus annuus*, sunflower rust and downy mildew. Wild sunflower species are native in North America and were brought to Europe by early explorers. However, the drastic improvement which made possible exploitation as a major crop is of recent date and took place in Russia since the nineteenth century. The sunflower rust, *Puccinia helianthi*, is prevalent wherever sunflowers are grown, but resistance has not been found in the domesticate. Downy mildew (*Plasmopara halstedii*) is a seed-borne and soil-borne disease of sunflower which causes heavy losses in crops in USA and elsewhere. It was recently found that a source of resistance to the rust derived from a natural cross of a cultivar with Texas Wild Annual, also confers resistance to downy mildew though through separate genes. Both resistances are derived from the wild stock.

Russian workers discovered downy mildew resistance in another species, *H. tuberosus* (Zimmer and Kinman, 1972).

Possibly more extensively than in any other economic plant, wild relatives have been used in breeding tomatoes resistant to the main pathogens – *Cladosporium, Fusarium*, nematodes, *Septoria, Stemphyllium*, tobacco mosaic virus and *Verticillium* – and, as in rice, cultivars with multiple disease resistance have been produced (for references see Rick, 1973).

Apart from resistance breeding, wild relatives of crop species have not been extensively used in plant breeding, although, as Harlan (1976) shows, there are many other opportunities. In recent years some new avenues have opened up which promise to extend significantly the boundaries of plant breeders' gene pools. One derives from the detailed knowledge of the genetic architecture which has resulted from work such as that of Kihara, Sears, Riley, Kimber and others on the wheat genomes. This has already led to sophisticated transfers, from chromosome or half-chromosome substitutions to gene transfers. Another prospect is the technique of cell hybridization which, however, at least for the time being, is likely to be restricted to closely related species, such as the cell hybridization of *Nicotiana glauca* and *N. langsdorffii* effected by Carlson, Smith and Dearing (1972).

The prospect of using closely related wild and weed species to extend the adaptive range of crops is as yet not widely explored. Zohary, Harlan and Vardi (1969) have drawn attention to 'the wild diploid progenitors of wheat [which] constitute large gene pools largely unexplored and untapped by plant breeders'. The three diploids differ in ecological adaptation and in many other respects, and incorporation by crossing (followed by back-crossing) should be easier and more readily exploitable than the more popular distant crosses. The authors urge that such species be extensively collected and studied. Nevo *et al.* (1979) suggest that genetic variation in wild barley, which has been shown to be at least partly adaptive, could be used in breeding for extreme environments (see p. 184). Autecological studies of wild tomato species by Rick (1973) have revealed a diversity of adaptations to extreme conditions which should be of considerable interest to tomato breeders. They include tolerance of an extreme range of conditions – from waterlogging to drought – in *Lycopersicon esculentum* var. *cerasiforme*; extreme resistance to desiccation in *Solanum pennellii*, thanks to the ability to store water in leaves, and similar drought resistance in *L. chilense* associated with an extensive root system; and near-complete freedom from insect predation in *L. hirsutum*, in its native habitat as well as under experimental conditions exposed to a range of insect parasites. Finally, we quote an attempt to widen the genetic base of the potato (*Solanum tuberosum*) by crossing with Neo-Tuberosum, which is a strain of *S. andigena* – the putative parent species of *tuberosum* – resulting from repeated mass selection for yield under Northern European conditions. F_1

hybrids of Tuberosum × Neo-Tuberosum have been encouraging, and the breeding potential of Neo-Tuberosum is being further explored (Simmonds, 1969; Holden, 1977).

It may be that in the context of ecological adaptations, especially to less favourable environments, the weedy relatives have a special place as genetic resources for future plant breeding. So far they seem to have been little considered in such a role, nor, with some exceptions, as sources of disease resistance. But having evolved under cultivation and in competition with crops, in many instances introgressed by crop and/or wild species, they are likely to have acquired adaptations possessed by neither.

Here the general observation is in place that most of the wild and weedy relatives of domesticates have been little collected (see Harlan, 1976), and collections which are made for specific purposes, such as studies of crop evolution, often are dissipated on their conclusion. For many of the principal crops – let alone minor ones – the number of wild accessions in major germplasm collections is pitifully small. While wild biota as a rule are not exposed to rapid extinction as are the land races, representative collections should be established for observation and use by plant breeders. After all, one cannot use what one does not know nor can readily obtain.

Availability and conservation of representative gene pools is one need, intensive study is another. We are beginning to study the population structure of wild relatives of domesticates as a guide to collection and utilization (see p. 184). Useful evidence comes from population studies on the introduced weed species, *Avena fatua* and *A. barbata*, in California, to which we have already referred (p. 199). Considerable differences between regions were shown to exist in the degree of polymorphism of both morphological characteristics and isozymes (Jain, 1969; Marshall and Allard, 1970). There is also a need for information on physiological characteristics of wild relatives, and some recent studies have yielded interesting results. Evans and Dunstone (1970) comparing wild and domesticated wheats at the three ploidy levels, found that evolution has involved a parallel increase in leaf and grain size, but coupled with a reduction in the rate of photosynthesis per unit leaf area. The wild species showed a fall in leaf photosynthesis during grain development, whereas in the domesticates it rose during that phase.

8.2.8 Wild species used by man

There are many wild species which have not been domesticated yet are extensively used (Fig. 8.1). Some are widely planted, though genetically and culturally in a near-wild state. Foremost among these are the forestry species which have their genetic reservoirs in primeval, or at any rate natural forests. Although continuously shrinking – in some parts of the world with increasing rapidity – for most species there is still scope for the salvage of a broad

range of genetic diversity, although perhaps not for very much longer where
pressure is most acute. Writing in 1967, Richardson (1970) could say that
the main threat to forest genetic resources came not from a rapid increase in
the tempo of utilization, but from population pressure on cultivable land.
This pressure has greatly intensified, especially in the tropics. In South-East
Asia the once vast resources of tropical moist forests are rapidly shrinking,
thanks to shifting cultivation with greatly reduced fallow periods, land
clearing for agriculture and plantation crops, clear-felling for pulpwood, and
intensive logging techniques which have replaced the selective practices of
the past. Today, foresters are increasingly concerned to renew timber
resources by planting fast-growing pioneer species. Quoting an ECAFE
report of 1974, Whitmore (1975a) predicts 'that both Malaysia and the
Philippines will have no remaining accessible virgin forests within a decade';
indeed, 'silviculture in natural forests has virtually ceased'. Whitmore lists
many South-East Asian and Melanesian species which are currently used for
industrial purposes, or as whole timber species. He names many others
worthy of silvicultural trial. The genetic reservoirs of all of these are under
threat. It is somewhat ironical that the rapid advance of plantation species
should coincide with the disappearance of their genetic resources on which
the improvement of productivity, pest resistance or timber quality will have
to depend. Guzman (1975) compiled a comprehensive list of forest species
in the Philippines which can be utilized for the production of food, including
fruits, oil, beverages, vegetables and spices. Indeed, tropical forests are the
main reservoir of as yet untapped resources of economic plants. For genetic
resources of fruit species see p. 218.

In some temperate zone forestry species genetic erosion has gone even
further than in the tropics. Two examples from California must suffice
(UNESCO, 1973). Douglas fir (*Pseudotsuga menziesii*) is an important
plantation tree in its native California and in other countries. It is protected
in reserves, but pollen from surrounding plantations, and aerial seeding of
non-Californian provenances threaten to erode the integrity and diversity of
the indigenous populations. The position is different in the Monterey pine
(*Pinus radiata*) of which there are only five small populations remaining in
its native habitat, which should be preserved *in situ*, considering the great
importance of radiata pine in plantations in New Zealand and Australia.
One may, however, wonder whether the secondary gene pools which must
have evolved in the vast areas planted in these countries are not perhaps
more useful genetic resources for tree breeding in Australia and New
Zealand than are the relic populations in California.

From the foregoing it is clear that indigenous forests have a dual function
as genetic resources. First, with regard to the species which are planted
either in monoculture, or – less frequently – into existing stands, the role of
indigenous forests parallels that of the land races of domesticated species as

sources of genetic diversity for adaptation to new environments (Guldager, 1975) and for selection and breeding. Second, indigenous forests, especially in the tropics, are a reservoir for potentially useful species. This will be discussed in the section which follows.

Wild species used by man may be endangered because in their native habitats they are over-exploited, rare, or both. This is the case with a number of medicinal plants native in India, including some of considerable economic and social importance. *Dioscorea deltoidea*, with a distribution in the western and north-western Himalayas, is an important source of diosgenin which is a precursor for the synthesis of cortisone and of steroidal sex hormones. The roots of *Rauvolfia serpentina* contain many alkaloids, notably reserpine, an antihypertensive drug. It occurs in hilly country in various parts of India, but also in Nepal, Sikkim and Bhutan. Pressure on these species is so great that they are likely to be seriously threatened unless successfully taken into cultivation. Another species requiring conservation measures is *Aconitum heterophyllum*, which contains an old aconite drug used to combat fever, and debility after malarial and other fevers (Dr B. K. Bhat, Regional Research Laboratory, Srinagar; and National Bureau for Plant Genetic Resources, New Delhi, personal communications).

Many wild species are used in pasture and range lands and as raw materials for chemical industries. In recent years some of these have become domesticated, usually in countries other than their own, as for example Mediterranean, African and South American pasture grasses and legumes in Australia. Large collections of such species have been established, but material continues to be collected in the countries of origin to obtain types adapted to new ecological conditions or for specific purposes. A large number of subtropical and tropical forage legumes, most of which had not previously been cultivated, were collected and introduced to Australia by the CSIRO Division of Tropical Crops and Pastures (Davies, 1960). Some of these, especially species of *Stylosanthes*, have been successfully established in Australia and are now being adopted by other tropical countries. Evaluation and further collecting are continuing. The major genera of interest include, amongst the legumes, ALYSICARUS, AESHYNOMENE, CENTROSEMA, DESMANTHUS, DESMODIUM, LEUCAENA, MACROTYLOMA, MACROPTILIUM, PHASEOLUS, STYLOSANTHES, VIGNA and ZORNIA. In the grass genera the emphasis is on ANDROPOGON, BRACHIARIA, CENCHRUS, CHLORIS, DIGITARIA, PANICUM and UROCHLOA.

On the whole such species are not under acute threat except in specific areas where rough grazing land is turned over to agriculture or forestry or other forms of land use. This is the case with ryegrass, *Lolium perenne*, an important forage grass, with ancient stands in pastures and meadows in various parts of Europe. These have served for years as the reservoirs from which modern high-producing strains have been developed by plant breeders at the Welsh Plant Breeding Station, Aberystwyth, and elsewhere. Now many of these old pastures have either been ploughed up or are threatened. One of historical interest, from which the first and most famous Aberystwyth strains were developed more than fifty years ago, was lost in the 1960s. Situated in the Monmouthshire moors of south-east Wales, it was mentioned in the Doomsday Book. It gave way to the buildings of a major steel works. Interesting populations with an extraordinarily high content of carbohydrates have been found in the Po valley of northern Italy. These also are now disappearing (Hayward, personal communication).

8.2.9 *Wild species of potential value to man*

Reference was made in the preceding section to wild species of prospective value. Such species are difficult to define. They stretch from plants which are now gathered as major vegetable foods or as condiments in many parts of the world, in the forests of southern Asia or the wasteland and scrub of Mediterranean countries, to potential raw materials for chemical and medico-chemical industries. One example of the latter must suffice. A great deal of research and development has been directed to the cultivation of the Australasian species *Solanum laciniatum* and *S. aviculare* which have a high content of the steroidal alkaloid solasidine, a raw material of steroidal contraceptives and other drugs. Production techniques have been advanced in several countries, including New Zealand (Fryer, 1972). In Australia, the heavy insect infestation has so far been prohibitive, but in countries of adoption in Europe and the USSR the parasite fauna does not seem to have adopted the immigrant species and large-scale production is in progress. Presumably there is scope for considerable genetic improvement. It should be noted that tribal people in south India are using the berries of some *Solanum* species for contraceptive purposes, those with the highest alkaloid contents of the berries (3–4.8%) being *S. incanum, S. indicum, S. xanthocarpum* and *S. trilobatum*. So far these have not been cultivated, and the productivity is not known (Viswanathan, 1976).

Indications that species are worthy of attention come either from past usage, from taxonomic, genetic, biochemical, or other associations with species now used, or from new requirements by industry. The history of crop production fails to suggest that new major crops are likely to be discovered (as against new plants deliberately engineered by man, such as *Triticale*). But there is reason for assuming that species may be taken into cultivation for specific ecological niches, or for requirements arising in nutritional, medical, chemical or other industrial research. Such leads should be taken up as they arise, but it hardly seems called for to collect and preserve a range of wild species on the basis of vague prospects. It would seem more purposeful to regard the preservation of potentially useful species as part justification for the conservation of ecosystems, with particular emphasis on tropical forests which appear to contain the most promising and diverse reservoirs of potentially useful plant species (see Evans, 1976).

8.2.10 Exploration

The term 'plant exploration' goes back to the era when plant species – mainly of horticultural interest – were explored and discovered in wild and rather inaccessible places in Asia and Africa. It is widely used as an alternative to 'collecting' and relates to materials gathered in fields, gardens, forests, wasteland, or any other sites where plants of actual or potential relevance to man are to be found.

In accordance with the emphasis throughout this chapter, plant exploration is intended for use, now or in the future, whether in research or in plant improvement, but not simply for conservation *per se*. Exploration must therefore be planned in the context of its purpose. Two objectives can be distinguished:

1. To collect *material which is endangered* and likely to disappear in the near future. Exploration of this kind can be regarded as a salvage operation. Targets are less specific than in the second category, although institutional or national interests will emerge in the preferential choice of target species or regions.

In earlier pages of this chapter we have referred to the current state of genetic resources in the field and in germplasm collections. On farmers' fields the land races which had been prevalent as recently as twenty-five years ago, have either been or are progressively being displaced by advanced cultivars. In germplasm collections, the representation of land race populations is far from adequate for most of the major and minor crops, but for many crops opportunities remain to complement available germplasm resources by exploration and collecting, albeit in remote and often rather inaccessible regions, and even there in the face of rapid erosion. Indeed, in some important areas the rate of displacement is such that only through most urgent action can one hope to salvage what remains. However, for wild species of economic significance there is no such widespread or urgent need for action, with the exception of species which are threatened by wholesale destruction or serious depletion of indigenous forests, by excessive exploitation of wild drug plants, by conversion of grazing land to agricultural or other development, or by overgrazing, often as a result of overpopulation.

2. To collect material expected to be *of value for current or proposed research or development*. Targets as a rule are more specific following up research leads, including earlier explorations, or filling gaps left by them. This may apply to crops which as yet have not been subjected to intensive breeding, such as many tropical vegetables and fruits, but more generally to

wild forestry and pasture species. In this case the purpose of collecting is *immediate use rather than conservation*, the main objective being the introduction of species of direct use in the country of adoption. The relationships between agronomic performance and the climate of origin have been the subject of several studies, most recently by Burt, Reid and Williams (1976) who used an extensive collection of the pasture legume *Stylosanthes* spp. There was a close relationship between groupings formed purely on morphological–agronomic classifications, and the climates of the countries of origin. This and further papers by the authors and their colleagues open the prospect of identifying the most promising sites for locating introductions, but also for indicating the climate, hence the region for obtaining promising introductions.

Continuing maintenance of such material is not aimed at its preservation as an irreplaceable resource but as an evaluated and available one likely to be useful in further work and to other institutions. A prominent example of such exploration and introduction of wild material for direct use and for plant breeding is the work on tropical forage legumes and grasses in Australia briefly mentioned on page 210.

(a) *Exploration targets – distribution and urgency.* This section is concerned with the geographical distribution of genetic resources, the priorities for exploration, and the methodology of exploration. The first question is where land races (and threatened wild species of economic relevance) are to be found, the second, what is the state of urgency, either on the grounds of an imminent threat to their continuing existence, or because of requirements for current plant breeding programmes. First thoughts turn to the centres, or regions, of origin where crops were domesticated from their wild progenitors, and where Vavilov and later plant explorers found concentrations of genetic diversity. Yet it had long been recognized – even by Vavilov himself – that there are many centres of diversity which are not centres of origin. Some such secondary centres, such as Ethiopia for barley, contain greater diversity than the centre of origin (Harlan, 1975a, p. 161) and, as has already been emphasized, they add greatly to the genetic diversity of crop species.

There is a good deal of rather general information on the geographical distribution of the areas of genetic diversity for the major and many of the minor crop species, and the recently published reference work by Zeven and Zhukovsky (1975) brings together useful information on the primary and secondary centres of diversity of a large number of plants used by man and of many of their wild relatives. The classification of the geographical distribution is based on Zhukovsky's twelve megacentres, which cover the greater portion of the earth's surface, each of them so large that even the additional information, where provided, gives little precise guidance where any species, less still its area of diversity, can be found. Valuable as such information is for general reference, more detailed information is required on the location and, if possible the main characteristics of crop genetic resources if they are to be identified and collected before it is too late.

Attempts to locate more precisely what remains to be collected have been made in the last ten years, but with only moderate success. A survey of genetic resources in their centres of diversity was organized in 1971/2 jointly by FAO and IBP, the International Biological Programme (Frankel, 1973; summarized in Frankel 1975a). It was conceived as an inventory and guide for plant collectors, but also as a call to action to scientists and governments. Limitations of time and finance prevented exhaustive surveys. Instead, information was requested from the best available sources – scientists working in the region, plant explorers, evolutionists, crop specialists and others. They encountered many difficulties such as sporadic distribution, isolated and remote location, or a lack of information on the varietal distribution in different parts of a country or region. The survey as published has many gaps and other shortcomings, yet some valuable information came forward such as accounts of the state of indigenous genetic resources of major crops in Africa by J. R. Harlan and in India by H. B. Singh, of fruits and grain legumes in Indonesia by Setijati Sastrapradja, on wheats in the Mediterranean basin and in Afghanistan by Erna Bennett, and many more. The publication gave encouragement and information for exploration, and, perhaps more importantly, it conveyed a sense of urgency. To this limited extent the purpose was achieved, and the attempt was not renewed. It brought forth, however, a clear lesson. Some of the most meaningful reports had come from active workers on genetic resources in the regions of diversity themselves, and it is on these that successful plant exploration would largely have to depend.

Valuable information may be derived from germplasm collections which have been adequately classified, evaluated and documented (see section 8.2.2). Leads for further exploration can come from observations on the extent of genetic diversity and on the occurrence of specific characteristics such as resistance to pathogens. An example, the discovery of resistance to BYDV and other barley diseases in collections from Ethiopia, has been mentioned (p. 191).

The general conclusion emerges that for most crops some valuable genetic resources are still in cultivation, though in some instances in a precarious condition. Of others a good deal remains, as of various pulse crops of great economic and nutritional importance (see reports by Van der Maesen, Kjellqvist and others in Frankel, 1973), but that changes may be so swift that none are really safe. A recent expedition to Himalayan valleys in Pakistan, isolated by high mountain ranges, found that even there the ancient local races of wheat are beginning to be replaced by cultivars from the south (Rao,

1977). In Nepal this process has gone even further. As a result of the successful introduction of selections from the semi-dwarf wheat cultivar Sonalika, the wheat area has expanded greatly, with 80% in improved varieties, and with corresponding displacement of indigenous land races. In the short time of seven years this has created an emergency situation in Nepalese genetic resources (IBPGR, 1978c). It is obvious that the scarce remaining genetic resources of major crops will be located and collected off the beaten track, mainly by workers with local knowledge and expertise.

This, it appears, can be the case even in the very centre of Europe. For some years Dr F. Kühn, of the Agricultural University of Brno in Czecho-slovakia, had drawn attention to indigenous land races and relic cultigens which had persisted in mountainous areas of northern Moravia and Slovakia. In 1974, Dr Kühn in collaboration with workers from the Zentralinstitut für Genetik und Kulturpflanzenzüchtung, Gatersleben, and others, conducted a collecting expedition which succeeded in obtaining 247 samples of cereals, grain legumes, vegetables and medicinal plants from fields, gardens and farmers' stores (Hanelt and Hammer, 1975). The collected material represents indigenous land races and relic cultigens, the latter in sporadic cultivation or as crop admixtures. The land races were local types, with great diversity within and between cultivars. The original collections and the first progenies have been botanically analysed (Kühn, Hammer and Hanelt, 1976), documenting the extensive genetic diversity. The authors point to the adjacent mountainous areas of southern Poland as other likely sites of persisting local land races, as indeed might be some other rather isolated mountainous regions in Europe. They emphasize, however, the rapid rate of disappearance of the remaining material in all the districts they visited, their effort probably being the last opportunity for salvage.

If detailed information in the main is confined to the notebooks and the memory of plant explorers and local scientists, and, of course, to the cultivators themselves, there is now fairly general agreement on the regional distribution of valuable genetic resources, and on priorities for collecting on the grounds of imminence of genetic erosion, or of requirements for current or projected breeding programmes. Lists of priorities had been worked out by the FAO Panel of Experts on Plant Exploration and Introduction since 1969, and have recently been revised and considerably extended by the International Board for Plant Genetic Resources (IBPGR, 1976a). Priorities among crops are allocated according to the risk of loss, the value of the materials in terms of current and potential economic and social importance, the recognized requirements of plant breeders, and the size, scope and quality of existing collections. Priorities among regions depend on the content of priority materials (see above), the rate of change affecting genetic materials, or crop failures endangering such materials. The priorities are considered as tentative and subject to continuing review. They are reproduced in Table 8.1. More recently, more specific exploration targets for some major crops were published by IBPGR (1978d); see also chapter 9, p. 245.

TABLE 8.1   *Regional priorities*

Region	Priority	Crop priority within each region
Mediterranean	1	1: wheat, barley and sugar beet 2: chickpea 3: maize, oats, rye, *Pisum* sp., *Vicia faba*, brassicas, olive and safflower
South-west Asia	1	1: wheat, barley and coffee 2: chickpea and sugar beet 3: oats, rye, *Pisum* sp., *Vicia faba*, brassicas, olive and safflower
South Asia	1	1: sorghum, millets and rice 2: chickpea, groundnut, *Vigna* sp., cassava, bananas, cotton and sugar cane 3: maize, pigeon pea, *Vicia faba*, yam, jute, brassicas and safflower
Ethiopia	1	1: wheat, sorghum, millets and coffee 2: chickpea, cowpea, soyabean, bananas and cotton 3: barley, maize, *Pisum* sp., *Vicia faba* and sunflower
Meso-America	1	1: *Phaseolus* 2: groundnut, cassava, potato, sweet potato, cotton, South American oil palm and cocoa 3: maize, pigeon pea, yam and sunflower
Western Africa	2	1: sorghum, millets and rice (*O. glaberrima*) 2: cowpea, groundnut, cassava and cotton 3: maize, yam and African oil palm
Andean Zone	2	1: *Phaseolus* 2: groundnut, potato, sweet potato and cotton 3: maize
Central Asia	2	1: wheat 2: rice, chickpea, *Vigna* sp., cotton, sugar beet and sugar cane 3: barley, maize, oats, rye, *Pisum* sp., *Vicia faba*, safflower and sunflower
South-East Asia	2	1: rice (*O. indica* and *O. javanica*) 2: soyabean, *Vigna* sp., cassava, sweet potato, bananas, cotton and sugar cane 3: maize, pigeon pea and yam

TABLE 8.1—*contd*

Region	Priority	Crop priority within each region
Brazil	2	2: groundnut, *Vigna* sp., cassava, sweet potato, cotton, South American oil palm, rubber and cocoa 3: maize and yam
Pacific Islands	3	2: sweet potato, sugar cane and bananas 3: yam
Far East	3	1: wheat, sorghum and millets 2: rice, groundnut, soyabean, *Vigna* sp., cassava, bananas, cotton and sugar cane 3: barley, maize, yam and brassicas
Eastern Africa	3	1: sorghum, millets and *Phaseolus* 2: rice, cowpea, groundnut, soyabean (*Glycine* sp.), cassava, bananas and cotton 3: maize, pigeon pea and yam
Southern South America	3	1: *Phaseolus* 2: groundnut, cassava, potato, sweet potato and cotton 3: maize

The designation of priorities may serve as a useful guide in the deployment of collecting effort and in allocating financial support. Among the reasons for priority ratings, rapid erosion through displacement or destruction is unequivocal and *should be indicated*. Other reasons are more subjective: what is a low-priority target in general may be a significant one in some circumstances. Statements of priorities on grounds other than imminent or likely danger may result in discouraging initiatives and in restricting national or international support for what could be valuable collecting efforts. Considering that the number of workers engaged in plant collecting is not large, and that germplasm collections in many of the lower priorities are inadequate, such priority ratings should not be regarded as more than a general guide.

It will be noticed that fruits and nuts are not included in Table 8.1 because of further consideration being given to these and some other crops. Here a brief comment on fruit species is in place, in view of their rapid disappearance on one hand and the restricted diversity of vegetatively reproduced cultivars on the other.

The ancient fruits and nuts of Near Eastern countries constitute rich, but all too little-appreciated and now fast-vanishing genetic resources.

Replacement by modern introduced varieties is widespread, and in some important areas old trees in home gardens may be the main residues of a rich heritage. This appears to be the position in Iran and Pakistan, where indigenous stocks of apples, pears, plums, grapes, apricots and almonds have all but disappeared and new plantings are confined to introduced varieties. However, in Turkey there still are considerable possibilities for salvage of important germplasm of a range of fruits and nuts (Zagaja, 1970; Sykes, 1972). Pistachio (*Pistacia vera*) is a nut with considerable market potential and economic prospects as an irrigation crop. Pistachio savannahs, and scattered trees which may be degraded savannahs, are widespread throughout the desert fringes of Asia (Maggs, 1972, 1973). Maggs states that the gene pool, though widespread throughout the Middle East, is being depleted by land clearance, kill-out grazing and the use of wild pistachio as rootstocks.

The cashew nut (*Anacardium occidentale*) is a native of tropical America, but East Africa and India account for 90% of the commercial production. In spite of the extensive cultivation in these areas, no cultivars have been formed. The existing germplasm collection in India has recently been extensively studied for a range of physiological and morphological characteristics, including the proportion of male and hermaphrodite flowers which may affect the yield. The conclusion was reached that the range of variability in the collection is rather low and that extensive exploration in the centre of origin could improve the utility of this valuable plantation crop (Kumaran, Nayar and Murthy, 1977).

The tree fruits of South-East Asia evolved from wild rain forest species, or were directly introduced into cultivation. Some of these are among man's oldest domesticates, and many will have been gathered for thousands of years prior to the advent of settled agriculture, as they are to this day. Ecologically and genetically there remains a close link between rain forest and village garden. In a locality in the Trengganu mountains in eastern Malaya Whitmore (1975b, p. 218) found twenty-six species of fruit trees being cultivated. Of these twelve are identical with the same species growing wild in the forest, six are improved selections from the wild, five are indigenous, but do not occur in the forest, and three are from the New World.

Many of the fruits make important dietary and economic contributions. Some fruit species are also useful timber trees, and some are attractive ornamentals. Selections and grafted varieties of the more important species are now available and are systematically introduced to farmers, replacing the traditional varieties. There is therefore an urgent need to collect and preserve traditional varieties before they disappear. A survey conducted by Sastrapradja (1973, 1975) in the countries of South-East Asia shows the extent of erosion which she estimated to have occurred. It is clear that with the widespread destruction of primary forests in many parts of the region the habitats of wild fruit species will be lost unless they are located and as far as is practicable protected. Yet their preservation will be essential for the future genetic improvement of these valuable food resources. It is to be hoped that the cooperation between countries with related problems, which is emerging will bring forth concerted action for conservation (Williams, Lamoureux and Wullijarni-Soetjipto, 1975, pp. 264–5; IBPGR, 1977a, 1978b).

(b) *Exploration – sampling procedures.* In view of the scarcity of remaining resources and, in many instances, the urgency of the situation there is a need for careful planning, preparation and execution in the collecting of primitive cultivars. Bennett (1970) describes in detail the facilities and preparations required, and recommends procedures during exploration. Site descriptions provide helpful guides in evaluating collected material, but the elaborate descriptions recommended by some authors are scarcely practicable in an inevitably busy collecting season. Nor is it clear in what manner detailed information on soil and vegetation features can be put to use, except in ecological studies of the locality rather than in plant breeding practice. Brief practical manuals for field collectors have been prepared by Chang, Sharma, Adair and Perez (1972) for rice, and by Hawkes (1976) for seed crops.

Inevitably, even in circumstances of acute scarcity, there occur problems of sampling since hardly ever is it possible to collect and retain even one entire population. As a rule there are limits to the size and number of samples that can be collected, evaluated and conserved, hence theoretical and practical considerations of sampling procedures have been examined by a number of authors (Whyte, 1958; Allard, 1970a; Bennett, 1970; Marshall and Brown, 1975). Whyte suggested a sample size depending on the breeding system, increasing from apomicts to cross-fertilizers to self-fertilizers. Allard presented a sampling model based on the distribution of variability in wild oats (*Avena fatua*) in California, which had been extensively studied by himself and his associates. For the size of the individual sample he suggested ten seeds of each of 200 plants. Bennett discussed the statistical and biological principles of sampling and the many practical considerations which enter into sampling procedures. Marshall and Brown's purpose was 'to formulate . . . a quantitative sampling theory for genetic conservation which permits the collection of the maximum amount of practically useful variability in the target species while keeping the number of samples within practical limits'. On the basis of these considerations they conclude that the aim should be to include 'at least one copy of each variant occurring in the target population with a frequency greater than 0.05'. In general the optimum strategy will be to collect 50–100 individuals per site, to sample as many sites as possible, and to include a representative range of environments within the target area.

Recently Marshall and Brown (1980) have extended their sampling theory to the collecting of forage plants. This calls for a consideration of sampling in natural populations. The basic sampling strategy is equally valid, but sampling procedures may be affected by the intrinsic ecological and genetic differences between cultivated and wild populations. Foremost among these is the greater population differentiation of natural populations, since they are not subject to the homogenizing influence of annual harvests and reestablishment on different sites. Plants adjacent to each other will tend to be more alike genetically than more distant ones, partly due to natural selection, partly to the finite size of interbreeding populations. Such considerations may affect sampling patterns, but not the principle of random sampling. In a highly variable site it is up to the collector to define the sampling site or sites according to his ecological assessment. But, as the authors point out, what is needed is a record of the sampling pattern used, rather than implying random sampling.

In both papers the authors discuss the issue of random versus biased sampling. They give reasons for adhering to the former resulting from the general principle of collecting *locally common alleles*, i.e. alleles favoured by natural (or deliberate) selection, as against rare alleles with conspicuous gene expression but ostensibly not associated with factors favoured in the particular environment. One may, however, recognize that a biased sample may be appropriate in forage species where a rare vegetative type may represent the 'ideotype' for the species in the mind of the agronomist-collector.

Qualset (1975), on the basis of character combinations he found in Ethiopian barley collections (see p. 198), suggested that to obtain 'interesting, but rare combinations of characters . . . the sample size should be five to ten times larger than that recommended by Marshall and Brown'. He further suggests that for evolutionary studies 'at least a portion of the populations, say ten per cent, should be sampled more extensively'. The first argument, as Qualset admits, has some validity only where a population has been extensively studied so as to reveal character combinations, which at any rate could be obtained by the breeder. Nor does it seem necessary, or even advisable to select a single population for evolutionary studies when, as Marshall and Brown have shown, a sample of similar size derived from pooling samples from separate sites provides a truer representation of the local genetic diversity.

Whatever the sampling strategy, experienced plant collectors never cease to emphasize the human factor. Especially in relatively unknown areas, information from local people frequently provides much needed ecological and social background to the crop diversity in a region. It can be invaluable in leading the explorer to atypical or rare strains which may be of evolutionary, genetic, or direct economic significance. We have previously given an example of such benefits derived from human contact (see p. 195).

(c) *Exploration – retention of the population structure of accessions.* The initial treatment of the sample collected in the field is an extension of the sampling strategy. The objective must be to safeguard the genetic integrity of the population that constitutes the accession and to avoid gene erosion at the very start. The earliest operation that exposes the population structure of an accession to possible erosion is the first progeny raised by the explorer (or a recipient of his collection). It is essential that at this stage the accession be treated as a *population sample*, i.e. that a sample consisting of aliquot parts of all components of the accession be used for multiplication, and, equally, for the sample set aside for conservation and future reference. In other words, the sample must be *randomized*, as against the not uncommon practice of taking out a few plants for multiplication and retaining – worse still, discarding – the balance. It is obvious that sampling on a population basis is rendered useless unless the randomizing procedure adopted during the collecting of the sample be retained through all subsequent processes of multiplication and regeneration.

It is obvious that this procedure refers only to what has been termed the conservation stream (chapter 9, p. 246). It does not restrict the freedom of the breeder or agronomist to establish and evaluate individual lines selected from the population sample, and to enter them into the conservation stream as derivatives from the original accession.

8.2.11 Induced mutations

While this book is concerned with natural diversity and its conservation, a review of variation in plants used by man would not be complete without reference to mutations induced by man and to their relationships to natural variation. Here we are not concerned with the various mutagens and mutagenic techniques, but with the nature of the mutants which have been or can be produced.

First, we can induce any mutation that has occurred naturally, and probably many which have either never occurred spontaneously or have been lost from the natural populations (Brock, 1971). Second, at the phenotypic level induced mutants are similar to the corresponding spontaneous ones, but there is no definite evidence of such correspondence at the level of the gene. Third, useful mutations may be, and often are, associated with unfavourable ones, which is the reason why natural sources for a desirable trait if available are preferred to induction of a mutation, in spite of the obvious advantage which the latter presents in modifying a specific gene in an otherwise well adapted and successful genotype. Yet among the 133 crop cultivars – 123 of them with generative propagation – developed by using induced mutations, 96 are the result of direct mutations, and only 37 are derived from crosses, with induced mutants as one of the parents (Micke, 1976).

Table 8.2, from Sigurbjörnsson and Micke (1974), indicates the kinds of induced mutations which have been found useful by plant breeders for incorporation in crop varieties. Another table in the same paper lists all varieties released by October, 1973, with their main improved characteristics. A number of the characteristics are qualitative and likely to be under single gene control. Others are quantitative and presumably due to random mutational change in genes affecting the character. Random mutation increases the variance of quantitatively inherited characters and shifts the mean away from the direction of previous selection, but increases the scope for selection in the desired direction. Selection pressure will be needed on other quantitatively inherited characters since they also are subject to random mutation.

TABLE 8.2　*Crop varieties developed through induced mutations*

Improved characters	No. of occurrences in released varieties			
	Cereals	Legumes	Other	Total
Higher yield	27	10	10	47
Lodging resistance	23	3	–	26
Disease resistance	13	9	2	24
Early maturity	19	9	8	36
Short stem	14	2	–	16
Quality (baking, malting, feeding, eating)	13	3	11	27
Winter hardiness	3	–	–	3
Higher protein	2	2	–	4
Shattering resistance	–	2	–	2
Improved plant type	3	3	3	9
Easier harvesting	1	2	–	3

Other improved characters include changes in awns, grain colour, 100-kernel weight, sprouting resistance, drought resistance, high lysine and better adaptability

From Sigurbjörnsson and Micke (1974).

In spite of many years' search for disease-resistant mutants the number of successful mutants has not been large (see Table 8.2). The need is greatest where known natural resistance is based on a single gene, as for example in barley yellow dwarf virus, or altogether absent. On the other hand a large number of mutants have been induced which modify specific physiological or morphological characteristics, or both, thus complementing the broad-spectrum variability of natural genetic diversity. As mutation research gathers momentum in an increasing range of crops, a number of mutants of interest to geneticists and plant breeders are emerging. These should be preserved as valuable genetic stocks for future reference and use (see section 8.2.6).

Is the induction of mutations likely to take the place of naturally evolved variation, thus rendering the conservation of genetic resources redundant? The answer lies in the complexity of adaptations and co-adaptations which we have already discussed (pp. 195–201). Brown (1978) lists various types of co-adaptation recorded in the literature. Accepting that an *a priori* assumption of widespread co-adaptation can be regarded as plausible, it must be admitted that the evidence in cultivated plants, reviewed earlier in this chapter (see above) is as yet scant. As Brown suggests, 'The finding of examples of coadapted complexes in populations of wild relatives and land races is of key importance in clarifying the need for genetic conservation.'

8.3 Summary

1. The genetic diversity of cultivated plants is derived from wild ancestral species, modified by adaptations in response to cultivation as distinct from natural conditions, and to the physical, biological, cultural and economic factors of the environment. 'Taxon-specific' adaptations, common to all members of a taxon (species, sub-species, etc.) are the result of domestication as such, whereas the exposure to environments in space and time resulted in a multitude of 'ecospecific' adaptations (pp. 175–9).

2. By comparison with the species used by hunter-gatherers, the number used under domestication declined, while infraspecific diversity greatly increased as a result of geographical expansion through migration and trade. Ecospecific adaptation resulted in the formation of land races suited to local environments (pp. 179–81).

3. There is good evidence, from direct observation and from germplasm collections, of extensive genetic diversity between and within land races, although their genetic structure has not been extensively studied; what information we have comes mainly from studies of wild relatives of crop species (pp. 181–4).

4. Germplasm collections have been used as research material in studies of the evolution and distribution of variation in crop species. Such studies suffer from the limitations of the collections themselves, deriving from inadequate collection and maintenance techniques, uneven representation, and inadequate documentation. Analyses on a modest geographical scale, with well-authenticated material tend to be more productive than surveys using world collections (pp. 184–91).

5. Collections have been extensively screened for characteristics of economic significance. Many have been identified, including resistance genes for various diseases and pests and adaptations to extreme environmental factors. There is as yet little evidence for co-adapted gene complexes. Land races continue to serve as genetic reservoirs but they are rapidly being replaced by advanced cultivars. Those remaining in the field should be preserved as a matter of urgency (pp. 191–203).

6. Advanced (current and obsolete) cultivars are resources which may have importance for future breeders. Those with known value characteristics should be preserved (pp. 203–4).

7. Wild and weedy relatives of crop species are important sources of disease resistance, and of physiological adaptations not possessed by their domesticated relatives. While not generally in need of preservation, they should be collected and studied to a greater extent than is now the case (pp. 204–8).

8. Many species are used in a wild or near-wild state. Among these the forestry species are the most important and also the most vulnerable, since (a) forest genetic resources can be seriously eroded or even destroyed through deforestation, and (b) *ex situ* conservation presents economic and technical difficulties. Forest genetic resources are at risk everywhere, but most significantly in South-East Asia where the immense wealth of forestry species is now seriously endangered, at a time when the value of the genetic resources of indigenous plantation species is beginning to be realized (pp. 208–12).

9. Land races still in cultivation are endangered almost anywhere and are likely to disappear in the very near future. For most crop species the holdings of land races in germplasm collections are incomplete to near-absent, hence collecting and preservation of what remains in the field is urgent. Target regions have been recognized, but in many instances specific targets still require to be identified for early action (pp. 212–19).

10. Collecting procedures need to be carefully planned and sampling methods defined with due consideration for population genetic principles (pp. 219–21).

11. When natural variation is scarce or absent or its transmission restricted by sterility barriers, induced mutations can usefully complement the natural genetic resources (pp. 221–3).

9

The conservation of plants used by man

9.1 Conservation in perspective

In this chapter we are concerned with the conservation of plant species, populations, genotypes which are actually or potentially useful to man. To this qualification we must add two others which derive from the first. Since we cannot determine what characteristics will be required in the future, or even which of the current economic species will continue to be used, a reasonable strategy of conservation would include (a) types likely to be useful in the near future, (b) representative samples of the genetic resources for use and study by future generations. Further, it follows from the qualification of usefulness that a *time scale of concern* (see Chapter 1) could be formulated in accordance with reasonable expectations of future uses. This has some tactical implications. For example, in view of the uncertainties of future requirements on one hand, the possibility of new methods for generating and using genetic variation on the other, a period of 50–100 years seems an adequate time scale of concern for the preservation of crop genetic resources. Such a commitment on behalf of the future is on a normal human scale, in contrast to long-term nature conservation which inevitably carries the burden of the awareness that commitments by any generation are subject to the vagaries of decision making by each subsequent one. Another implication of a relatively short time scale of concern is the realization that the regeneration of stored seeds which are losing vitality – perhaps the most burdensome operation in seed preservation – may possibly be avoided altogether (see section 9.4.2).

In the preceding chapter we have recognized four classes of genetic resources with distinctive genetic and ecological characteristics (section 8.2). In germplasm conservation we are concerned with only two categories with quite distinctive requirements for conservation – wild species and domesticates. *Wild species* are best conserved in their natural habitats within the communities of which they form a part, and this applies to those wild species which are regarded as actual or potential genetic resources. In general the principles of nature conservation (Chapter 5) apply, with some qualifications discussed below in section 9.2. Only when such communities are threatened, or some individual species within them as, for example, the wild relatives of vegetable or fruit species are at present, then some form of protection may be in place, either under natural conditions as in forestry reserves or in specially designed 'genetic reserves' (see section 9.2; or *ex situ*, section 9.3). However, the genetic resources of the great majority of wild species used by man can be regarded as reasonably safe in at least a proportion of their natural habitats, although in some instances there is a need for protection, in others for continuing watchfulness.

By way of contrast, *all domesticates* require positive measures for their maintenance and preservation. Few if any of the most vulnerable genetic resources, the land races still remaining in cultivation, can be regarded as safe unless they are adequately represented in a well-conducted germplasm collection, i.e. *ex situ*. All genetic resources maintained in whatever form of *ex situ* preservation require continuing supervision, maintenance and protection, involving in some instances considerable expenditure. Hence efforts to limit the burden, by cooperation on one hand (section 9.7), rationalization on the other (section 9.5), are to be supported.

The life cycle, the size and cost of the unit of preservation, and the method of preservation, largely determine the scale and effectiveness of operations. For example, fruit species, preserved as trees, are expensive to maintain and are subject to virus and insect attack, restricting both scale and effectiveness. Preservation as seeds, pollen or tissue culture would allow to increase the scale, reduce costs and expedite distribution.

Here we make a special plea for the preservation of accessions or stocks which have been *extensively used or evaluated in investigations* of any kind, overriding differences of use, purpose or biological status. They include materials used in cytogenetic, evolutionary, physiological, biochemical, pathological or ecological research on one hand, accessions evaluated for their agronomic or breeding propensities on the other. An example of the latter group are the wild species collected for their potential as forage species in other parts of the world. It is not unusual for such accessions to be examined and evaluated for a number of years, only to be rejected and lost if not found entirely suitable. Yet such material might be of use in other environments. While it is true that if sites are well documented it could be collected again, the expense of exploration is considerable, to which must be added the value of previous evaluation which strictly applies only to the material on which it was obtained. We urge that material that can be regarded as reasonably promising be offered to others who might be interested, either directly or through the IBPGR secretariat (see section 9.7).

9.2 Conservation in situ *(Fig. 9.1)*

Both domesticates and wild plants can be preserved *in situ*, i.e. in a natural environment where their evolution has been a continuing process. *In situ* conservation is therefore described as *dynamic* (Fig. 9.1). In domesticates *in situ* conservation is meaningful only for land races cultivated in their home environment (see section 8.2). Once land races are displaced by modern cultivars as part of a change in the agronomic and social system, *in situ* conservation of land races, as a deliberate measure, would meet with difficulties. Attempting to retain 'primitive farms' in a radically changed social and technological climate would, except in the short term, be self-defeating. However, based on long experience in Iran and Turkey, Kuckuck (in Bennett, 1968, pp. 32 & 61) proposed the establishment of what might be called 'crop reservations' – areas of $\frac{1}{2}$ to 1 ha in size where a local crop variety would be maintained under the supervision of a local agricultural officer. The areas would be subject to changes in the environment brought about by agricultural development – fertilizers, cultivation, etc. – and to genetic change mediated by natural hybridization, mutation and natural selection. They would thus form 'mass reservoirs' with opportunities for gradual adaptation to changing environments and with genetic self-renewal through mutation and introgression.

Frankel (1970a) questioned the practicability of maintaining a number of small isolates, with all the inherent tactical difficulties, let alone the

		Domesticated	Wild
dynamic	*in situ*	Land races in their areas of cultivation	In natural communities
	ex *situ*	Mass reservoirs (Simmonds, 1962) Genetic reserves (Dinoor, 1976)	Forest provenances
static		seeds plants cell, tissue, meristem culture	

Fig. 9.1 Conservation of domesticated and wild species of economic significance.

inevitable loss of identity. Proper maintenance might be available in experiment stations, i.e. technically *ex situ*. This, however, would remove the land races from the environments and cultural conditions to which they are adapted and would limit the number of sites.

In situ conservation of *wild species used by man* (see section 8.2.8) provides the relative stability of multi-species diversity within a co-adapted community. It also tends to preserve genetic polymorphisms and to provide opportunities for recombination. Thus the conservation of economic species is akin to the conservation of natural ecosystems (see Chapter 5), but the economic end purpose introduces considerations which have bearings on strategy and tactics. In ecosystem conservation the emphasis is on representativeness of particular communities within the ecosystem as a whole (UNESCO, 1974), and in long-term reserves, such as the biosphere reserves, the time scale of concern is without a notional limit. In the conservation of economic species the emphasis is on the preservation of particular species or associations of species, and a notional time limit can be formulated in accordance with their life cycle and the anticipated economic purposes – perhaps a century for wild relatives of crops, in accordance with the time scale for the crops themselves (see section 9.1), and perhaps at least 500 to 1000 years for forest tree species which are likely to be needed for a very long time to come.

These considerations have some social and tactical consequences. First, conservation with an economic purpose and with a notional end point is more likely to receive public support than nature conservation for its own sake. Second, legal safeguards need not be as comprehensive as in long-term nature conservation. Third, ecological safeguards of size and organization can also be more relaxed, but there is a greater requirement for ecological diversity so as to encompass a representative range of genetic diversity. Fourth, control and management need not be as stringent as in strict nature reserves. Roche (1975), while recognizing that strict nature reserves are the prototype of forest genetic conservation, advocates a relaxation of control and recommends multi-purpose reserves in which genetic conservation is only one of the objectives. This would seem particularly appropriate for the tropical rain forests of South-East Asia where economic and social pressures impose limits on the number and size of strictly maintained nature reserves. However, forest reserves can be maintained under near-natural conditions, with less stringent controls allowing for a limited extraction of timber, of other forest products such as fibres, poles, fruits, leaves which are customarily obtained by villagers from neighbourhood forests, and for the enjoyment and recreation of both country and towns people (Frankel, 1978a). Indeed, flexibility of a system of reserves would be the best guarantee for continuing public support.

We conclude that for purposes of genetic conservation of economic species, size, diversity, multiplicity and security of conservation sites are likely to be more important for the long-term success of genetic conservation than is the strictness of management.

In this discussion national parks or other long-term nature reserves are considered as the prototype for the conservation of 'wild' genetic resources. Many of the existing nature reserves all over the world contain genetic resources of a number of species, although these may not all be on record. The biosphere concept includes provision for inventories to be obtained of all species represented in a biosphere reserve (UNESCO 1973, 1974), and it has been suggested that these could be indexed for species of actual or potential economic or scientific relevance (Frankel, 1978c). Guidelines for the population sizes required for stable conservation are discussed in Chapter 4.

Special reserves are established for the conservation of particular species or groups of species because they are particularly important, or because they are rare and not adequately represented in existing nature reserves. The most prominent example of the first kind are the forestry reserves established in areas rich in forest resources (see also section 9.3). These serve not only as sources of seed, but as permanent repositories of forest genetic resources. Indirectly, they help to protect a host of associated species.

Attempts are made – mainly, but not exclusively, in the United States – to protect endangered plant species, usually in restricted sites, where they are given a measure of protection from predators and competitors. This could be extended to species of economic significance, though more often than not *ex situ* preservation may be more economical. In a greatly stimulating and informative chapter on the conservation of wild species useful to man Jain (1975) envisages the establishment of 'genetic reserves' in which the population biology of individual species can be studied, and wild relatives of crop species can be protected.

9.3 Populations ex situ *(Fig. 9.1)*

When populations of domesticated or wild species are maintained in habitats other than those in which they evolved or became adapted, they are exposed to selection pressures prevailing in their adopted site. Such exposure may be incidental, as in many introductions of crops, forages, or forestry species which are regularly reintroduced from the country of origin (pasture species from New Zealand, eucalypt species from Australia). On the other hand exposure to, and selection in new environments may be the main purpose of the operation, resulting in populations adapted to different habitats. Some authors regard such an approach as a valid form of genetic conservation and this is the reason why it must be considered here.

Simmonds (1962) proposed the use of mass reservoirs – composite crosses of large numbers of diverse parental types (see e.g. Harlan, 1956) as more appropriate for long-term conservation than what he called 'museum collections' of individual accessions kept separately. His reasons were the high rate of attrition and absence of continuing natural selection in the latter. Attrition has been largely overcome by current technology (see section 9.4). That mass reservoirs have a place as 'an adjunct to plant breeding' (Simmonds, 1962) has been asserted by some, questioned by others. Here we are concerned with their role in conservation. The question is whether genetic variation is maintained more effectively than in collections of individual accessions. Marshall and Brown (1975), using the loss of from 50 to 70% of the original genetic variation for height and heading date in barley Composite Cross V as evidence, argue that mass reservoirs 'are of little value in *preserving* variation, potential or expressed'. Nor would, as has been suggested, growing mass reservoirs in many different places – to counteract genetic depletion – reduce the effect by comparison with maintaining individual lines in storage. Further, they quote evidence to show that different composite cross populations, far from maintaining their differences in allelic frequencies, were approaching the same end point after a number of years, thus tending to retain the same fraction of variation. Frankel (1970a) questioned whether exposure to current selection pressures may adversely affect the frequency of specific characteristics (see chapter 8), and even more so of co-adapted complexes (*ibidem*) which may be needed in the future, or even currently in other environments. However, collections maintained in the form of individual entries make it possible to study separate gene pools and to relate their characteristics to their original environments. The pragmatic anonymity of bulk methods is in itself an argument against their exclusive application.

Dinoor (1976) suggested that land races (or possibly composites of land races?) be grown in what he terms 'nature reserves' (rather a misleading term since they would require cultivation) or 'genetic reserves', under conditions of continuing evolution and, more significantly, of coevolution with pathogens. By comparison with the *in situ* preservation of land races which have been, or are about to be, displaced (see section 9.1), the populations proposed by Dinoor can be raised wherever there are expectations of host–pathogen evolutionary changes, presumably in the centres of diversity of the pathogen (see Leppik, 1970), or wherever a concentration of pathogen diversity has been ascertained. Such an experiment could be an aid in plant breeding or in phytopathology, but could not be regarded, as Dinoor suggests, as a method of genetic conservation, if for no other reason than that the number of strains that could be included would of necessity be small (Frankel, 1978b).

Ex situ populations of wild species are widely used in silviculture where they are called provenances. Seeing that most species are outbreeders, variation between and within provenances is often intense. Provenances with high breeding value are therefore worth preserving (Maini, Yeatman and Teich, 1975). *Ex situ* conservation, apart from being a convenience, becomes a necessity if the indigenous resources of a species are threatened. This is an increasingly common problem in the tropics and also among some temperate zone species (see section 8.2.8). However, it will be realized that *in situ* conservation cannot be fully replaced in this way. As self-renewing communities, natural forests have the capability of retaining levels of genetic diversity, adaptively evolved in time, which no sampling pattern can attempt to approach. Security from predators or from catastrophes such as fire or land slides can be obtained by a multiplicity of reserves, no less needed for *ex situ* plantings which are less likely to recover from drastic damage than are natural areas. We therefore believe that foresters, and especially forest geneticists, should make every effort to secure adequate reserves of forest genetic resources in natural forests, rather than rely on *ex situ* conservation which in the longer term may be no more secure than natural sites and lacks the evolutionary and social values of natural communities.

Guldager (1975) distinguishes three kinds of *ex situ* conservation:

Static conservation, retaining as far as possible the population structure of the original population. The author distinguishes between conserving 'genotype frequency' and 'gene frequency', the former being possible only in vegetatively propagated material. The aim is to prevent the loss of genetic information.

Evolutionary conservation, which implies the initiation of 'new natural evolution' (see below), i.e. natural selection in the environment of adoption.

Selective conservation, which implies the application of selection pressure against clearly undesirable characteristics, without undue depletion of the available diversity.

In the context of dynamic conservation the prospects of *ex situ* 'evolutionary conservation' of silvicultural species call for comment. In agricultural crops which are grown for their seed, reproductive capacity as a rule is positively correlated with economic productivity. In silvicultural species a correlation between reproduction and economic value is bound to be weaker if it exists at all, possibly through the selective elimination of the weakest producing individuals. It may be that selection for timber productivity is most effective in the region where a species or strain has evolved, possibly because of the more extensive variation, an opinion which is not unusual among some breeders of livestock. Yet, as in livestock breeding, selection in the environment where the introduced material is to survive and produce, has patent advantages. However, in both cases natural selection may have to be supplemented by deliberate selection.

9.4 Static conservation (Fig. 9.1)

In contrast to the dynamic conservation of populations discussed in the preceding sections, the conservation of seeds or of individual plants aims to preserve as far as is possible the genetic integrity of individuals (or populations). This means that while frequencies of genotypes or alleles may undergo inevitable changes, gene and allele erosion are minimized and recombination with genetically or ecologically alien material is avoided. The purpose is to retain as far as is practicable genes and gene assemblies which have evolved in land race populations (see Chapter 8) and in modern varieties and various genetic stocks which are or may be required for research or breeding purposes.

It should be emphasized here that static and dynamic forms of conservation are in no way mutually exclusive. As we have seen they have different purposes and may be pursued independently of each other.

9.4.1 Conservation of seeds

Conservation is both safest and cheapest if life processes are reduced to a low level. This is the case in species which can be preserved in the form of seeds. Some species have seeds which, by currently used procedures, can only be kept alive for short periods, and some plants produce no seeds. However, the majority of seed-reproduced crop species can be stored over long periods of years without substantial loss of vitality and, as far as existing evidence goes, without substantial genetic damage.

The advantage of long-term storage over the alternative procedure of regular regeneration at intervals of three to five years, is the avoidance or reduction of natural selection, genetic drift in small populations, natural hybridization, destruction by parasites or climatic rigour, or loss through human error. Seed conservation is also substantially less expensive.

Roberts (1975) reviewed the problems of long-term storage of seeds which can be maintained under conditions of low moisture content and low temperature. Such seeds have been called 'conventional' or 'orthodox', in contrast to 'recalcitrant' seeds in which a decrease in moisture content below relatively high levels (between 12 and 31% depending on the species) decreases viability. The vast majority of species have orthodox seeds, but many forestry and fruit species and tropical crops have recalcitrant seeds.

Roberts (1975) emphasizes that the long-term storage of orthodox seeds presents no serious problems, although the principles are not yet fully understood. The quantitative relationships between seed viability, temperature and seed moisture content are reasonably clear, but those of oxygen pressure and seed viability are not as well defined. Roberts developed what he has termed the three basic viability equations which describe the relationships between temperature, moisture content and seed viability. Oxygen pressure is ignored since in open storage it is that of the atmosphere, in sealed conditions it soon drops to zero due to respiration. The equations have been used to develop viability nomographs for a number of species for which the required experimental evidence is available.

Chromosome breakage in ageing seeds, which has been known for many years, has been found to be closely associated with viability, hence the conditions for seed survival – time, temperature, moisture and oxygen – are also the conditions for chromosome stability (Roberts, Abdalla and Owen, 1967; Abdalla and Roberts, 1968, 1969). That gene mutation is subject to similar conditions is reasonably well established (Abdalla and Roberts, 1969).

The predictions for storeability under different combinations of temperature and seed moisture can be applied in the design of installations for long-term storage. On this basis the FAO Panel of Experts on Plant Exploration and Introduction proposed standards and procedures for storage installations used for long-term seed conservation (FAO, 1975). The 'preferred', or desirable, conditions are storage at −18 °C or less in air-tight containers at the seed moisture content of 5 ± 1%. Since only a small proportion of the existing storages attain this level, an 'acceptable' standard was set at 5 °C or less, either in air-tight containers at 5–7% seed moisture, or in unsealed containers at not more than 20% relative humidity. Under these conditions the multiplication interval, i.e. the number of years taken for viability to drop by 5%, should be 390 for wheat, 33 500 for barley, and 1600 for broad bean (*Vicia faba*) (Roberts, 1975), but the author adds that the barley cultivar used may have a very much longer viability period than other cultivars. The Panel specified operating procedures, including entry procedures, accession size, routine germination tests, regeneration and documentation. None of these could be found burdensome by a well-run storage laboratory. Roberts (1975) supplied useful information on the economics of seed storage.

To reduce the burden of periodical germination tests, the Vavilov Institute of Plant Industry routinely tests 5% of accessions in any one group (place of origin, year of collection or last regeneration etc.) resorting to testing other accessions in the group when test samples show a drop in vitality (Budin, personal communication).

A recent report prepared by an expert committee under the chairmanship of Professor Roberts, though based on the recommendations of the FAO Panel of Experts, added much practical and detailed information on design and cost aspects of storage facilities (IBPGR, 1976b). It shows how storage facilities of dimensions commensurate with the size of any collection should be designed, equipped, housed, maintained and managed, what service facilities are needed and how they should be equipped, what safety precautions are required, and how much all this is to cost, including transport. Clearly the installation of a storage facility is no longer in the pioneering stage, nor need the cost be astronomical.

Storage at the temperature of liquid nitrogen has yielded encouraging results with orthodox seeds, and the difficulties, especially in cooling and thawing, have proved less than anticipated (Sakai and Noshiro, 1975). Very low temperature storage would further reduce life processes, and, as Roberts (1975) suggests, may have a place in the preservation of recalcitrant seeds.

Recalcitrant seeds, as already mentioned, cannot be dried without loss of viability. Storage in carbon dioxide may delay senescence, but usually not beyond one year (Harrington, 1970). Villiers (1975), followed up the well-known experience that seeds can remain viable for long periods buried in soil in a fully imbibed state. He found that seeds maintained at normal temperatures in a hydrated state retained their viability without loss, that chromosome damage was virtually absent, and that malformations suggesting mutations were absent, against loss of viability and severe genetic damage in the air-dried controls kept at the same temperatures, but with moisture contents between 5 and 10%. Seeds damaged in storage, or by irradiation, actually recovered in viability when imbibed. Villiers suggests that the explanation may lie in repair systems remaining active in imbibed seeds, but being inactivated in dry seeds. Research on imbibed seeds is as yet in an early stage of development. The approach has implications for the maintenance of recalcitrant seeds, and for seeds of many wild species which present storage problems.

The conservation of wild species other than those of economic significance is the subject of chapters 3 and 4 but it is convenient to discuss the preservation of seeds of wild species in the context of seed preservation in general.

Preservation of *wild species* in the form of seeds has, as Thompson (1975) pointed out, some advantages over maintaining living specimens. It is difficult to preserve an adequate representation of a species as living plants, especially of large and long-lived ones, and seed preservation may be at least a useful adjunct in a conservation programme. Further, seed preservation is relatively safe from environmental hazards and from changing policies of successive administrations. Distribution also is easier in the form of seeds. Last but not least, it is cheap in terms of space and maintenance.

What information there is on storage requirements of wild plants points to a considerable diversity. Harrington (1970) lists a large number of species in which seed germinated after many years of burial in soil, and others which remained viable for more than fifty years under herbarium conditions. Many annual or short-lived species with dry seeds can be stored like seeds of orthodox crop species. Others, as we have seen, are short-lived when dehydrated. For many species, as Thompson (1975) remarks, little is known about storage conditions and germination requirements, which are complicated by dormancy, hard-seededness, or requirements for a post-ripening period. Accordingly, the seed bank of the Royal Botanic Gardens, Kew, provides a variety of seed treatments and of conditions for germination tests, including a range of temperatures, diurnally fluctuating temperatures, chilling, and treatment with gibberellic acid. Storage conditions also are varied (Thompson and Brown, 1972). Recently Thompson (1976) listed criteria which may help in evaluating the suitability of taxa for conservation in seed banks. These are available information on taxonomy and reproductive biology, the accessibility in the field, storage responses, and amenability to cultivation. Applying these criteria, he finds *Clarkia* highly suitable, *Gentiana* and *Ulmus* intermediate, and *Shorea* (Dipterocarpaceae) quite unsuited for seed conservation.

9.4.2 *Entry into storage and regeneration of stored seed (see Fig. 9.2)*

When an entry is received in a base collection (see section 9.7) it may be composed of clearly discernible kinds of seed, e.g. of different colours. These may be separated as subsamples. But neither at this nor at any subsequent stage, such as multiplication or regeneration, must selection be exercised. Indeed, *avoidance of selection is a cardinal principle of conservation.*

The first step must be a germination test to ascertain whether the entry is fit for storage. The FAO Panel of Experts on Plant Exploration and Introduction (FAO, 1975) recommended a minimum germination of 80% for storage acceptance. An entry with a germination at or above this level and of sufficient size – the Panel report specifies storage requirements – is directly entered into storage, the balance of seed passing to an 'active centre' for multiplication. If germination is less than 80%, it is passed to the active centre which will grow it and return the required amount to the base collection for storage. The principles for multiplication are identical with those for regeneration (see below).

The viability of stored seed is monitored by periodic germination tests. When stored seed has to be replaced because supplies are nearing exhaustion or viability has dropped, then regeneration, i.e. the production of new seed, becomes necessary. The stage at which the drop in viability makes regeneration necessary is subject to judgement and argument. The FAO Panel recommends regeneration when germination has dropped below the initial value by 5% ('preferred' standard) or by 10% ('acceptable' standard). Some curators may regard this as an unnecessarily high standard, and K. J. Symes (personal communication) considers a 20% drop a reasonable limit, presumably in view of the risk to genetic integrity caused by shorter

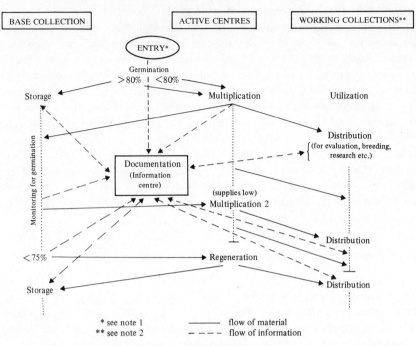

Fig. 9.2 Organization chart of conservation and utilization system. See text for explanation

(1) An 'entry' into the system can be an original collection sample, an introduction, or a breeding line. It is presumed to be a population.

(2) Working collections of plant breeders, pathologists, entomologists, etc. have no responsibility for maintaining germplasm, but, as users, have opportunities for contributing information derived from evaluation and use of material. This should be supplied to the documentation and information system for the benefit of others.

regeneration cycles. Allard (1970a) observes that after a drop in germination to half the original level 'the genetic composition of the entry is drastically changed due to differential survival of genotypes'. Unfortunately there is little precise information on this phenomenon, but on balance the Panel's recommendations seem well founded, especially if regeneration is conducted at a reasonably safe standard.

The purpose of regeneration is to restore an appropriate amount of seed for continuing the 'conservation stream' and at the same time to provide seed for the 'utilization stream' conducted by the 'active centres' (see section 9.7) which are charged with multiplication, distribution and evaluation of accessions. The aim should be to maintain intact, as far as possible, the genetic structure of the population, whatever its composition, i.e. whether a 'pure line', a land race population, a composite of such populations, etc. (see section 8.2). Two requirements need to be met. First, the breeding system must be controlled, which involves the prevention of outcrossing with *other* entries. Further, in cross-fertilized species fertilization *within* entries needs to be controlled, whether by isolation of the entry, or by controlled crossing.

The second requirement is to reduce the effects of natural selection in an environment other than the original one. Hence it is often claimed that regeneration should take place in the environment in which the entry had evolved. This is often impracticable, always difficult and expensive, and altogether unnecessary *provided that survival is maximized*, i.e. if all or most of the components of a population survive and reproduce. This may require the collaboration of a number of centres in different climate zones, in extreme cases (such as some wild species) cultivation under controlled environment conditions, procedures such as vernalization where appropriate, and protection from parasites and pests. It is, of course, necessary that approximately aliquot contributions of seed are obtained from each component of a population, e.g. in the cereals one (or more) medium-sized heads per plant.

Active centres as a rule conduct evaluation and are engaged in plant breeding. They also are responsible for distribution of seed to plant breeders and to other genetic resources centres. As seed stocks of entries in their care become exhausted or lose viability, a new multiplication is conducted. In the first place, if germination is still high, the centre's own seed can be used, otherwise seed should be called up from the base collection. This should be done in any case after some two or three multiplications. Some of the new seed should be returned to the base collection to supplement its stocks.

Allard (1970a) draws attention to the powerful effect of genetic drift in small populations, whether the breeding system is selfing or intercrossing, leading towards fixation and genetic erosion, less powerful only than 'the most stringent of selection of heterozygotes'. The population will increasingly consist of a random sample of the original genes as homozygotes, including, due to the absence of selection, deleterious genes causing a weakening of the population. This reasoning is derived from model populations of ten to twenty individuals (Allard and Hansche, 1964) which would be smaller, perhaps by a factor of 10^2 to 10^3, than populations likely to be used for the regeneration of collection entries, since the combined requirements for the conservation and utilization streams (see section 9.7) would be such as to obviate the risk of genetic drift. If there is no requirement of seed for purposes other than replacement of stocks in long-term storage, then the population size should correspond to that recommended for sampling in the field, i.e. a minimum of 100 plants (cf. section 8.2.10).

In a large germplasm collection regeneration is a continuing task, but not at any one time need it be a heavy burden. With an expected viability of between thirty and fifty years, regeneration of any one accession would be a rare event within the notional time scale of concern of one century or thereabouts. Hence the number due for regeneration, if reasonably spaced out, need never be unduly large.

9.4.3 Pollen

The long-term storage of pollen is far less worked out than that of seed. Yet it would have the practical advantage of making pollen directly available for crossing purposes. This could be of importance in the breeding of tree species which take some years to flower if raised from seed. Long-term storage of pollen would reduce the need of maintaining living collections of primitive and wild types which would be mainly or exclusively used for cross-breeding.

In discussing the prospects of developing methods for long-term storage of pollen, Roberts (1975) concludes that at present pollen storage appears to be a much less promising method of preservation than the storage of seeds because, under similar conditions, pollen has a far shorter period of viability. Yet he points out that the range of conditions available for pollen storage has not been fully explored, especially very low temperatures and moisture contents. He suggests that the promising preliminary results should be followed up with longer-term storage experiments.

Long-term pollen storage would be particularly useful in species with seed storage problems. Promising results come from recent work by Akihama, Omura and Kozaki (1978) who kept freeze-dried pollen of peach for nine years and of peas for six, without a marked loss in germination or in fruit setting by comparison with fresh pollen. Kobayashi, Ikeda and Nakatani (1978) maintained citrus pollen stored at −20 °C for three years.

9.4.4 Vegetative propagation

There are many economic plants which either do not produce seeds, or which are not normally reproduced from seeds so as to maintain intact a highly heterozygous genotype. To either of these belong many short-lived crops which are propagated from tubers, roots, bulbs, etc., and many long-lived shrubs and trees.

Storage organs of short-lived plants have a storage life which is measured in months rather than years. Potatoes, for example, can be kept at 4 °C to the next spring, or even for a further 12 months (Hawkes, 1970); but most other root and tuber crops have a storage life of less than twelve months. However, the effect of respiratory inhibitors such as maleic hydrazide, while applied to stimulate sprout production, suggests that they may have a potential role in the regulation of metabolic processes and in extending the storage life (Martin, 1975). Other vegetative parts of plants present similar storage problems. Unrooted or rooted woody cuttings held between −2 °C and 2 °C have been stored for up to five years (Howard, 1975; Akihama and Nakajima, 1978, pp. 80–111). Thick tea roots have been kept at 5 °C with wet sphagnum moss in vinyl bags for five years, and, judging by their carbohydrate metabolism, can be expected to survive for another ten years (Sakai, Doi and Nakayama, 1978). Longer term storage possibilities should be explored. These would include the most suitable type of material, the storage conditions and the regenerative procedure after storage (Howard, 1975).

Species which are normally propagated asexually may produce seeds which are more readily preserved than are vegetative organs, such as tubers or roots. This is the case in many species related to the potato, *Solanum tuberosum*, many of which produce storable seeds. However, the majority of fruits, both tropical and temperate, have seeds with a short storage life.

Germplasm of species requiring vegetative forms of reproduction is being preserved in living collections maintained in experiment stations, research institutions, botanic gardens, and by commercial enterprises such as nurseries. The maintenance of living collections presents biological and economic difficulties. The biggest single biological difficulty comes from the likelihood of virus attack. But the organizational and economic problems of space (especially for trees and large shrubs) and maintenance are considerable, with the further consequence that numbers and sizes of samples are likely to be much reduced by comparison with what is regarded as desirable in seed-reproduced crops. Against this it must be recognized, first, that vegetatively reproduced crops have uniform vegetative progenies; that multiple grafts can reduce space requirements; and that most of them are highly heterozygous, so that seeds they produce – or can be induced to produce by various means – reproduce highly diverse progenies.

The problems of maintaining live collections of germplasm are formidable, and every effort should be made to reduce the burden involved. It must be recognized, however, that in several countries live collections – of tuber, root or tree crops – are being maintained, especially in some tropical countries where land and labour are still available. Such efforts should be encouraged and supported. Indeed, there is an urgent need to establish collections of germplasm of great value which is in danger of impending extinction, as is the case with the many temperate fruit species that have their gene centres in Iran, Turkey and Pakistan, and tropical fruits in the countries of South-East Asia. Comprehensive national or regional germplasm collections are a most urgent need, and in view of the value to the rest of the world, should be internationally supported.

9.4.5 Tissue culture

The term plant tissue culture generally includes a range of systems such as organ, meristem, cell or protoplast cultures. For genetic conservation, tissue culture has clear advantages over the maintenance of live plants, and even over the preservation of seeds. Space requirements and cost of maintenance are low in comparison with live plants, and tissue cultures can be maintained free from pathogens. In outlining these advantages, Henshaw (1975) points to difficulties in maintaining tissue cultures. These include (1) difficulties in establishing cultures, which in some species requires a great deal of research; (2) regeneration in some plants is difficult and has so far not been solved; and even in plants which can be regenerated in fresh cultures, the morphogenetic potential appears to be lost after long periods of *in vitro* cultivation; (3) plant tissue cultures tend to be chromosomally unstable.

The most common change in non-meristem tissue cultures is to polyploidy, which apparently is favoured under *in vitro* conditions. As a consequence of polyploidization, aneuploidy is tolerated, and structural changes are accumulated. In many species there is great variation in chromosome numbers and constitution (D'Amato, 1975).

It is therefore of great importance that there appear to be good prospects of maintaining the chromosomal integrity of tissue cultures stored at the temperature of liquid nitrogen (-196 °C). Several authors (Nag and Street, 1973; Bajaj, 1976; Bajaj and Reinert, 1977; Sugawara and Sakai; 1978) found that tissue cultures of a number of species, including *Daucus carota, Acer pseudoplantanus, Atropa belladonna, Datura stramonium*, were stored from 100 days to 2 years without undue loss in cell viability. In one experiment, for example, the optimal cell viability was 65%. The essential components of the system were the age and nature of the cells, the cooling rate, the storage temperature and the thawing rate. The highest rate of survival is obtained in cell cultures at the lag phase or early exponential growth. Cooling at a rate of 2 °C per minute, or alternatively prefreezing at -40 °C or -70 °C for $\frac{1}{2}$ hour have been successful prior to storage at -196 °C. The thawing rate is critical, rapid thawing in water at 35 °C–40 °C being recommended. A programmed freezer is available for controlling and monitoring the cooling process (Bajaj and Reinert, 1977). A storage temperature of liquid nitrogen is far superior to higher temperatures of, for example, -20 °C, -40 °C or -78 °C. A protectant is an essential part of the system and its nature has been explored by several authors. DMSO (dimethylsulphoxide) and glycerol seem to be the best protectants. The morphogenetic potential is unimpaired by the freezing preservation technique and, accordingly, normal plants were reproduced as from unfrozen tissue culture, producing flowers and seeds in the normal way. Withers (1978) provided a useful review of techniques and achievements of freeze-preservation and indicated problem areas in need of further research.

The only form of tissue culture developed so far which can be, and is used for germplasm conservation is meristem culture. Apical meristems of many plants can be readily cultured in normal culture media with added gibberellic acid and raised potassium level. Plantlets are raised in test tubes and can be maintained indefinitely by transferring stem cuttings to new medium every six to twelve months. The advantages are small size, very high speed of multiplication, and absence of pathogens (Morel, 1975). Indeed the method has been used successfully to raise virus-free stocks of potatoes and other plants. Meristem culture has been used extensively in grape cultivation and the maintenance of stocks. Morel (1975) notes that 800 cultivars can be maintained in six replications on a space of 2 m^2, and that after fifteen years of experience, with one transfer every year, it is apparent that stocks can be maintained indefinitely. He observes that the technique can be applied to many plants, and might be useful in the preservation of species with 'recalcitrant' seeds (see p. 234).

9.5 Evaluation and documentation

Conservation of genetic resources of domesticated plants – or animals – has little purpose except for actual or potential utilization in agronomy, forestry, or plant or animal breeding, or in research in genetics, physiology, evolution, microbiology, biochemistry, nutrition, or processing technologies. The utilization of a germplasm collection for any purpose whatever necessitates scanning, or evaluation for characteristics relevant to the particular purpose.

Evaluation is an essential preliminary to utilization. It may consist of no more than a site description of the places of origin and a few morphological or phenological descriptions, or it may extend to any of a range of tests such as physiological, genetic, biochemical or plant pathological examinations, and many more. At whatever level, they have, and indeed should have, the common motivation of being related to *utilization*, whether in some area of research, or in plant introduction or plant breeding. To add a few examples to those already given in Chapter 8 – large collections of wild wheat and *Aegilops* species have been analysed for their protein characteristics in order to discover the ancestor that contributed the B-genome of the common wheat, *Triticum aestivum* (Johnson, 1975). Rick assembled representative collections of *Lycopersicon* species for studies of the evolution of the tomato. A systematic search was made for wheats rich in protein and lysine; and, as is evident from sections 8.2.2 and 8.2.3, the search for resistance to an enormous range of parasites is never ending. Extensive collections of wild legume species from Africa and South America have been examined, under controlled environment conditions and in the field, for their suitability for different regions of Australia. The evaluation of these pasture species was aided by numerical analysis of variation patterns (Burt *et al.*, 1971), and by further statistical techniques developed in recent years (see e.g. Burt and Williams, 1980).

Yet the effect of the best-organized evaluation can be ephemeral unless the observations or assays are systematically recorded in a form which can be read and used not only by the persons who initiated or conducted the examinations, but by others concerned with the same species. A systematic descriptive documentation system is the key to continuing use of collections, and even to their continuing existence. It is no overstatement that poorly conducted records are responsible for the limited use which has been made of the accessions contained in many germplasm collections, as well as for the neglect and the abandonment of valuable collections. Deficiencies of documentation restrict the usefulness of a collection even within the institution owning it, and much more so to other potential users. The difficulty arises from the vast amount of information, and from the lack of agreed standards and procedures which would facilitate the communication of information between collections, and between collections and users. What is needed is a system to collect, record, classify, analyse and communicate information about accessions, and to keep it up to date. If such a system is adopted by a number of cooperating institutions the information obtained by any of them can be transmitted to others so that experience gained in evaluation and utilization is shared, potentially world wide.

A programme to develop such a system was initiated in 1973. Based on research carried out over a number of years by information specialists at the University of Colorado at Boulder, Colorado (see Hersh and Rogers, 1975), a project was established jointly by FAO, the Consultative Group on International Agricultural Research (CGIAR) and the University of Colorado, under the title Genetic Resources Communication, Information and Documentation System (GR/CIDS). From an earlier system, TAXIR, a system called EXIR was developed specifically for the requirements of the various phases of genetic resources work. In the course of the last four years it has undergone exhaustive tests on a number of collections representing a diversity of crops and conditions, and, according to IBPGR (1977b), EXIR or TAXIR are now in use in a number of institutions, mostly on IBM or CDC, but also on some other machines. A users' manual and a manual on the organization of data are obtainable from FAO or the University of Colorado.

Recently, IS/GR, the Information Sciences/Genetic Resources Program of the University of Colorado, working under contract with IBPGR, has assembled large quantities of information from some of the world's largest germplasm collections of wheat and maize, to enable the crop advisory committees for these crops (see section 9.7) to formulate standard lists of descriptors for use by plant breeders and genetic resources centres. Data bases were also set up for a number of other crops, including sorghum, bulrush millet, *Phaseolus* and other grain legumes. Many particular tasks were undertaken, such as selecting from a collection a set of entries representing the diversity in the collection, or methods for identifying duplicates (see section 9.6 below) (IBPGR, 1978a).

It must be realized that during this period – and preceding it – a number of institutions, both national and international, developed their own information systems which they found satisfactory for their purposes and, in some instances, more convenient to operate than EXIR. While a large degree of international coordination of information is highly desirable, it is obvious that this cannot be achieved by a universal acceptance of a single program, but by means of transmission of information between systems.

Indeed, recent developments in computer technology and in program design have rendered the big and expensive programs of recent years largely unnecessary. Minicomputers can fulfil most of the tasks required by genetic resources centres, and information can freely flow between centres using different systems. Computer facilities are becoming freely available, and IBPGR sponsored training courses are held in different parts of the world.

9.6 Size of germplasm collections

How large need plant germplasm collections be? Theoretically, collections should be representative of variation which is, or could be of interest or use to scientists within the compass of an institution, a country, a region, or the world, depending on the function and responsibility of a particular collection. This may be specific – for example, the cytogenetic stocks of a species, or sources of resistance to specific pathogens. However, as will have become evident in this and the preceding chapter, many germplasm collections have broader responsibilities and functions, for example a representative collection of Indian rice or of Near Eastern stone fruits, or a world collection of sorghum. In such collections the scale of operations, i.e. the size of the collections, is not limited by their function. This raises the issue of the 'economy of scale', which can be formulated like this: *how can one organize a germplasm collection to be as representative as possible, yet not to exceed manageable limits?* In the conservation of livestock genetic resources, as we shall see in Chapter 10, this issue is even more acute because the greater unit value and size of animals imposes a rigid 'economy of scale'.

In principle, the larger a collection, the better the chance that it will include genetic information which at some point in time may be found useful. Since the nature of such information cannot be foreseen, this argues for the largest possible representation of available genetic diversity. But the 'economy of scale' imposes limits on what otherwise might be an indiscriminate assembly of materials. The cost of storage and maintenance of seed collections rises in proportion with numbers once a basic size is exceeded; for vegetative collections this is the case even without the latter qualification. But the largest burden, in terms of effort and investment, comes from evaluation, in which a number of institutions may be involved. 'Redundancy' i.e. near-identity or close similarity for all identifiable genes or characters, is an expensive, though unavoidable fault of collections assembled without restraint. How is it to be avoided or reduced?

We can make a distinction between *named varieties*, i.e. recognized types, and *populations which are collected in the field*, representing land races or wild species. In both, some steps to reduce redundancy can be taken even prior to inclusion in a collection. In cultivars as a rule some information is available on origin and parentage, phenology, resistances, performance, adaptation and a varying number of morphological characteristics. This information can be used to establish the extent of correspondence between cultivars. In field collecting, characterization of the target population and of its environment are valuable aids in avoiding redundant collecting, as can be the names and farmers' assessments of local land races.

Once varieties or populations are entered into a collection, they can be subjected to evaluation for a wide range of directly relevant characteristics. As suggested by Brown (1978), electrophoretic surveys of proteins can provide an objective measure of diversity between and within plant populations. Similarly, they may help in assessing genetic differences between closely related cultivars. A multivariate analysis of all available information should then provide presumptive evidence of redundancy (or diversity).

Once such evidence is available, what positive steps are open? Four possibilities suggest themselves:

1. The entire material is retained, in spite of evidence suggesting redundancy. This is appropriate in collections obtained in primitive areas subject to rapid development, where there is little likelihood of repeat collecting; or in collections of great scientific or historical interest.
2. A random selection of apparently redundant entries is retained.
3. Redundant entries are combined, and the bulk sample becomes a single accession with consequent saving in handling and in storage space. Depending on sample size, land for regeneration is also reduced. However, recognition of some characteristics may be less efficient in bulks than in separate entries.
4. The sample size may be kept to a minimum, while keeping entries apart.

Each of the options 2–4 incurs a loss of genetic information compared with 1. In an appendix to this chapter Dr A. H. D. Brown discusses the nature of this loss.

9.7 Organization

It remains to outline the institutional context in which genetic resources are assembled, evaluated and conserved. Plants have been, and will continue to be collected, used, stored, and in many instances discarded by individuals and institutions. Yet in the precarious position in which the world's genetic resources are placed, the idea has gained ground that an organized effort is needed to salvage as much as possible of what is left in the field, and to insure that both the existing collections as well as land races which are still cultivated by farmers are preserved in properly organized and equipped gene banks or genetic resources centres, and further that germplasm collections are described, evaluated, recorded and made available for use.

The idea of a genetic resources centre (GRC) emerged from the vast and still unparalleled collections assembled by N. I. Vavilov and his colleagues, the tangible result of their collecting activities in many parts of the world. The USSR Institute of Plant Industry became in the 1920s and 1930s the most highly developed and successful GRC ever established, embodying all aspects, from worldwide exploration, through extensive study and evaluation, to long-term conservation. It provided the inspiration to corresponding activities in USA, Great Britain, Japan, Australia and elsewhere, though most of these were mission-oriented 'plant introduction services' whose task was to obtain materials currently required by agronomists or plant breeders rather than a representation of the existing diversity. However, large collections were established for all the major and many of the minor agricultural crops, and these are maintained by various institutions engaged in crop improvement. References to germplasm collections are found throughout Chapter 8.

The broader concepts of genetic resources work, clearly visualized by Vavilov but in jeopardy in the USSR during the twenty-five years of Lysenko's dominance, began to be recognized at a conference convened by FAO in 1961 (Whyte and Julén, 1963) and were widely accepted after a further conference organized by FAO and the International Biological Programme (IBP) in 1967 (Frankel and Bennett, 1970b), where, incidentally, the term 'genetic resources' was first used.

The decade of 1965 to 1974 saw the emergence and growth of what one might call the genetic resources movement. The initiatives came largely from the FAO Panel of Experts on Plant Exploration and Introduction (which had merged with the IBP committee for plant gene pools), a group of scientists associated with it, and the small genetic resources unit of FAO. Their tangible contributions were two volumes which explored and clarified many of the scientific and technical issues involved in the various aspects of genetic resources work (Frankel and Bennett, 1970b, and Frankel and Hawkes, 1975). However, of equal importance was the impact the group or its members made on FAO itself, on the Stockholm Conference on the

Human Environment, 1972, on administrators in various countries, and, last but not least, on the scientific public. Members of the group played a prominent part in a meeting convened at Beltsville in 1972 by FAO and the Consultative Group on International Agricultural Research (CGIAR), which made recommendations for a global network of GRCs and for its coordination, which ultimately led to the establishment of the International Board for Plant Genetic Resources (IBPGR) by CGIAR. The events leading up to the current international activities and organization were well described by Harlan (1975b).

With growing interest in national and international activities it was found necessary to define the areas of activity and their structural relationships. This was particularly needed with regard to conservation on one hand, evaluation and multiplication on the other. Conservation, whether of seeds or plants, requires major installations in relatively few sites, whereas evaluation gains from a multiplicity of sites. The FAO Panel of Experts proposed definitions for germplasm collections as 'base collections' or as 'active collections'. Base collections of seeds are responsible for the long-term conservation of what essentially are reserve stocks, the 'conservation stream' (see section 9.4.2), to be called upon by the corresponding active collections when required, Active collections are responsible for the multiplication and distribution of entries (the 'utilization stream'), hence must possess medium-term storage facilities. They are also responsible for regeneration of stocks when required (see section 9.4.2). All centres participate in documentation. Functions of both base and active collections may be performed in the same or in separate institutions, but, as we have seen (p. 234) a base collection may need to collaborate with several active centres to secure appropriate environmental conditions for the regeneration of a wide range of entries. Existing GRCs represent examples of various combinations of this kind. Evaluation should be as broadly based as possible, with not only active centres, but also experiment stations and plant breeding institutions participating, whose collections are called 'working collections'. Information from all sources, including base, active and working collections, should be entered in the documentation system. For a more detailed discussion of organization and function of GRCs see Frankel (1975b).

The FAO Panel of Experts on Plant Exploration and Introduction was discontinued after the establishment of the International Board for Plant Genetic Resources with near-complete change of membership. The FAO Panel of Experts on Forest Gene Resources, however, has continued its useful activities.

The entry of the international institutes for agricultural research into the area of crop genetic resources has brought much needed new strength to the international effort. CIMMYT, the International Centre for maize and wheat, maintains a large collection of maize assembled over many years by the Rockefeller Foundation and others. IRRI, the International Rice Research Institute, has assembled over the last ten years a collection representative of most of the rice growing areas, which is still being enlarged in cooperation with workers in Asia and Africa, but is defective in wild species material. CIP, the International Potato Institute, is well on the way to complete the assembly of potato germplasm (J. G. Hawkes, personal communication). Other Institutes in the dry and humid tropics are establishing collections of a range of important crops. Storage facilities are being provided for all those which need them.

By comparison, support for national efforts in the developing countries in Asia, Africa and Latin America, where the centres of genetic diversity are situated, is slow to get off the mark. In its first three years of activity, the main achievement of the International Board – and one of importance – has been the advance in the field of documentation (see section 9.5). Yet in the crucial top-priority areas of exploration and conservation – the main purposes for which the Board was set up – activity had been slow to start. Much time was spent on the establishment of priorities (IBPGR, 1976a) resulting in an order which, though more elaborate, differed little from one prepared by the FAO Panel of Experts years earlier. Concerted and massive support was given by IBPGR to the development of the information system for genetic resources (see section 9.5); and valuable support has been given to training, and some emergency exploration was supported.

Recently, activities sponsored or supported by IBPGR have been intensified (IBPGR, 1978a). Following upon the appointment of crop germplasm advisory committees, collecting activities were expanded. In 1977, grants for collecting activities were received by seventeen institutions in fifteen countries. The first four grants for storage facilities were allocated, and one for upgrading of a collection of live plants (bananas). Publications include descriptors for germplasm information systems (see section 9.5) for wheat, maize, banana and potato, a bibliography (with a supplement) of plant genetic resources, and monographs on plant quarantine in genetic resources transfer, and on tropical vegetables and their genetic resources. Regional cooperative activities, perhaps the most difficult of the proposed projects, are well under way in South-East Asia, where cooperative planning and action are supervised by an intergovernmental committee (IBPGR, 1977a).*

As the Report points out, 'important national activities will no doubt be started or continue to progress in many countries', and, one may add, they will continue to hold the oldest and most extensive germplasm collections of many economic plants. The Board's business, continues the Report, 'is not to control these activities: it exists to promote them, serve them, and help to articulate them into a network', i.e. to promote collaboration and the widest interchange of materials and information. Indeed, as in all scientific

* Since then, activities in all areas have been greatly expanded.

activities, encouragement of initiative and individual enterprise is more likely to be productive than rigid centralization of planning and support. One may hope that, in spite of set priorities, national or international encouragement may not be wanting for individual effort, taking advantage of the diversity of interest and expertise of individuals, as long as they are concerned with the broad objectives of what one may recognize as the 'genetic resources movement'.

9.8 Summary

1. From the qualification of 'usefulness to man' derives a *time scale of concern* which is commensurate with economic and social predictabilities of usefulness (section 9.1).
2. Conservation is *'dynamic'* under conditions facilitating continuing evolution. For land races this is meaningful only under continuing (commercial) cultivation in their home environments, i.e. *in situ*. Wild species of economic significance are most effectively conserved within a co-adapted community, as are wild species in general (see chapter 5). However, diversity and multiplicity of conservation sites are likely to be more important than is strictness of management (sections 9.1 and 9.2).
3. *Dynamic conservation ex situ* in the form of mass reservoirs is a plant breeding rather than a conservation method. *Ex situ* populations of wild species are widely used in silviculture where they are called provenances. *Ex situ* conservation is useful for research, for the selection of locally adapted types, or for the conservation of material which is threatened in its native habitat (section 9.3).
4. *Static conservation* aims to preserve as far as is possible the genetic identity of individuals or populations. This can be achieved safely (and cheaply) in the form of seed stored under appropriate conditions of temperature and moisture content. Seeds which can be stored at low temperature and humidity can be stored for decades or centuries without loss of vitality and without genetic damage. Appropriate techniques and procedures are described. Seed preservation of species with 'recalcitrant' seed and of many wild species requires further study. (Section 9.4).
5. *Regeneration* of stored seed which begins to lose vitality should take place in an environment where survival is maximized (section 9.4.2).
6. Storage of *pollen*, and the preservation of *vegetative material* and of *tissue cultures* are briefly outlined (sections 9.4.3–9.4.5).
7. *Evaluation*, extensively dealt with in Chapter 8, is the essential preparation for the utilization of germplasm, but is of little practical value without adequate *documentation* which can be understood by and is accessible to all potential users. A documentation system especially

designed for genetic resources has been developed and is beginning to be widely used (section 9.5).

8. Large germplasm collections should be subject to an 'economy of scale' so as to keep the effort and cost involved in maintenance, conservation, and especially in evaluation within manageable limits. Ways are discussed in which the burden of redundancy (i.e. of related and/or closely similar entries) can be reduced, and the loss of genetic information which is incurred is the subject of an appendix (section 9.6 and appendix below).

9. *Genetic resources centres* are institutions engaged in long-term conservation ('base collections'), in multiplication, distribution and evaluation ('active collections'), or both. All centres participate in documentation. The role and contributions of international organizations and institutions are outlined (section 9.7).

Appendix: efficient handling of apparently redundant entries (A. H. D. Brown)

The reduction of a collection of samples to a manageable size, either for the purpose of testing, or for those of regeneration and maintenance, raises a number of issues. These issues can be conveniently studied using simple probability models. In these models we will make the simplifying general assumption that the numbers involved are sufficiently large that 'sampling with replacement' models are adequate.

Let the collection originally consist of k_1 entries each containing n_1 propagules (seeds), making a total of $N_1 (= k_1 n_1)$ units altogether. Next we assume that these k_1 entries have no known differences in origin, nor in results from evaluation. Thus the user only knows that the entries probably differ from one another genetically, but lacks any evidence of such differences. His problem is to reduce the numbers from this group of entries to be processed on some rational basis. Three basic options are considered (cf. p. 244):

2. The entry-deletion option. He may simply discard at random a number of the entries in entirety, so that the new collection consists of k_2 entries ($k_2 < k_1$), each still with n_1 propagules to amount to a new total of N_2 units ($N_2 = n_1 k_2$).

3. The sampled bulk option. He combines equal quantities from each of the k_1 entries into one large bulk, from which $N_2 (= n_2 k_1)$ gametes are drawn.

4. The truncated separate sample option. Under this scheme, equal quantities (again $n_2 < n_1$) are kept separate as k_1 individual entries of reduced size ($N_2 = n_2 k_1$).

The consequences of these options can be evaluated in the extreme case in which only one of the k_1 original entries, unknown to the user, contains a desirable allele with frequency p in that sample but all the other samples lack this allele. Further we can introduce a recognition factor, denoted as r ($0 < r < 1$). This factor quantifies the relative difficulty that might arise in recognizing that a tested plant possesses the desired allele when assayed on a single plant basis, compared with a progeny row test. Thus when $r = 0.1$, the user might detect the fact that a single plant in a bulk population carries an allele for mild resistance to a disease at an efficiency of 10% compared with its detection when all the plants for the relevant entry are tested in the field contiguously.

We consider the probability of recovery of the desired allele under three options. In the case of the first or deletion option, this probability is

$$P_d = k_2 [1 - (1 - p)^{n_1}]/k_1 \simeq k_2/k_1$$

For the second, or bulk option

$$P_b = 1 - (1 - rp/k_1)^{n_2 k_1}$$

Finally, for the separate entries

$$P_s = 1 - (1 - p)^{n_2}$$

Some general conclusions can be established. First, P_s always exceeds P_b, although in most situations the difference is marginal. Thus for example, consider a strategy (or value of n_2) designed to recover all alleles with frequency greater than 0.05 (our arbitrary target frequency), with a certain minimum probability of success in the situation of separate entries. An equivalent level of testing in the bulked situation would recover all alleles with frequency greater than 0.0513, at the same or higher probability, assuming that there was no recognition problem ($r = 1$). Alternatively, the sample size (n_2) in the bulk situation could be increased to compensate for the dilution effect of bulking. For a gene frequency of $p = 0.05$, the value of n_2 should be increased by a factor of 1.026 times its value in the separate entries situation. It is clear that the decision of whether or not to bulk would be most frequently made on pragmatic, rather than on probabilistic grounds.

Second, values of P_b and P_s are not greatly affected by k_1. Thus we can choose a value of $k_1 = 20$ entries and compute the value of n_2 required to achieve certain probabilities of recovery, for a given gene frequency and values of r. These values of n_2 and comparisons can be used irrespective of the actual original number of entries. Some examples of k_2 (the parameter to be varied under the first deletion option) and n_2 (for the second and third option) are given in Table 9.1. To interpret these figures, we consider the last column, pertaining to our basic arbitrary strategy. Let us assume that each entry originally consisted of 100 seeds ($n_1 = 100$). The original total collection is thus 2000 seeds, which, we assume, must be cut. The first deletion option, however, only allows a reduction to $19 \times 100 = 1900$ if our basic strategy is still to be achieved. Further reduction is only achieved at

TABLE 9.1 *Values of k_2 or n_2 required to achieve a given probability of recovery for a desirable allele present in only one of (k_1 =) 20 entries*

Allele frequency p	0.50	0.05	0.05
Probability of recovery	0.95	0.50	0.95
k_2 values for –			
2. Deletion option	19	10	19
n_2 values for –			
3. Bulk options $r = 1.0$	6	14	60
$r = 0.1$	60	139	599
$r = 0.01$	599	1386	5991
4. Separate entries option	5	14	59

Symbols: k_1 = number of apparently redundant entries in the original collection; k_2 = number of the original k_1 entries represented in the modified collection; n_2 = number of propagules (seeds) per entry in the modified collection; r = probability of detecting the desirable allele when screening the bulk population as compared with testing separate entries.

greater genetic cost. The second strategy, when there is no recognition problem ($r = 1.0$) and the third strategy, indicate that the total collection size may drop to about $20 \times 60 = 1200$. A further drop to about $20 \times 14 = 280$ (or 14% of the original size) means that about half the localized alleles of frequency $p = 0.05$ will be lost. Note that the deletion option allows only a drop to 50% of the original total size for an equivalent genetic cost. The decision between the second and third options would, on these figures, largely depend on the severity of recognition problems compared with the labour of manipulating several virtually indistinguishable entries.

Third, the probability of recovery in the deletion option is independent of p. This is shown by comparing the first and last columns of the Table. Both common and rare alleles give the same probability of recovery for a given amount of deletion. These results indicate that deletion would rarely, if ever, be an appropriate option.

10

Conservation of livestock genetic resources

10.1 Genetic resources of crops and livestock – analogies and contrasts

In considering the genetic resources of animals, comparisons with those of plants can serve as a useful introduction, since a great deal more has been written – and done – about the gene resources of plants than those of animals. There are significant biological differences which are reflected in the adaptation and population structure of plants and animals. For example, plants are earthbound, hence can be expected to be more closely adapted to environmental niches than the peripatetic animals. Many plants are, or can be, self-fertilized, whereas all livestock species are heterosexual. Livestock species and domesticated plants differ greatly in their reactions to diseases and parasites.

Yet there are significant analogies, which are clearly reflected in the evolution of domesticated plants and animals, as are dissimilarities, and both are reflected in the structure and function of present-day genetic resources. As in plants, three categories of genetic resources can be recognized: ancestral wild species, unimproved breeds corresponding to land races, and improved breeds corresponding to advanced cultivars (see Chapter 8). In several animal species the progenitor can be no more closely defined than it can in a number of crops, especially since several related forms may have been involved. In the sheep the main ancestral types, the moufflon, the urial and the argali, are not only closely related but are linked by intermediate forms (Zeuner, 1963).

With the exception of the dog, which undoubtedly is the oldest domesticate, it is uncertain which came first, plant or animal domestication, partly because it is not easy to determine when hunting and herding of animals merged into domestication. It appears that the domestication of reindeer, goats and sheep approximately coincided with that of the oldest crop plants, barley and wheat, while cattle and horses followed later. Thus crops and livestock are of comparable age, but whereas the ancestors of crops have survived because wild plants are highly resilient, wild animals subject to exploitation by man are more vulnerable. With the advent of domesticated derivatives they are exposed to introgression from domesticates, to progressive habitat restriction, and to the tender mercies of the hunter. Thus the last specimens of the aurochs, *Bos primigenius*, the wild progenitor of European cattle, died in a Polish forest some 350 years ago. The only remaining wild cattle are the White Park cattle in England, though it is rather hard to credit that they have altogether escaped introgression from domesticated cattle. Indeed, some of the herds have been preserved from inbreeding depression by deliberate crossing with domesticated breeds (J. M. Rendel, personal communication).

In general the evolutionary distance between wild and domesticated species is less in animals than in plants. Many crop plants have evolved characteristics such as the solid rachis (spindle) of the ear of wheat or the single, large cob of corn, which would prevent their survival in nature, having passed the point of no return; whereas Heck (1952) succeeded in breeding a 'reconstituted aurochs' from crosses between domesticated breeds exhibiting individual characteristics of the aurochs, the process taking less than thirty years. Seeing that the aurochs became extinct hundreds of years before, the similarity may have been more superficial than real. Yet an analogous reconstruction would be hard to envisage in most of the crop species.

Experimental evidence on processes of domestication comes from the work of D. K. Belyaev and his colleagues at the Institute of Cytology and Genetics at Novosibirsk. Belyaev started from the hypothesis that drastic changes to the reproductive system, including increased fertility and loss of strict reproductive seasonality took place under domestication as correlated responses to selection for behaviour. Reproduction in wild species is regulated by the length of day, acting upon the hypothalamus which plays a vital role in controlling reproduction as well as affecting the behaviour pattern. Silver foxes derived from stocks which had been reared on Siberian farms over seventy years without exhibiting behavioural or reproductive changes, were selected for 'tranquil behaviour', i.e. lack of the aggressiveness characteristic of the species. After fifteen years of selection, and in the absence of any form of training, the behaviour of the selected group was rapidly approaching the behaviour pattern of the dog against the unchanged control group, and the tranquil behaviour was found to be genetically associated with earlier reproduction within the reproductive season and with higher fecundity. Genetic change in behaviour has led to a change in the neuro-endocrinal mechanisms which control the hormone status of the organism. Physiological and biochemical research, including both normal and castrated animals, indicated the involvement of the pituitary–adrenal system and the sex glands in the selection for behaviour. There is an apparent disorganization of the physiological system evolved over long periods of natural selection, including even an increase in the number of microchromosomes in animals selected for either tranquillity or super-aggressiveness. It will be of great interest to follow the effect of stabilizing selection when the selection effects for tranquillity level off (Belyaev, 1969; Belyaev, Volobuev, Radjabli and Trut, 1974; Naumenko, Trut, Pavlova and Belyaev, 1976).

With the possible exception of the sheep and goat, closely related wild species have a great deal less to contribute to the improvement of domesticated animals than to that of crop plants. The wild progenitors of livestock species either no longer exist as in the case of the cattle or horse, are uncertain, as in dog or pig, or the domesticate is hardly removed from the wild as in elephant or camel. In sheep, attempts have been made to transfer adaptation to high altitudes and genes for size and vigour from wild to domesticated species, but the inaccessibility of their habitat, which protects the wild sheep, also inhibits their utilization. A symposium paper by Foote (1976) relates these and other experiences of crosses of domesticated and wild species of sheep and of goats obtained in Russia, Israel and at the University of Utah. In general, however, there is a marked contrast between plants and animals in the availability and use of wild genetic resources, at least until the time that genetic components of distant species, or even orders, become available for transfer. The gene pool of livestock species is thus almost entirely domestic, and this emphasizes the need to preserve and manage it to best advantage.

As in plants, local breeds of livestock evolved wherever animal husbandry took roots; and unimproved breeds, like land races of crops, exhibit characteristic differences between breeds. Livestock species spread throughout the Old World and acquired adaptations to a great range of environments. Livestock lend themselves a great deal more than crops to the selective impact of the breeder since characteristics given a selective value are more readily appreciated and mating can be easily controlled. While in crop evolution we have stressed the predominant impact of the environment, hence of natural selection, man may be presumed to have been an influential partner with nature in the formation of livestock breeds, starting from the early stages of domestication: social and cultural impacts supplemented, and in some instances may have outweighed, the impact of the environment.

The resultant diversity of animal breeds, which so greatly stimulated Darwin, differs from that of land races of crops in some relevant respects. First, land races of crops have acquired specific characteristics of adaptive significance, such as resistance to parasites, many of which are simply inherited. This is less common in animals. But thanks to the universal outbreeding and prevalent polygenic inheritance, disease resistance polymorphisms are widespread in livestock breeds exposed to pathogens (see section 10.2.1). Second, unlike crop cultivars, most animal breeds can be readily distinguished by external characteristics. Indeed, the retention of such characteristics is fostered by selection. In some instances one may wonder whether such breed characteristics resulted from individual and social preferences becoming traditionalized. Even so, such breeds, though closely related populations, are co-existing in reproductive isolation from each other and are likely to evolve genetic (and physiological) differences which affect productivity. Yet the question arises – and will be discussed further – whether any but major differences between related breeds are sufficient to justify conserving, for example, do we need *all* the breeds of Zebu cattle? This question arises in similar ways in plants, but with the significant difference that breeds of livestock are a great deal more difficult and costly to maintain than are land races of crops. Operationally this may be the most powerful factor in determining the relative scope of conservation which is practicable in animals and in plants. Indeed, it may be easier and cheaper to conserve thousands of wheat or barley cultivars than one breed of cattle or sheep. Finally, quarantine restrictions impose far more stringent constraints on the movement, hence on the use, of germplasm of animals than of plants.

Genetic resources of a large number of crops have been assembled in germplasm collections which contain thousands of entries derived from different parts of the world. They can be preserved by routine procedures for long periods of time, the stocks they contain are readily available for research or breeding. Analogous collections of livestock species can only be realized when both semen and ova, or embryos, can be safely preserved and animals can be produced from them (see section 10.4). Until such time animal breeds can only be preserved either in agricultural use, or in some form of controlled breeding reserves. The nearest analogy in plants are those tree crops which cannot be preserved in the form of seeds, because their seeds are 'recalcitrant' (see section 9.4.1) or because they have no seeds at all. Many fruit species are in this category. The difficulties and expense of maintaining fruit trees restricts the number and the size of collections. Hence in fruits, as in livestock, there is a real danger that modern high-producing breeds will displace the wealth of genetic diversity which the breeder will continue to need. This is happening in the orchards of Turkey, Iran and Pakistan, as in the dairy herds of Europe. Cosmopolitan apples and pears, peaches and apricots are replacing the ancient diversity of fruits in the Near East, and Friesian dairy cattle the diversity of the local dairy breeds which had evolved all over Europe. The prospect that similar patterns will develop the world over is all too patent.

From these comparisons we draw some preliminary conclusions. In animals as in plants the main concern is for the primitive and local breeds which are exposed to displacement by improved, higher-producing breeds, whether cosmopolitan introductions or upgraded locally adapted breeds. The result, already widespread, is the extinction of local land races of crops and breeds of livestock. Reasons for the preservation of the former were given in Chapter 8, and, as shown in section 10.2, there is also a case for the preservation of genetic diversity in livestock. Yet, compared with plants, animal germplasm conservation needs to be highly selective. In plants which can be preserved as seeds, selection for germplasm conservation can be on a random basis modified by ecological considerations, and subject to further selection on the basis of subsequent evaluation. In animals the cost of preservation imposes far more stringent restrictions. *Selection is inevitable*, but on what premises or evidence? It may be that the failure to find answers to this question is the main cause for the lack of concerted policies, let alone action programmes for the preservation of animal genetic resources.

10.2 The state of animal genetic resources and the case for their conservation

There are three reasons for the conservation of livestock genetic resources – agricultural, scientific and cultural (Mason, 1974). As in domesticated plants, the justification rests on current or potential use, interest or enjoyment. Hence it is wholly anthropocentric.

10.2.1 Agricultural

Genetic resources of livestock can be characterized in three fairly distinct categories: the primitive local breeds, mainly in less developed parts of the world, the improved local breeds which were prevalent in Europe until the middle of the present century and remain so in large parts of Asia (especially cattle and buffalo), and the high-producing, largely cosmopolitan breeds of cattle, pigs and poultry and the specialized breeds of sheep bred in Europe, Australia, USA and New Zealand (Table 10.1). Progress towards increased productivity is invariably associated with reduced genetic diversity, *between breeds*, because of the selective elimination of primitive and of improved local breeds, *within breeds*, through the reduction in the number of distinctive strains within the increasingly dominant breeds of today, and *within populations*, as a result of selection or genetic drift which follows a fall in the effective number of stud flocks and herds and a fall in the effective number of sires that accompanies commercial exploitation of the development of techniques such as artificial insemination (AI) (see Maijala, 1974).

Lauvergne (1975) conducted a survey of disappearing cattle breeds in Europe and the Mediterranean basin. In this region 115 indigenous breeds are threatened with extinction – many of them differentiated into distinctive strains – and only 30 are holding their own. Nearly all the threatened breeds were confined to, and, presumably, adapted to their local environment. The standardization has progressed farthest in dairy breeds: only two or three of these supply the milk for most Europeans and Americans. While the loss of

TABLE 10.1   Categories of livestock breeds and their uses as genetic resources

Category	Examples	Actual or potential use
(i) Primitive local breeds	Sheep, goat and cattle in Africa and Asia, e.g. the Kuri cattle of Lake Chad (Mason, 1975a)	(a) Upgrading by selection and/or cross-breeding, for low-input or climatic stress conditions (b) Transfer of specific characteristics to improved breeds, e.g. high fecundity of Dahman sheep in Morocco (see Turner, 1977), or of the Finnish Land race of sheep (see Mason, 1975b)
(ii) Improved local breeds	European local breeds of cattle, sheep and poultry; breeds of Zebu cattle and buffalo in India and Pakistan	(a) Further improvement for use in special areas (specific or disadvantaged environments), by selection and/or cross-breeding with improved breeds (b) Cross-breeding among local breeds, or with breeds of category (iii) for production of new breeds (c) Potential future use with changed consumer demand, new industrial processes, etc. (e.g. current interest in Simmental cattle) (d) Actual or potential use in breed crosses, e.g. in poultry (van Albada, 1964)
(iii) Improved high-producing breeds, and distinctive strains within breeds	High-producing largely cosmopolitan breeds of cattle, pigs and poultry, bred in Europe or USA; sheep bred in Europe, Australia or New Zealand	Recognized strains within breeds, e.g. Friesian dairy cattle, Rhode Island Red and White Leghorn fowl, Merino sheep. Such strains constitute remaining genetic diversity in the pre-eminent breeds used for intensive livestock production

some breeds by fusion with similar ones, as has been effectively carried out in France, is commended, Lauvergne (1975) deplores the 'tidal wave which is engulfing the local populations and resulting in the disappearance, or irreversible dilution, of their genes'.

A worldwide survey by Mason (1975b) for cattle (other than European), sheep, goats and pigs provides a general view of the erosion of local breeds – advanced in Europe and USA, incipient or gradual in Asia and Africa. But even in European countries there remain a number of local breeds of sheep, though some in small pockets, whereas pigs have been reduced to two dominant breeds, with primitive breeds remaining only in the extensive pastures of Spain, Portugal and Corsica. By way of contrast, there remain a large number of Zebu cattle breeds in India and Pakistan, though some with declining numbers; and local breeds of sheep still prevail in most parts of Asia and Africa.

The gene pool in use in poultry production in the Netherlands is a good example of rapid erosion and severe restriction (Albada, 1964). The large number of traditional country breeds going back to the seventeenth century are now relegated to the backyards of fancy breeders. Towards the end of the last century they were replaced by a range of imported breeds and new breeds produced in Holland, but this diversity was gradually reduced to two breeds, White Leghorn and Rhode Island Red becoming dominant. The need for replenishment after World War II and the replacement of 'pure breeds' by 'breed crosses' (i.e. F_1 hybrids) led to the exclusive use of three American breeds, White Leghorn, Rhode Island Red and New Hampshire, itself a derivative of Rhode Island Red. Within breeds distinctive strains have been developed which now constitute the effective gene pool. However, the economics of poultry production inevitably led to a reduction in the number and an increase in the size of poultry farms, and in consequence to a reduction in the number of strains in use, further reducing the genetic diversity available to breeders to what may be regarded as minimum levels.

A group of experts in poultry breeding discussed arguments for and against efforts to conserve a greater genetic diversity than appears likely to survive the drive towards increasing homogeneity on a worldwide scale (FAO, 1973). Village type production under low-input, labour-intensive and relatively primitive conditions was likely to continue for a long time, to which local types may make distinctive contributions (Table 10.1 i (a), ii (a), (b)), and, in parenthesis, this is surely the case in other species such as sheep or pig. It was doubted, however, whether such breeds had much to offer towards the improvement of modern, high-producing breeds, since the latter were likely to incorporate the productive potential of more primitive breeds. New diseases were liable to be controlled by husbandry or veterinary methods, nor was there extensive evidence of greater resistance among primitive breeds. Yet there is surely a case for preservation because of possible genotype–environment interactions which could necessitate genetic readjustment in case of environmental or economic change.

It is evident that in all species of livestock there is a severe and accelerating loss of genetic diversity. It takes different forms in different species and areas, but there is a common trend at all three breed levels – unimproved, improved and high-producing. Is there a case for preservation of vanishing genetic resources based on their potential usefulness? What ideas are emerging as a basis for a conservation strategy?

In plants, as we saw in Chapter 8, germplasm collections serve as sources of specific characteristics and/or of adaptations to specific conditions. The characteristics or objectives can be specified and screening methods devised not only for collection stocks, but for hybrid generations derived from them. In animals the evaluation of potential parents and hybrids presents technical and economic difficulties of an altogether different order. In consequence the discussion about the potential use of genetic resources has tended to give emphasis not, as in plants, to potential donors of specific characteristics, but to breeds as such which might find a useful place in changed circumstances in the future. Concern has also been expressed for the maintenance of an adequate number and diversity of strains within the predominant breeds which are becoming progressively more homogeneous. This is of special importance for breed crosses in poultry (Albada, 1964).

A round-table conference on the conservation of animal genetic resources (FAO, 1974) concluded that breeds which had been or were now being displaced could again find a useful or even prominent place as a result of changes in the environment (feeding, housing, management, etc.) or in the demand for animal products. In recent years some previously little known local breeds have proved their usefulness for special purposes, such as the fast-growing Charolais beef cattle, the Cornish game-cock as a meat breed of poultry, and the extremely meaty Piétrain land race of pigs from Belgium. Three local breeds of sheep, the Romanov (USSR), the Finnish Landrace, and the Dahman sheep from Morocco, and also the Booroola Merino, a strain selected in Australia, have well-documented high fertility (see section 10.2.2).

Local hardy breeds of sheep and cattle, still in use in many parts of Asia, Africa and even Europe, are capable of exploiting unfavourable or extreme environments or extensive conditions of management. Such breeds may have a continuing or even expanding role in the future when the exploitation of such environments may be enhanced by population pressure. Indeed, as H. N. Turner (personal communication) remarks, local breeds have an important and continuing role in countries like Indonesia, where the environment is not likely to be drastically improved for many years. She stresses that such breeds should be guarded against indiscriminate crossing with introduced breeds until the value of the crossbreds is adequately assessed.

There is a real risk that valuable Indonesian breeds of sheep, two of which are highly prolific, are swamped by the introduction of Australian sheep so that the identity of the Indonesian breeds may be lost altogether. Nevertheless, in the long term perhaps the most general case can be made for the preservation of threatened breeds on the grounds of their role as potential parents in cross-breeding, either for the production of (F_1) breed crosses, or in breeding programmes for the production of new breeds.

Genetic resistance to diseases and parasites of livestock has come into prominence only in recent years, though it is unlikely to assume the major importance it has in crop plants where the prospect of finding new sources of resistance ranks highly among grounds for the conservation of genetic resources. There are many reasons for this difference. Animals have the potential to develop immunity and to retain it for some time, though genetically susceptible to the causative organism. Indeed, immunity to many diseases can be induced by vaccination. Various diseases are controlled by other veterinary techniques, and many parasites are held in check by chemicals, though in some instances more and more precariously in the face of rapid adaptation of the parasite. Resistance in plants is prevailingly effected by hypersensitivity, through jettisoning of invaded cells, tissues, or sectors, whereas animals are not organized for such auto-amputations.

At the genetic level, plant cultivars as a rule are a great deal more uniform than animal breeds, hence more vulnerable to virulent races of a pathogen. Nor are they able to store genetic information at low frequencies within a population, whereas in animals resistance to many diseases appears to be widespread even though at low levels. There is a tendency for relatively resistant hosts and relatively benign disease agents to co-exist in fairly stable equilibrium, although it is likely that continued exercise of control, with a lessening or lack of challenge, will result in selection for resistance being weakened (Halpin, 1975). Twenty years ago Hutt (1958) stated that for every disease of domesticated animals which had been adequately investigated evidence for genetic resistance had been found where the disease was prevalent. Yet while sources of resistance may be more readily available than had been thought, selection for resistance is fraught with logistic and economic difficulties. Large populations of crop plants for selection under severe infestation can be readily available. In animals the cost, and in many instances also the risk of spreading infection, inhibit or altogether preclude selection procedures which have been so successful in plant breeding.

It is therefore not surprising that research into genetic resistance is further advanced in poultry than in the large mammals. A high degree of resistance to two virus-induced respiratory diseases, lymphoid leucosis and Marek's disease was obtained through selection within two strains of Cornell White Leghorn, combined with selection for increased egg production. In one of the strains mortality from all causes was reduced from 51.5% to 12.7%, from neoplasms (mainly leucoses and Marek's disease) from 14.2% to 3.7%, while egg production to 500 days, per hen completing the test, increased from 169 to 243, demonstrating that disease resistance could be combined with increased production (see below) (Cole and Hutt, 1973). It has recently been shown that resistance to Marek's disease, a herpesvirus-induced lymphoma, is associated with the B^{21} allele of the major histocompatibility locus B. B^{21} is thus 'the first genetic marker for resistance to herpesvirus-induced neoplastic disease' (Longenecker *et al.*, 1976).

In mammals, genetic resistance has been established mainly at the level of breeds or strains, although the pattern of resistance within breeds is likely to follow the general model discussed previously. Breed resistance to trypanosomiasis has been well established for a long time. An important source of resistance is the West African Shorthorn, an indigenous, unhumped breed of cattle which is widespread in West Africa. It is extremely heterogeneous, hence its productivity could be improved (Stewart, 1951). Another well known source of resistance is the N'dama breed of southern Nigeria, which, however, is susceptible to rinderpest; whereas the Fulani breed, which is used outside the tse-tse belt, is susceptible to trypanosomiasis, but resistant to rinderpest (Halpin, 1975). A bacterial pathogen, *Dermatophilus congolensis*, causes cutaneous streptothricosis, a contagious dermatitis, in cattle, mycotic dermatitis (or lumpy wool) in sheep, and a range of disorders in many other animals. It debilitates or kills the animals and damages hides and wool. It is prevalent on cattle in parts of Africa. Oduye (1975) reports that some local breeds, the Muturu, the N'dama and the Baole – are resistant, whereas Chad Brahma and Arab Zebu are repeatedly infected without gaining immunity. Observations of this kind are multiplying with growing interest in genetic resistance.

One of the most important instances of resistance, because of the wide distribution of the parasite and its serious effects on susceptible (European) breeds, is the resistance of Zebu cattle and of Zebu-derived crossbreds to ticks. Bonsma (after Hutt, 1958) found that Afrikander cattle, a South African Zebu breed, was highly resistant to the bont tick, *Amblyomma hebraeum*, the carrier of a rickettsia which causes heartwater disease. Among Afrikanders from endemic areas mortality from heartwater was nil. Mortality was higher among cattle imported from a tick-free area, and in herds which normally are dipped in acaricide when dipping was withheld. Crossbreds with European breeds had resistance roughly proportional to their Zebu content. In Australia, cattle ticks (*Boophilus microplus*) are the carriers of blood parasites the most important of which is *Babesia bigemina*, to which Brahman and Afrikander cattle are highly resistant. In two recent experiments derivatives of Afrikander and Brahman crosses with Hereford × Shorthorn were compared with Hereford × Shorthorn (all in F_3–F_4). When uncontrolled, ticks reduced the weight gain in Hereford × Shorthorn by 40–54 kg a year, and in Afrikander crosses by 20 kg, but had no effect on the Brahman cross. Tick resistance accounted for 90% and 45% respectively of the superiority of Zebu over European (Seifert, 1971; Turner and Short, 1972). In one of these experiments, control of gastrointestinal helminths resulted in substantial liveweight increases in Hereford-Shorthorn and the Afrikander cross, against virtually none in the Brahman cross, demonstrating the helminth tolerance of the latter.

The problem of evaluation for resistance to infectious diseases has already been mentioned. The induction of epidemic conditions required for identification of resistance may be costly and even dangerous, and naturally occurring epidemics or endemic diseases, like foot and mouth disease, are likely to be dealt with by veterinary measures or destruction of stock. No wonder, therefore, that characters associated with the incidence of a disease, which can be used as indicators or markers, have evoked much interest. Cancer eye in Hereford cattle is a hereditary affliction, yet selection is difficult because of the late appearance of the disease. However, resistance has been found to be associated with a protective ring of pigment around the eye, and to the extent that disease incidence depends on absence of pigment round the eye, it is easy to select against susceptibility (Bonsma, 1949, and others quoted by Hutt, 1965).

Genetic resistance of the White Leghorn breed to pullorum (or white diarrhoea) disease is associated with superior ability to raise the body temperature rapidly during the first few days after hatching. Plus and minus selection for rectal temperature followed by exposure to *Salmonella pullorum* resulted in mortality rates of 8.6% and 40.7% respectively. Resistance could thus be achieved by selection in the absence of the pathogen (Hutt and Crawford, 1960).

What Halpin (1975) calls 'liability' to a disease should be regarded as distinct from resistance, though also under genetic control. Skin folding is by far the most important determinant of susceptibility to blowfly strike in sheep, and is controlled by a number of genes (Morley, 1953). It has long been known that the incidence of mastitis in cattle has an hereditary component. Some characters, such as peak flow rate, pendulosity of the udder, and teat shape are known to be associated with mastitis and have high heritabilities (see Spooner, Bradley and Young, 1975). Halpin (1975) notes that there is evidence that serum transferrin genotype may be a biochemical marker for liability to mastitis; and absence of haemoglobin B appears to be a marker for the resistance of some cattle breeds to trypanosomiasis.

H. N. Turner (personal communication) notes that in the Gangetic plane of India temperate sheep breeds are liable to become affected with lung trouble, whereas local breeds do not. South Australian Merino sheep, selected in a low rainfall area, suffer more from 'fleece rot' when exposed to high rainfall than do fine-wool Merinos.

There are indications that liability to disease may be associated with morphological or physiological characters which condition or are a consequence of high productivity. This appears to be the case in mastitis of cattle where there are indications that genetic resistance may be associated with reduced yield potential (less stress on the organism) and increased tightness of the streak canal. However, in poultry, as we have seen, combined selection for disease resistance and egg production was highly successful.

Having stressed the difficulties which are encountered in the quest for genetic resistance, it is necessary to view its significance in the context of the growing concern with the effects and dangers of the application of drugs and chemicals due to loss of effectiveness, adverse affects on the animal and/or its products, environmental pollution, and in some instances increasing toxicity. New chemicals are increasingly difficult and expensive to devise and test. These are powerful reasons for economy in the use of drugs, and for supplementing or replacing them by genetic resistance and by management practices. Yet the dependence on preventive or curative measures, the destruction of affected stock, and the streamlining of the genetic composition of livestock species, far from encouraging the emergence of resistance, are wholly conducive to deprive present and future generations of the sources of genetic resistance which natural selection has succeeded in preserving, with little help or encouragement from animal breeders. On their preservation and use may, to a large degree, depend the future of our livestock industries.

Before concluding the discussion of agricultural uses of animal genetic resources, reference must be made to an analytical approach in animal improvement which may be significant for the future use of genetic resources. Workers at the Tropical Cattle Research Centre in Rockhampton, Australia, have studied the reasons for the different growth rates of European, Zebu, and European × Zebu cattle (Fig. 10.1). Growth rates of British breeds (*Bos taurus*) are lower under tropical than under temperate conditions; Zebu breeds (*Bos indicus*), though attuned to the tropics, have growth rates which are lower than temperate standard rates. The factors limiting growth rates are, in the European Hereford-Shorthorn cattle, their susceptibility to environmental constraints – susceptibility to ticks and worms, to the 'pinkeye disease' (Frisch, 1975), to heat (Colditz and Kellaway, 1972), and, it seems, to fluctuating food supplies. All these constraints result in reduced feed intake. American Brahmans, on the other hand, are relatively resistant to all of these, yet their growth rate is only marginally better than that of Hereford-Shorthorn. This is due to a substantially lower voluntary intake of food. The Brahman and to a lesser extent its cross with Hereford-Shorthorn, have slower growth and lower appetite in optimum conditions, but bigger reserves – because of smaller requirements – in terms of protein and energy. Since they have a lower heat output they have a

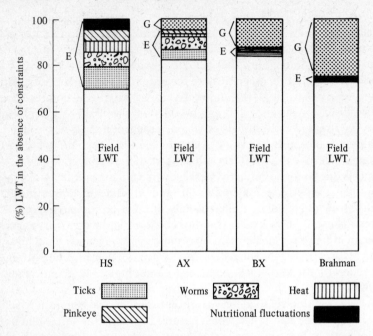

Fig. 10.1 A model of reasons for differences in growth in various breeds. G = difference in liveweight caused by genetic differences in appetite. E = difference in liveweight caused by genetic differences in resistance to environmental constraints: ticks, worms, heat, pinkeye, nutritional fluctuations (see key). LWT = liveweight at fifteen months under field conditions. HS = Hereford × Shorthorn. AX = Afrikander × HS. BX = Braham × HS. From Frisch and Vercoe (1978).

smaller heat load hence are less affected by stress, apart from their better heat dissipation. Afrikander and Brahman crosses with Hereford-Shorthorn exceed the parent breeds in liveweight increase, though, as will be seen in Fig. 10.1, for different reasons: in resistance and food intake Brahman crosses with Hereford-Shorthorn (BX) are nearer to the Zebu, the corresponding Afrikander crosses (AX) to the European parent (Frisch and Vercoe, 1978, also for references). Though this study was greatly assisted by the multiple resistance of the Zebu, an analytical approach to production constraints could be similarly successful in using a series of genotypes carrying resistance to different constraints to maximum production. Indeed, it should now be possible to improve performance by introducing parental material excelling the original stocks in one or more of the characters which have been studied. There is a wide field for the application of genetic resources – subject to availability and to access across quarantine barriers.

10.2.2 Scientific

Scientific reasons for genetic conservation merge into those reviewed in the preceding section. In the interest of gaining a fuller understanding of life processes on which productivity is based, it is desirable to have available as wide a spectrum of genotypes differing from what might be seen as the normal state so as to broaden the scope for physiological and genetic studies. The distribution of blood groups, the physiology and genetics of environmental adaptations, differences in productivity between and within breeds, the nature of resistance to diseases and pests – these and many other problems will be in the centre of interest for many years to come, and will require a wide range of experimental material. High-fecundity breeds of sheep (see section 10.2.1) have been used in experiments assessing the progress of selection for high reproductive rate, and have provided material for research into the contribution to fecundity of ovulation rate and the role of pituitary and ovarian hormones (Bindon and Piper, 1976). Vanishing breeds with distinctive characteristics which render them uneconomic today, such as dwarf cattle of the Himalayas, the Sinhala Zebu of Sri Lanka, the humpless Tibetan cattle, and the dwarf breeds of East Africa may represent adaptations to extreme environments which should be preserved for study since otherwise they are likely to vanish (Mason, 1975b). Last but not least, 'rare breeds are often missing links in the history of breeds and a study of their blood groups, protein polymorphisms and coat characters, can throw light on the origin and relationship of breeds' (Mason, 1974). Similar views were expressed by Kidd (1974) and also by Manwell and Baker (1976) who, on the evidence of published data on protein polymorphisms in several species, suggested that a detailed study of multiple protein polymorphisms may unravel hybridization pathways in the evolution of domesticated animals. Manwell and Baker (1977) attempted to estimate the genetic distance between two breeds of sheep (Australian Merino and Poll Dorset) by comparing genetic variation at thirty loci for blood proteins and estimating from these data the time of their divergence; this led to somewhat equivocal results. Yet such an approach could be extremely useful in providing an objective measure of the genetic relationships between breeds which are under consideration for long-term preservation (see section 10.3).

10.2.3 Cultural

Cultural reasons for the conservation of extinct or vanishing breeds of livestock relate to the association of man and the animals which have served him throughout the ages. 'Historical breeds' are of interest not only to the man and woman on the land, but to townspeople who delight in the colourful diversity of ancient breeds of cattle or sheep, horses or pigs, fowl or duck, from home and abroad. Domesticated animals and their wild relatives are popular exhibits in animal parks or zoological gardens. Members of the Rare Breeds Survival Trust in Britain maintain herds of rare and vanishing breeds, and, as we have seen earlier in this chapter, perform a social service which in later generations may yet bear economic fruits. Relic breeds of poultry which delight us in pictures by the great Dutch painters of the seventeenth and eighteenth centuries, are still maintained in backyards in the Netherlands for their beauty and historical interest.

10.3 What to preserve: strategies of conservation

As we have seen, selection among breeds is inevitable since it is not possible, and perhaps not even desirable, to attempt an adequate representation of every breed in existence. In an introductory paper to the Round Table on the Conservation of Animal Genetic Resources, Mason (1974) lists the following groups which should be considered:
(1)   indigenous breeds uniquely adapted to their environment or showing hybrid vigour when crossed with exotic breeds;
(2)   local productive breeds little known outside their home country (e.g. high-fecundity sheep, see section 10.2.1);
(3)   genetically unique breeds;
(4)   bizarre or beautiful breeds (e.g. St Kilda sheep with four horns, or piebald Jacob's sheep);
(5)   historically important breeds (e.g. Shorthorn cattle in England, French Rambouillet sheep, Galla cattle in Ethiopia).

These are circumscribed and moderate proposals which do not make excessive demands for resources beyond those currently applied, especially in categories 3–5. Other animal geneticists would wish to go further. Turner (1977) considers that 'the need for conserving genes to provide future genetic variation is as great for animals as for plants'. She regards the lack of information about the particular characteristics of many breeds as the main problem in designating conservation targets. However, catalogues and descriptions of breeds from many parts of the world have been published by FAO and the Commonwealth Bureau of Animal Breeding (for references see Mason, 1974); and the latter collects, stores and disseminates information on breeds of livestock, which is obtained from the conventional literature and from less conventional publications and reports. Turton (1974) states that 'a great deal of information is available on the breeds that are economically the most important in the developed world', and that efforts are being made to collect information on rare and perhaps endangered breeds. But this is clearly not the case in many of the developing countries where most of the remaining unimproved breeds of livestock are to be found. Accordingly, there is an urgent need for a survey and characterization of local breeds, as a basis for decision and action leading to conservation and utilization.

How comprehensive need it be? There have been insistent demands for performance tests, including productivity tests in identical environments. Yet, in many instances, reliable observations on specific characteristics, such as comparative resistance to parasites, or high fecundity, may constitute relevant and adequate information. Comprehensive tests are undoubtedly desirable for rational decisions on conservation, and are essential for utilization, yet the dynamics of extinction, combined with lack of resources, may force the pace of decision making. Miller (1977) supports these views. He considers that 'the only rational, feasible criteria . . . are the likelihood of extinction of a breed and the degree to which it possesses unique genes', and concludes that 'the gathering of still more performance data is likely to be a fruitless exercise that will delay undertaking any concrete action'.

Whatever the level and extent of available information, decisions on what to preserve should result from the integration of all relevant information on identifiable groups – whether breeds, strains, populations or genetic stocks – which are being considered. Such decisions involve a judgement of the presumptive value, and often a process of selection among stocks related by origin, adaptation, genetic constitution, etc. The nature of the information which will be required is summarized in Table 10.2. It should include the presumed function or functions in animal husbandry or breeding (Table 10.2, a), the current condition and location (b and c), the type of use (d) and any specific characteristics of the material (e). The notes which follow will further explain the table and the way in which its contents can assist in developing conservation strategies.

(a) Function of genetic stock(s). The actual, or anticipated use and function of a breed or stock considered for conservation should be specified, as a basis

TABLE 10.2 *Information required for the formulation of conservation strategies*

(a) Function of genetic stock (breed, strain, population, etc.): continuing use; donor of specific characteristics or genes; breed crosses (F_1); parent for new breeds; research

(b) Population status: increasing; steady; declining; rare; near-extinct; introgressed; oligo- or monotypic[a]

(c) Distribution (geographical, ecological): global; regional ecological range; ecological specialization

(d) Type of use: milk; meat; draught; eggs; wool etc. and combinations

(e) Specific characteristics: morphological; physiological; adaptations; resistances etc.

---

[a] See pages 258 and 272.

for decisions on conservation. It will also influence the choice of conservation methods.

(b) Population status. Conservation is automatic when numbers are steady, but specific steps are needed if numbers are declining towards the danger point. A danger signal would be a drastic reduction in the number of stud breeders or of government studs of a breed, depending on the organization of the industry. But neither may exist, as is the case in most of the developing countries, or a breed may be confined to few and isolated habitats, as, for example, the highly fecund Dahman sheep which has a small population in an oasis in Morocco; many breeds in Asia and Africa would fall into this category (H. N. Turner, personal communication). In such circumstances, as so often in conservation, action may depend on local, national or international organizations, but in particular on the initiative of individuals who have become alerted to the value and significance of genetic resources.

Rare breeds in specialized ecological niches have some analogies with rare species of plants (section 7.1), which occur 'in widely separated, small subpopulations . . . or [are] restricted to a single population'. They have 'developed the ability to exploit an extreme habitat'. Rare breeds of livestock occur in isolated areas and/or in extreme environments to which they are adapted, such as high mountains or deserts. Adaptations which such breeds have acquired over long periods of selection may be of great interest for research and of use in constructive breeding (see Mason, 1975a). Under modern conditions they tend to be progressively endangered, and their continued survival should be closely watched.

(c) Distribution. In some livestock species breeds have evolved which have a near-global distribution. This is the case with breeds which are maintained at high levels of nutrition and protection, such as the American poultry breeds which dominate intensive poultry production (see section 10.2.1), or with Friesian dairy cattle, and could be the case, under the obverse conditions of climatic and nutritional stress, with the Australian merino sheep. Under both intensive and extensive conditions, distinct strains evolve in response to differences in the environment, in breeding objectives or methods, or in market requirements. Such strains are the residual gene pool of the breed, and as such are important for the maintenance of productivity and health, and, in poultry, swine, beef and dairy cattle as well as sheep breeding, of hybrid vigour in breed crosses (section 10.2.1). However, while the maintenance of adequate genetic diversity in a predominant breed is ostensibly the responsibility of breeders, experience has shown that they cannot necessarily be relied upon (see section 10.2.1).

A regional or local distribution may reflect adaptations to ecological conditions, land use patterns, social and cultural systems, or combinations of these. As we saw in section 10.2, many breeds of this kind have been or are being displaced by others with higher economic productivity, though this may not be the case in the country of adoption, for example when Australian sheep are introduced into the wet tropics for which they are not really suited. In some countries of the developing world considerable wealth of traditional strains or breeds is rapidly eroding. In many instances there is a multiplicity of breeds with different though often overlapping distribution and function. The problem arises which breeds should be taken care of. Should it be those which are still popular but which in the near future may be as exposed to erosion of diversity and/or to introgression by imported breeds, as are now the endangered ones? Or those which are now failing, presumably for reasons of lesser efficiency in present ciircumstances? To what extent do local breeds differ, making their continued separate identity worthwhile? Here the comparison of allelic frequencies in polymorphisms could be of some help (see section 10.2.2).

(d) Type of use. In all livestock species there are breeds which are specialized for particular uses. We have encountered chicken breeds excelling in egg production, and others selected as meat breeds. There are beef and milk cattle, fat lamb, mutton, fine wool and carpet wool sheep, milking and angora goats, race horses and draft horses. However, there are multipurpose breeds which perform well in more than one role such as beef or milk and draught, and, of course, some types of livestock perform badly in all directions such as much of the South-East Asian poultry, except for cock fighting. Type of use must be an important component in the selection of breeds which are to be preserved. But it is perhaps salutary to remember that recently dairy cattle such as the Friesian, and even the Jersey, have demonstrated their substantial potential for beef production.

(e) Specific characteristics will assume prominence in conservation strategies oriented towards breed improvement or the production of new breeds. This area has already been discussed, with special reference to disease resistance, in section 10.2.1. In each case the question arises whether a breed (or genetic stock) with a specific useful characteristic is to be used as such, or as source of a gene, or genes, for transfer to other breeds. This will on one hand depend on the general suitability of the stock, on the other on the type of inheritance of the characteristic, whether based on one or a few major genes or on multiple genes, as, presumably, in high fecundity in sheep.

(f) Productivity. We have refrained from giving prominence to productivity, or measured economic performance, beyond the performance level implicit in the relative prominence of a breed. The detailed assessment of productivity, which is the integration of all factors impinging upon the economic return of the animal, is relevant for animal husbandry or breeding. Genetic conservation is concerned with future conditions as much as with present ones, with the potential in recombination as much as with direct use, in short with the genetic content as much as with the organism as a whole. While genetic conservation is oriented towards animal breeding, there is a

difference due to their respective 'time scales of concern' (see Chapter 1). Animal breeding is concerned with adaptation to current or predictable conditions, genetic conservation is equally concerned to safeguard genetic materials for future, hence less predictable, needs.

Conservation strategies. On the basis of information of the kind outlined in the preceding sections and summarized in Table 10.2, it should be possible to develop *conservation policies and procedures* for specific situations. An example may serve to illustrate this. Let us assume that a conservation strategy is to be devised for *Zebu cattle in India.* In 1953, Joshi and Phillips described twenty-eight Zebu breeds in India and Pakistan. The number may now be somewhat larger or smaller. Each recognized breed will be characterized for its expected function as a genetic resource, its population dynamic state, distribution, type of use, and specific characteristics. Some of these are relatively straightforward, such as the type of use – whether milk, draught or dual-purpose. But the current condition of breeds, for example the degree of introgression from exotic breeds, may be hard to ascertain and subject to change. However, a rapid survey of available information would indicate (a) the most popular breeds, (b) the geographical and ecological distribution, and (c) the representation of breeds in larger herds under government or private control.

The results of the survey would identify those breeds which remain reasonably intact and prominent, those which are intact but endangered, and those which are introgressed beyond redemption. The attempt to reconstitute such virtually extinct breeds from mixed herds, according to some kind of ideotype of the breed, seems as futile as it is purposeless. The survey will further identify breeds with broad adaptability as distinct from those adapted to specific environments. In the Indian context one may perhaps anticipate this to lead to three groupings – one adapted to the humid tropics, another to dry tropics, and a third to mountain conditions. If this is the case, we are faced with a 3 × 3 matrix of use categories (milk, draught and dual-purpose) and environments (humid tropics, dry tropics and mountains).

A conservation strategy must include a number term. In the example in question one might visualize a maximum of two breeds in each of the 'ecological' and 'use' categories, to which could be added a small number of breeds with special characteristics, such as resistance to disease, should they not already be included. This results in a formidable number, but it may be found that a much smaller one might adequately represent the essential information contained in Indian cattle breeds once the breed characteristics are closely specified in relation to use and function.

Here we come to the real question of the justification for the conservation of breeds which have similar ecological and evolutionary backgrounds, but which have been separated by differences in environment and use, but perhaps even more by social history and tradition. Modern developments lead to ascendancy for some, obsolescence for others. *Should obsolete breeds be preserved*, and why? While the proposed procedure may be of help in presenting the evidence, it cannot make decisions which must rest on biological, economic and social grounds.

The arguments in favour of the preservation of obsolete breeds have already been discussed in section 10.2. In brief, breeds which are now obsolete may again be wanted in the future; they may contain characteristics or adaptations which may be required for cross-breeding; they may be needed in research; they have historical and social connotations. The first argument is open-ended and, without support from one of the others, points to a representative sample as an alternative to the preservation of *all* obsolete breeds. The second argument is substantive, and the evidence in each case must be considered on its merits; the same evidence will also be decisive on the third. The fourth can only be considered in the social context.

Breeds or genes. In the preceding discussion we have given prominence to the preservation of breeds or strains as the operative units in animal husbandry. But, as we have recognized, breeds also represent gene pools assembled in long-term selection, hence potentially valuable resources for constructive breeding. In this context it is the genetic materials they contain rather than the breeds as such which are the object of conservation. This distinction is important when considering the alternatives of conserving individual breeds *versus* gene pools of grouped breeds. This question is discussed more fully in the section which follows, but here it is relevant to observe that if the maintenance of representative or prominent breeds is regarded as justified in the face of gradual displacement and/or introgression, this is neither practicable nor necessary for all the endangered or obsolete breeds, many of which have no more than local significance. If the genetic information they contain is deemed to be worth preserving, pooling of such breeds seems the only practical alternative to their rapid extinction.

General conservation strategy. From this discussion the following general strategy of genetic conservation emerges:
(a) *conservation of one or more significant breeds* representative of the principal ecological conditions and uses,
(b) conservation of *recognized or likely sources of adaptations or other characteristics* which may be of use in breeding or research. This may lead to the preservation of breeds, or of breed composites (or gene pools) as discussed in the following section. The conservation of *genetic resistance to diseases and parasites* in breeds or gene pools should be located where the pathogen or parasite is endemic.

Monotypic livestock? In concluding the discussion of conservation strategies, we must consider to what extent the conservation of genetic resources is likely to be put to productive use. So far we have explored the biological opportunities for, and restraints on the use of genetic resources, but it remains to discuss the relevant economic and social factors which can be recognized in the light of current trends. Indeed, this examination would be unrealistic without a consideration of the impact that the commercial organization of the industry makes on the structure of existing breeds, resulting in drastically reducing the genetic diversity between and within existing breeds in a growing number of livestock species (see section 10.2.1). J. M. Rendel (personal communication) comments: 'It is the trend in all livestock for a very few studs to provide all breeding stock; bulls in cattle, day old chicks in poultry, rams in sheep. The tendency is for production livestock to have the same genetic source all over the world . . . For reasons of finance, advertising and commercial drive in addition to excellence in performance, American egg laying poultry dominates the world.' This situation is spreading to developing countries wherever intensive production methods are introduced. It can also be expected to extend to new breeds should any be able to become established.

As a counter to the commercial – hence biological – monopolies which stifle genetic diversity, Rendel suggests that ways should be explored 'to limit by legal measures the range of a person or company providing breeding stock', thus ensuring that local differentiation is maintained, say, at the level of a state in terms of the USA, and correspondingly elsewhere. In addition to countering the drive towards oligotypic or even monotypic breeds, this would give a chance for the emergence of new breeds which under the conditions which are now evolving would have little chance of penetrating the protective screens of highly effective commercial networks.

10.4 Conservation methods

Following Maijala (1964), three main approaches are discussed. They are:
(1) maintenance of pure breeds or strains,
(2) establishment of one or several gene pools,
(3) banks of frozen semen, embryos, or gonadal tissues.

The *maintenance of pure breeds* has the advantage of making animals with known pedigree and performance records readily available for observation and measurement, for research, and for breeding purposes, including (F_1) breed crosses. In less developed countries, as Mason (1974) points out, it may be necessary to bring local breeds into a national breeding plan. Presumably this would result in establishing superior yet typical strains.

In breed conservation, foremost attention must go to the preservation of genetic diversity; this has already been discussed. J. M. Rendel (personal communication) suggests that the best way for preserving genetic variation may be to rationalize breeding practices. Artificial insemination (AI) centres 'should be partial isolates, breeding for local adaptation and introducing only a small percentage of stock from outside'. Obviously there must be a number of conservation centres for any breed, preferably representing ecological variation within the breed area. The scope and the need may be less in the case of rare breeds in isolated areas.

By comparison with the conservation of individual breeds or strains, *gene pools* consisting of a number of breeds are simpler and cheaper to maintain. As we have seen, they have special significance in the preservation of endangered and obsolete breeds (section 10.3), but there are biological reasons for a more general application of breed pooling in conservation procedures. F. H. W. Morley (personal communication), in discussing the choice between many small and few large gene pools, favours the latter, for the following reasons:

(1) the probability of loss of genes by random drift, especially of those genes with heterozygous advantage, would be reduced,

(2) the pool could be added to as unrepresented types seemed to be endangered. If a gene pool of 2000 breeders existed, the inclusion of 200 females from an endangered group would have a fair chance of incorporating important genes for preservation, without necessarily causing major changes in gene frequencies in the gene pool,

(3) avoidance of inbreeding would reduce problems of management as well as random losses,

(4) the effective size of the small groups would be difficult to maintain because of practical problems restricting numbers of male parents,

(5) new combinations of genes could arise. Some might be recognized and be of value in the development of new breeds. There seems to be no good reason why the development of new breeds should continue to be largely from selection among the offspring of the crossing of only two breeds.

Pooling of related breeds has the advantage of reducing the number of conservation units, yet preserving to a considerable extent the genetic information they possess. To what extent this would be the case should breeds which diverge in adaptation or function be combined in one pool, is open to question. One may visualize pooling separately the milk, draught and dual-purpose breeds among the twenty-eight Zebu cattle breeds which Joshi and Phillips listed in 1953 (see p. 270). But it may be doubted whether in jointly pooling *all* breeds the specific adaptations of the three groups might not be dissipated. In essence the argument against global mass reservoirs of plant cultivars (see section 9.2) applies, though to a lesser degree: in a gene pool of all Indian Zebu breeds there would be a better prospect of preserving the integrity of environmental adaptations than there would be in a global mass reservoir.

The population size of gene pools must receive attention. (For a general discussion of population size see section 3.1.) Morley (personal communication) suggests that the following should be considered:

(a)   Conservation of rare but unrecognized genes. If their frequency were, say, p (< 0.001) then populations of $4/p = (4000)$ or more would be necessary to give a reasonable chance of including the gene in the sample. Even there the probability of extinction by chance, especially if aided by some selection, is alarmingly high. Genes of such low frequency are of little importance and are being maintained by mutation.

By comparison a sample of 100, and a little care in arranging random selection, would be sufficient for unrecognized genes with $p \simeq 0.1$, a sample of, say, 1000 for genes with $p \simeq 0.01$, combined with considerable care to counteract natural selection.

(b)   Conservation of rare but recognized genes should not present major difficulties. If they are recognized and rare are they likely to be important?

(c)   Reduction of inbreeding (and random drift) to levels of say less than $\Delta F/dt < 0.001$ per generation, where F is Wright's inbreeding coefficient, would require effective populations of 250 males per generation. Perhaps a $\Delta F/dt$ of 0.01 would be acceptable in which case at least 25 males per generation would be needed, assuming that breeding is more or less at random (Chapter 3).

Morley concludes that populations of 2000 each at a minimum of two sites, with random selection of breeding stock, would be adequate for the conservation of gene pools. Rendel comments that if the number in each sex is kept high, 500 of each would be sufficient. A herd of this size could hold on to its relatively rare, neutral genes for many generations. Here, the devil's advocate might argue that much smaller herds would suffice if we opted to ignore neutral genes because natural selection would aid in preventing beneficial genes from becoming lost. Superficially, this criticism makes sense, until one considers that many genes might be selectively neutral in some environments but not in others. For example, genes for disease resistance would be neutral or even deleterious in the absence of the disease, but would be strongly favoured during an epidemic. The problem is that in small herds such genes would have a high rate of loss because of genetic drift.

Breeds or strains retained for *research* purposes, or obsolete breeds retained for *cultural or aesthetic* reasons cannot be subjected to pooling. The former are the concern of research institutions or zoological gardens, the latter of breed societies, organizations such as the Rare Breeds Survival Trust in Britain, or of individual supporters of breed conservation.

The *storage of semen* at low temperatures ($-196\ °C$) is well established for cattle, but for sheep it has not been equally successful, and in pigs it is still problematical. What is still in doubt, even in cattle, is the length of the fertile life of spermatozoa under storage. Some observers report a drop in fertility after two years' storage, others that the storage life of bull and ram semen can be extended for up to five years without an effect of ageing. Yet Salisbury and Hart (1975) in a review of the functional integrity of sper-matozoa after storage, draw 'the inescapable conclusion that spermatozoan senescence is a universal phenomenon which decreases the fertility of the spermatozoa and increases the mortality in embryos fertilized by aged sperms'. The authors further discuss the effect of ageing on the sper-matozoan genome and report on experiments which examine the suggestion that senescence could result in an alteration in the deoxyribonuclear protein complex (DNP) which is translated into aborted development. So far the evidence has failed to establish a change in DNA in the mammalian gamete during senescence but the authors point to the well-established evidence of the mutagenic effect of senescence on spermatozoa in *Drosophila melanogaster* and conclude that much more research is needed before very long-term storage of semen can be regarded as entirely safe.

Egg transfer is an essential condition for the *storage of fertilized eggs*, and superovulation – multiple ovulation by a donor animal – is useful for increas-ing available numbers for storage and for evaluation. Techniques for superovulation and transfer are well developed in cattle, sheep and pigs. In cattle, up to eight ova have been obtained from one donor in one season, with fertility rates on transfer of up to 70–90% (for techniques and results, see review paper by Gordon, 1975). Eggs of rabbits, pigs, sheep and cattle have been transported over large distances. This should be greatly facilitated by freezing techniques which are rapidly being developed, and there are 'hopes that cattle embryos may soon be shipped around the world in flasks of liquid nitrogen in much the same way as bull semen'. Gordon (1975) points out that this may cause quarantine requirements to be modified in line with those now applying to semen.

Low-temperature storage of mammalian embryos has also made good progress in recent years. It was found that embryos are extremely sensitive to the rates of freezing and thawing which have to be relatively slow. Further, species differ in the stages at which storage was successful, for example, in the mouse 1-cell, 2-cell, 8-cell and blastocyst, in cattle blasto-cyst, in sheep morula and blastocyst stages. Storage periods of up to one year have been successful (Whittingham, 1975). These results justify the hope that low-temperature storage of embryos will be developed to the stage of reliability reached now with storage of semen. The open question is still the extent of the storage period. For the long-term conservation of genetic resources a period of even twenty years would lead to notable economies in conservation. Indeed, 'semen preservation, associated with egg storage and simple techniques for egg transplantation, could provide the cheapest way of maintaining animal populations over the years' (FAO, 1967). It would also greatly extend the scope for genetic conservation of livestock.

10.5 Conclusions and summary

(1) Genetic resources of plants and animals have many analogies in their evolutionary history, and in the formation of local primitive breeds and their recent erosion. Wild progenitors of most livestock species no longer exist, the main exception being sheep and poultry. The cost of conservation of animal breeds, compared with cultivars, necessitates a more selective attitude to conservation of animal than of plants (section 10.1).

(2) Selection for increased productivity is inevitably associated with reduced genetic diversity both between and within breeds. This has led to many local breeds becoming rare and endangered and to calls for their preservation, because they may again become useful (as some of them have in recent years), and, more importantly, because they may be of value as sources of resistance to diseases and parasites, or of other characteristics such as high fecundity. Breeds adapted to specific environments may be useful as material for breeding and for research. Historical breeds are part of man's heritage (section 10.2.1).

(3) Conservation strategies provide guidelines for the selection of breeds, strains, genetic stocks, etc. which are to be preserved. A conservation strategy can be formulated using the following elements of information (see Table 10.2): the function of a breed in animal husbandry or breeding; the current condition of the breed and its distribution; the type of use; and any specific characteristics it may possess. On the basis of this information for the breeds of a region it should be possible to formulate a strategy for the region as a whole and to make specific decisions relating to breeds, strains etc. This is illustrated using Indian Zebu cattle as an example (pp. 265–70).

(4) A basic strategy of conservation is proposed, providing for (a) the conservation of one or more significant breeds representative of the principal ecological conditions and uses, (b) the conservation of sources of adaptations or other characteristics of potential use in breeding or research, including obsolete breeds so far as they satisfy above qualifications. Significant or prominent breeds or strains [(a) above] are conserved in use, though they may have to be protected from introgression, especially by exotic breeds. However, breeds or stocks in category (b) will require specific measures for their preservation (pp. 270–2).

(5) The worldwide tendency of livestock species to become oligotypic or monotypic as a result of commercial concentration in the production of

breeding stocks, threatens the remaining diversity within and between breeds. It could possibly be countered by legal measures to regionalize the production and sale of breeding stocks (p. 272).

(6) While breeds preserved for current and future use must be preserved as such, the question arises whether breeds (or other stocks) in category (b) are to be preserved as distinct entities, or alternatively as components of gene pools. Pooling of related breeds, while reducing the number of conservation units, can be expected to preserve the genetic information they contain. To what extent this may be dissipated should *all* breeds of a region be pooled without discrimination, is open to question. Arguments in favour of gene pools are that constructive breeding is not concerned with breeds as such but with the genes or gene combinations they contain, that gene pools are cheaper and simpler to maintain, and under appropriate conditions of population management should hold gene frequencies at adequate levels (pp. 272–4).

(7) While low-temperature storage, at $-196\,°C$, of cattle semen is a routine operation for shorter periods, long-term storage is still problematical, as it is for other species. Storage of fertilized eggs has been successful for periods up to one year. Progress in these areas would greatly extend the scope for genetic conservation of livestock (pp. 274–6).

References

Abdalla, F. H. and Roberts, E. H. (1968). Effects of temperature, moisture and oxygen on the induction of chromosome damage in seeds of barley, broad beans, and peas during storage. *Ann. Bot.* **32,** 119–36.

Abdalla, F. H. and Roberts, E. H. (1969). The effects of temperature and moisture on the induction of genetic changes in seeds of barley, broad beans, and peas during storage. *Ann. Bot.* **33,** 153–67.

Abdou, Y. A. M., Gregory, W. C. and Cooper, W. E. (1974). Sources and nature of resistance to *Cercospora arachidicola* Hori and *Cercosporidium personatum* (Beck and Curtis) Deighton in *Arachis* species. *Peanut Sci.* **1,** 6–11.

Abplanalp, A. A. (1974). Inbreeding as a tool for poultry improvement. In *First World Congress on Genetics Applied to Livestock Production*, pp. 897–908. Graficas Orbe. Madrid, Spain.

Ajayi, S. (1975). *Domestication of the African Giant Rat* (Cricetomys gombianu, *Waterhouse*), Department of Forest Resource Management. University of Ibadan, Nigeria.

Akihama, T. and Nakajima, K. (eds.) (1978). *Long Term Preservation of Favourable Germ Plasm in Arboreal Crops*. Fruit Tree Research Station, M.A.F. Fujimoto, Japan.

Akihama, T., Omura, M. and Kozaki, I. (1978). Further investigation of freeze-drying for deciduous fruit tree pollen. In *Long Term Preservation of Favourable Germ Plasm in Arboreal Crops*, ed. T. Akihama and K. Nakajima, pp. 1–7. Fruit Tree Research Station, M.A.F. Fujimoto, Japan.

Albada, M. van (1964). *Conservation of a Gene Pool. The Situation in the Netherlands.* World Poultry Association, Sezione Italiana, 2ª conferenza avicola europea, pp. 409–27. Academia nazionale di agricoltura, Bologna.

Allard, R. W. (1970a). Population structure and sampling methods. In *Genetic Resources in Plants – Their Exploration and Conservation,* ed. O. H. Frankel and E. Bennett, *IBP Handbook No. 11*, pp. 97–107. Blackwell Scientific Publications, Oxford.

Allard, R. W. (1970b). Problems of maintenance. In *Genetic Resources in Plants – Their Exploration and Conservation,* ed. O. H. Frankel and E. Bennett, *IBP Handbook No. 11,* pp. 491–4. Blackwell Scientific Publications, Oxford.

Allard, R. W., Babbel, G. R., Clegg, M. T. and Kahler, A. L. (1972). Evidence for coadaptation in *Avena barbata. Proc. Natl. Acad. Sci. USA* **69,** 3043–8.

Allard, R. W. and Hansche, P. E. (1964). Some parameters of population variability and their implications in plant breeding. *Adv. Agron.* **16**, 281–325.

Allard, R. W., Kahler, A. L. and Weir, B. S. (1971). Germ plasm sources and world collections of genes. In *Barley Genetics*, vol. II, ed. R. A. Nilan, pp. 1–13, Proceedings of the 2nd International Barley Genetics Symposium, 1969. Washington State University Press, Pullman, Washington.

Allard, R. W., Kahler, A. L. and Weir, B. S. (1972). The effect of selection on esterase allozymes in a barley population. *Genetics* **72**, 489–503.

Andrews, D. J. (1970). Breeding and testing dwarf sorghums in Nigeria. *Expl. Agric.* **6**, 41–50.

Anon. (1972). World EQ index. *Int. Wildl.* July/August 1972, 21–36.

Anon. (1977). Inbreeding and behavior. *Oryx* **14**, 113.

Anon. (1978). Anthrax in Namibia. *Oryx* **14**, 309.

Anon. (1979). Washington symposium on wildlife regulations. *Syst. Zool.* **28**, 119–121.

Antonovics, J., Bradshaw, A. D. and Turner, R. G. (1971). Heavy metal tolerance in plants. *Advanc. Ecol. Res.* **7**, 2–58.

Armstrong, R. A. and McGehee, R. (1976). Coexistence of species competing for shared resources. *Theor. Pop. Biol.* **9**, 317–25.

Aschmann, H. (1959). *The Central Desert of Baja California: Demography and Ecology.* Manessier Publishing Co. Riverside, California.

Ashri, A. (1971a). Evaluation of the world collection of safflower, *Carthamus tinctorius* L. I. Reaction to several diseases and associations with morphological characters in Israel. *Crop Sci* **11**, 253–7.

Ashri, A. (1971b). Evaluation of the world collection of safflower, *Carthamus tinctorius* L. II. Resistance to the safflower fly, *Acantophilus helianthi* R. *Euphytica* **20**, 410–15.

Ashri, A. (1973). *Divergence and evolution in the safflower genus* Carthamus L. Final research report, P. L. 480. USDA.

Ashri, A. (1975). Evaluation of the germplasm collection of safflower, *Carthamus tinctorius* L. V. Distribution and regional divergence for morphological characters. *Euphytica* **24**, 651–9.

Ayala, F. J. (1965). Relative fitness of populations of *Drosophila serrata* and *Drosophila birchii*. *Genetics* **51**, 527–44.

Ayala, F. J. (1969). Evolution of fitness. V. Rate of evolution in irradiated populations of *Drosophila*. *Proc. Natl. Acad. Sci. USA* **63**, 790–3.

Ayala, F. J., Tracey, M. L., Barr, L. G. and Ehrenfeld, J. G. (1974), Genetic and reproductive differentiation of the subspecies *Drosophila equinoxialis caribbensis*. *Evolution* **28**, 24–41.

Ayala, F. J. and Valentine, J. W. (1974). Genetic variability in a cosmopolitan deep-water ophiuran, *Ophiomusium lymani*. *Mar. Biol.* **27**, 51–7.

Ayensu, E. S. (1976). International co-operation among conservation-oriented botanical gardens and institutions. In *Conservation of Threatened Plants*, ed. J. B. Simmons, R. I. Beyer, P. E. Brandham, G. Ll. Lucas and V. T. H. Parry, pp. 259–69. Plenum Press, New York.

Ayensu, E. S. and De Fillips, R. A. (1978). *Endangered and Threatened Plants of the United States.* Smithsonian Institution Press, Washington DC.

Bader, R. S. (1965). Heritability of dental characters in the house mouse. *Evolution* **19**, 378–84.

Bajaj, Y.P.S. (1976). Regeneration of plants from cell suspensions frozen at −20, −70 and −196 °C. *Physiol. Plant* **37**, 263–8.

Bajaj, Y. P. S. and Reinert, J. (1977). Cryobiology of plant cell cultures and establishment of gene banks. In *Applied and Fundamental Aspects of Plant Cell, Tissue and Organ Culture*, ed. J. Reinert and Y. P. S. Bajaj, pp. 757–89. Springer-Verlag, Berlin.

Barnett, S. A. and Stoddart, R. C. (1969). Effects of breeding in captivity on conflict among wild rats. *J. Mammalogy* **50**, 321–5.

Basilevsky, P. *et al.* (1972). *La Faune Terrestre de l'Isle de Sainte Helène*. Deuxième Partie, II. Insectes 9, Coleoptera. *Ann. Mus. Royale Afrique Centrale, Série Octavo Zoologie*, No. 192.

Bateson, P. (1978). Sexual imprinting and optimal outbreeding. *Nature* **273**, 659–60.

Batten, P. (1976), *Living Trophies*. The Crowell Co., New York.

Bawa, K. S. (1974). Breeding systems of tree species of a lowland tropical community. *Evolution* **28**, 85–92.

Belyaev, D. K. (1969). Domestication of animals. *Sci. J.* January 1969, 47–52.

Belyaev, D. K., Volobuev, V. T., Radjabli, S. I. and Trut, L. N. (1974). Investigation of the nature and the role of additional chromosomes in silver fox. II. Additional chromosomes and breeding of animals for behaviour. *Genetika* (Moscow) **10**, 83–91.

Benado, M. B. F., Ayala, F. J. and Green, M. M. (1976). Evolution of experimental "mutator" populations of *Drosophila melanogaster*. *Genetics* **82**, 43–52.

Benirschke, K., Lasley, B. and Ryder, O. (1980). The technology of captive propagation. In *Conservation Biology: An Evolutionary–Ecological Perspective*, ed. M. E. Soulé and B. A Wilcox, pp. 225–42. Sinauer Associates, Sunderland, Mass.

Bennett, E. (ed.) (1968). *Record of the FAO/IBP Technical Conference on the Exploration, Utilization and Conservation of Plant Genetic Resources*, 1967. FAO, Rome.

Bennett, E. (1970). Tactics of Plant Exploration. In *Genetic Resources in Plants – Their Exploration and Conservation*, ed. O. H. Frankel and E. Bennett, *IBP Handbook No. 11*, pp. 157–79. Blackwell Scientific Publications, Oxford.

Berger, E. (1976). Heterosis and the maintenance of enzyme polymorphism. *Amer. Natur.* **110**, 823–39.

Bermant, G. and Lindburg, D. G. (1975). Introduction and overview. In *Primate Utilization and Conservation*, ed. G. Bermant and D. G. Lindburg, pp. 1–4. John Wiley and Sons, New York.

Berndt, R. and Sternberg, H. (1969). Über Begriffe, Ursachen und Auswirkungen der Dispersion bei Vögeln. [On the forms, origins and efforts of dispersion in birds.] *Vogelwelt* **90**, 41–52.

Berry, R. J. (1971). Conservation aspects of the genetical constitution of populations. In *The Scientific Management of Animal and Plant Communities for Conservation*, ed. E. Duffy and A. S. Watt, pp. 177–206, *Symposium of the British Ecological Society, 11*. Blackwell Scientific Publications, Oxford, London.

Berwick, S. (1976). The Gir forest: endangered ecosystem. *Am. Sci.* **64**, 28–41.

Bindon, B. M. and Piper, L. R. (1976). Assessment of new and traditional techniques of selection for reproduction rate. In *Proceedings of the International Sheep Breeding Congress, Muresk, 1975*, ed. G. J. Tomes, pp. 357–71.

Bingham, R. T., Hoff, R. J. and McDonald, G. I. (1971). Disease resistance in forest trees. *Ann. Rev. Phytopathol.,* **9**, 433–52.

Black, F. L. (1975). Infectious diseases in primitive societies. *Science* **187**, 515–518.

Bock, W. J. (1970). Microevolutionary sequences as a fundamental concept in macroevolutionary models. *Evolution* **24**, 704–22.

Bodmer, W. F. and Cavalli-Sforza, L. L. (1976). *Genetics, Evolution and Man.* W. H. Freeman and Co., San Francisco.

Bonnell, M. L. and Selander, R. K. (1974). Elephant seals: genetic variation and near extinction. *Science* **184**, 908–9.

Bossert, W. (1967). Mathematical optimization: Are there abstract limits to natural selection? In *Mathematical Challenges to the Neo-Darwinian Interpretation of Evolution*, ed. P. S. Moorhead and M. M. Kaplan, pp. 35–46, *The Wistar Institute Symposium Monograph No. 5*. Wistar Institute Press, Philadelphia.

Bouman, J. (1977). The future of Przewalski horses in captivity, *Inter. Zoo Yearb.* **17**, 62–8.

Bowman, J. C. and Falconer, D. S. (1960). Inbreeding depression and heterosis of litter size in mice. *Genet. Res. Camb.* **1**, 262–74.

Brambell, M. R. (1977). Reintroduction. *Inter. Zoo Yearb.* **17**, 112–16.

Bretsky, P. W. and Lorenz, D. M. (1970). An essay on genetic-adaptive strategies and mass extinction. *Geol. Soc. Amer. Bull.* **81**, 2449–56.

Brock, R. D. (1971). The role of induced mutations in plant improvement. *Radiat. Bot.* **11**, 181–96.

Brown, A., Nevo, E. and Zohary, D. (1977). Association of alleles at esterase loci in wild barley *Hordeum spontaneum*. *Nature* **268**, 430–1.

Brown, A. H. D. (1978). Isozymes, plant population genetic structure and genetic conservation. *Theor. Appl. Genet.* **52**, 145–57.

Brown, J. H. (1971). Mammals on mountaintops: nonequilibrium insular biogeography. *Amer. Nat.* **105**, 467–78.

Brown, J. H. and Kodric-Brown, A. (1977). Turnover rates in insular biogeography: effect of immigration on extinction. *Ecology* **58**, 445–9.

Bruce, A. B. (1910). The mendelian theory of heredity and the augmentation of rigor. *Science,* **32**, 627–8.

Brücher, H. (1968a). Südamerika als Herkunftsraum von Nutzpflanzen. In *Biogeography and Ecology in South America* vol. 1, pp. 251–301.

Brücher, H. (1968b). Die genetischen Reserven Südamerikas für die Kulturpflanzenzüchtung. *Theor. Appl. Genet.* **38**, 9–22.

Bunting, A. H. and Curtis, D. L. (1968). Local adaptation of sorghum varieties in northern Nigeria. In *Agroclimatological Methods. Proceedings of the Reading Symposium*, pp. 101–16. UNESCO, Paris.

Burt, R. L., Edye, L. A., Williams, W. T., Grof, B. and Nicholson, C. H. L. (1971). Numerical analysis of variation patterns in the genus *Stylosanthes* as an aid to plant introduction and assessment. *Aust. J. Agric. Res.* **22**, 737–57.

Burt, R. L., Reid, R. and Williams, W. T. (1976). Exploration for, and utilization of, collections of tropical pasture legumes. I. The relationship between agronomic performance and climate of origin of introduced *Stylosanthes* spp. *Agro-Ecosystems* **2**, 293–307.

Burt, R. L. and Williams, W. T. (1980). Strategy of evaluation of a collection of tropical herbaceous legumes from Brazil and Venezuela. III. The use of ordination techniques in evaluation. *Agro-Ecosystems* (in press).

Camin, J. H. and Ehrlich, P. R. (1958). Natural selection in water snakes (*Natrix sipedon* L.) on islands in Lake Erie. *Evolution* **12**, 504–11.

Campbell, S. (1978). *Lifeboats to Ararat*. Times Books, New York.

Campbell, S. (1980). Is reintroduction a realistic goal? In *Conservation Biology: An Evolutionary–Ecological Perspective* ed. M. E. Soulé and B. A. Wilcox, pp. 263–70. Sinauer Associates, Sunderland, Mass.

Carlquist, S. (1965). *Island Life, a Natural History of the Islands of the World*. The Natural History Press, New York.

Carlson, P. S., Smith, H. H. and Dearing, R. D. (1972). Parasexual interspecific plant hybridization. *Proc. Natl. Acad. Sci USA* **69**, 2292–4.

Carson, H. L. (1955). The genetic characteristics of marginal populations of *Drosophila*. *Cold Spring Harbor Symp. Quant. Biol.* **20**, 276–87.

Carson, H. L. (1964). Population size and genetic load in irradiated populations of *Drosophila melanogaster*. *Genetics* **49**, 521–8.

Case, T. O. and Gilpin, M. E. (1974). Interference competition and niche theory. *Proc. Natl. Acad. Sci. USA* **71**, 3073–7.

Cavalli-Sforza, L. L. and Bodmer, W. F. (1971). *The Genetics of Human Populations*. W. H. Freeman and Co., San Francisco.

Chaisson, R. E., Serunian, L. A. and Schopf, T. J. M. (1976). Allozyme variation between two marshes and possible heterozygote superiority within a marsh in the bivalve *Modiolus demissus, Biol. Bull.* **151**, 404.

Chang, T. T. (1976). The rice cultures. *Phil. Trans. R. Soc. Lond.* **B275**, 143–57.

Chang, T. T., Ou, S. H., Pathak, M.D., Ling, K. C. and Kauffman, H. E. (1975). The search for insect and disease resistance in rice germplasm. In *Crop Genetic Resources for Today and Tomorrow*, ed. O. H. Frankel and J. G. Hawkes, *International Biological Programme 2*, pp. 183–200. Cambridge University Press.

Chang, T. T., Sharma, S. D., Adair, C. R. and Perez, A. T. (1972). *Manual for Field Collectors of Rice*. IRRI, Los Baños.

Chasen, F. N. (1940). A handlist of Malaysian mammals. *Bull. Raffles. Mus.* **15**, 1–209.

Chen, K., Gray, J. C. and Wildman, S. G. (1975). Fraction I protein and the origin of polyploid wheats. *Science* **190**, 1304–6.

Cicmanec, J. C. and Campbell, A. K. (1977). Breeding the owl monkey (*Aotus trivirgatus*) in a laboratory environment, *Lab. Anim. Sci.* **27**, 512–17.

Clarke, C. H. D. (1954). The bob-white quail in Ontario. *Tech. Bull. Fish & Wildl. Ser. No. 2*. Ontario Dept. Lands and Forests, Ontario Ministry of Natural Resources, Toronto, Ontario.

Clayton, G. A., Knight, G. R., Morris, J. A. and Robertson, A. (1957). An experimental check on quantitative genetical theory: III. Correlated responses. *J. Genet.* **55**, 171–80.

Clegg, M. T. and Allard, R. W. (1972). Patterns of genetic differentiation in the slender wild oat species *Avena barbata*. *Proc. Natl. Acad. Sci. USA* **69**, 1820–4.

Clegg, M. T., Allard, R. W. and Kahler, A. L. (1972). Is the gene the unit of selection? Evidence from two experimental plant populations. *Proc. Natl. Acad. Sci. USA* **69**, 2474–8.

Coe, M. (1980). African wildlife resources. In *Conservation Biology: An Evolutionary–Ecological Perspective*, ed. M. E. Soulé and B. A. Wilcox, pp. 273–302. Sinauer Associates, Sunderland, Mass.

Colbert, E. H. (1951). *The Dinosaur Book: the Ruling Reptiles and Their Relatives.* McGraw-Hill, New York.

Colditz, P. J. and Kellaway, R. C. (1972). The effect of diet and heat stress on feed intake, growth, and nitrogen metabolism in Friesian, F_1 Brahman × Friesian, and Brahman heifers. *Aust. J. Agric. Res.* **23**, 717–25.

Cole, R. F. and Hutt, F. B. (1973). Selection and heterosis in Cornell White Leghorns: a review, with special consideration of interstrain hybrids. *Anim. Breed. Abstr.* **41**, 103–18.

Comstock, R. E. (1977). Quantitative genetics and design of breeding programs. In *Proceedings of the International Conference on Quantitative Genetics*, ed. E. Pollack, O. Kempthorne and T. B. Bailey, pp. 705–18. Iowa State University Press, Ames.

Conway, W. G. (1980). An overview of captive propagation. In *Conservation Biology: An Evolutionary–Ecological Perspective*, ed. M. E. Soulé and B. A. Wilcox, pp. 199–208. Sinauer Associates, Sunderland, Mass.

Converse, P. J. and Williams, D. R. R. (1978). Increased HLA-B heterozygosity with age. *Tissue Antigens,* **12**, 275–8.

Coon, C. S., Garn, S. M. and Birdsell, J. B. (1950). *Races, a Study of the Problems of Race Formation in Man.* Charles C. Thomas, Springfield, Illinois.

Cousins, D. (1978). Gorillas – A Survey. *Oryx* **14**, 374–6.

Coyne, J. A. (1976). Lack of genic similarity between two sibling species of *Drosophila* as revealed by varied techniques. *Genetics* **84**, 593–607.

Crow, J. F. (1970). Genetic loads and the cost of natural selection. In *Biomathematics: 1, Mathematical Topics in Population Genetics*, ed. K. Kojima, pp. 128–77. Springer-Verlag, Berlin.

Crow, J. F. and Kimura, M. (1970). *An Introduction to Population Genetics Theory.* Harper and Row, New York.

Dale, W. L. (1959). The rainfall of Malaya I. *J. Trop. Geogr.* **13**, 23–37.

D'Amato, F. (1975). The problem of genetic stability in plant tissue and cell cultures. In *Crop Genetic Resources for Today and Tomorrow*, ed. O. H. Frankel and J. G. Hawkes, *International Biological Programme 2*, pp. 333–48. Cambridge University Press.

Dana, T. F. (1975). Development of contemporary Eastern Pacific coral reefs. *Mar. Biol.* **33**, 355–74.

Darlington, C. D. (1969). The silent millennia in the origin of agriculture. In *The Domestication and Exploitation of Plants*, ed. P. J. Ucko and G. W. Dimbleby, pp. 67–72. Duckworth, London.

Darlington, C. D. and Mather, K. (1949). *The Elements of Genetics.* Allen and Unwin, London.

Darwin, C. (1859). *The Origin of Species.* John Murray, London.

Davies, J. G. (1960). Pasture and forage legumes for the dry sub-tropics and tropics of Australia. In *Proceedings of the VIII International Grassland Congress*, ed. C. L. Skidmore, pp. 381–5. Reading.

Dempster, J. P. (1977). The scientific basis of practical conservation: factors limiting the persistence of populations and communities of animals and plants. *Proc. R. Soc. Lond.* **B197**, 69–76.

Denniston, C. D. (1978). Small population size and genetic diversity: implications for endangered species. In *Endangered Birds: Management Techniques for Preserving Threatened Species*, ed. S. A. Temple, pp. 281–9. University of Wisconsin Press, Madison.

de Wet, J. M. J., Harlan, J. R., Stalker, J. T. and Randrianasolo, A. V. (1978). The origin of tripsacoid maize (*Zea mays L.*). *Evolution* **32**, 233–44.

Diamond, J. M. (1972). Biogeographic kinetics: estimation of relaxation times for avifaunas of southwest Pacific islands. *Proc. Natl. Acad. Sci. USA* **69**, 3199–203.

Diamond, J. M. (1975a). Assembly of species communities. In *Ecology and Evolution of Communities*, ed. M. L. Cody and J. M. Diamond, pp. 324–444. Belknap Press, Cambridge, Mass.

Diamond, J. M. (1975b). The island dilemma: lessons of modern biogeographic studies for the design of natural reserves. *Biol. Conserv.* **7**, 129–46.

Diamond, J. M. (1976). Island biogeography and conservation: strategy and limitations. *Science* **193**, 1027–9.

Diamond, J. M. and May, R. M. (1976). Island biogeography and the design of natural reserves. In *Theoretical Ecology: Principles and Applications*, ed. R. M. May, pp. 163–86. Blackwell Scientific Publications, Oxford.

Dickerson, G. E., Blunn, C. T., Chapman, A. B., Kottman, R. M., Krider, J. L., Warwick, E. J. and Whatley, J. A., Jr., in collaboration with Baker, M. L., Lush, J. L. and Winters, L. M. (1954). Evaluation of selection in developing inbred lines of swine. *Univ. Missouri Coll. Agr. Res. Bull.* **551.**

Dinoor, A. (1975). Evaluation of sources of disease resistance. In *Crop Genetic Resources for Today and Tomorrow*, ed. O. H. Frankel and J. G. Hawkes, *International Biological Programme 2*, pp. 201–10. Cambridge University Press.

Dinoor, A. (1976). Germplasm and the phytopathologist. *Plant Genetic Resources* **32**, 36–8.

Dobzhansky, Th. (1967). *The Biology of Ultimate Concern: Perspectives in Humanism*, ed. R. N. Ansben. The New American Library, New York.

Dobzhansky, Th. (1970). *Genetics of the Evolutionary Process*. Columbia University Press, New York.

Dodson, C. H. (1975). Coevolution of orchids and bees. In *Coevolution of Animals and Plants*, ed. L. E. Gilbert and P. H. Raven, pp. 91–9. University of Texas Press, Austin.

Dorofeev, V. F. (1975). Evaluation of material for frost and drought resistance in wheat breeding. In *Crop Genetic Resources for Today and Tomorrow*, ed. O. H. Frankel and J. G. Hawkes, *International Biological Programme 2*, pp. 211–22. Cambridge University Press.

Drummond, D. C. (1970). Variation in rodent populations in response to control measures. In *Variation in Mammalian Populations*, ed. R. J. Berry and H. N. Southern, pp. 301–20. Academic Press, London, New York.

Drury, W. H. (1974). Rare species. *Biol. Conserv.* **6**, 162–69.

Durrell, G. (1975). Foreword. In *Breeding Endangered Species in Captivity*, ed. R. D. Martin. Academic Press, London, New York.

Eanes, W. F. (1978). Morphological variance and enzyme heterozygosity in the monarch butterfly. *Danaus plexippus. Nature* **276,** 263–4.

Eberhart, S. A. (1977). Quantitative genetics and practical corn breeding. In *Proceedings of the International Conference on Quantitative Genetics*, ed. E. Pollack, O. Kempthorne and T. B. Bailey, pp. 491–502. Iowa State University Press, Ames, Iowa.

Ehrlich, P. R. (1961). Has the biological species concept outlived its usefulness? *Syst. Zool.* **10,** 167–76.

Ehrlich, P. R. (1980). The strategy of conservation, 1980–2000. In *Conservation Biology: An Evolutionary–Ecological Perspective*, ed. M. E. Soulé and B. A. Wilcox, pp. 329–44. Sinauer Associates, Sunderland, Mass.

Ehrlich, P. R., Ehrlich, A. H. and Holdren, J. P. (1977). *Ecoscience: Population, Resources, Environment.* W. H. Freeman and Co., San Francisco.

Ehrlich, P. R. and Feldman, S. S. (1977). *The Race Bomb: Skin Color, Prejudice, and Intelligence.* Quadrangle/New York Times Book Co., New York.

Ehrlich, P. R. and Raven, P. H. (1969). Differentiation of populations. *Science* **165,** 1228–32.

Ehrlich, P. R., White, R. R., Singer, M. C., McKechnie, S. W. and Gilbert, L. E. (1975). Checkerspot butterflies: a historical perspective. *Science* **188,** 221–8.

Eidt, R. C. (1968). The climatology of South America. In *Biogeography and Ecology in South America*, ed. E. J. Fittkau *et al.*, pp. 54–81. W. Junk, the Hague.

Eisenberg, J. F. (1980). The density and biomass of tropical mammals. In *Conservation Biology: An Evolutionary–Ecological Perspective*, ed. M. E. Soulé and B. A. Wilcox, pp. 35–56. Sinauer Associates, Sunderland, Mass.

Eldredge, N. (1971). The allopatric model and phylogeny in Paleozoic invertebrates. *Evolution* **25,** 156–67.

Eldredge, N. and Gould, S. J. (1972). Punctuated equilibria: an alternative to phyletic gradualism. In *Models in Paleobiology*, ed. T. J. M. Schopf, pp. 82–115. Freeman, Cooper & Co., San Francisco.

Ellingboe, A. H. (1975). Horizontal resistance: an artifact of experimental procedure? *Aust. Plant Path. Soc. Newsl.* **4,** 44–6.

Elton, C. S. (1958). *The Ecology of Invasions by Animals and Plants.* Chapman & Hall, London.

Elton, C. S. (1975). Conservation and the low population density of invertebrates inside neotropical rainforests. *Biol. Conserv.* **7,** 3–15.

Endler, J. A. (1973). Gene flow and population differentiation. *Science* **179,** 243–250.

Endler, J. A. (1977). *Geographic Variation, Speciation and Clines.* Princeton University Press, Princeton, NJ.

Enfield, F. D. (1977). Selection experiments in *Tribolium* designed to look at gene action issues. In *Proceedings of the International Conference on Quantitative Genetics*, ed. E. Pollak, O. Kempthorne and T. B. Bailey, pp. 177–90. Iowa State University Press, Ames, Iowa.

Eriksson, H., Halkka, O., Lokki, J. and Saura, A. (1976). Enzyme polymorphism in feral, outbred and inbred rats (*Rattus norvegicus*). *Heredity* **37,** 341–9.

Evans, G. C. (1976). A sack of uncut diamonds: the study of ecosystems and the future resources of mankind. *J. Appl. Ecol.* **13**, 1–39.

Evans, L. T. and Dunstone, R. L. (1970). Some physiological aspects of evolution in wheat. *Aust. J. Biol.* **23**, 725–41.

Falconer, D. S. (1960). *Introduction to Quantitative Genetics*. Oliver and Boyd Ltd, London.

Falconer, D. S. (1977). Some results of the Edinburgh selection experiments with mice. In *Proceedings of the International Conference on Quantitative Genetics*, ed. E. Pollack, O. Kempthorne and T. B. Bailey, pp. 101–16. Iowa State University Press, Ames, Iowa.

FAO (1967). *Report of the FAO Study Group on the Evaluation, Utilization and Conservation of Animal Genetic Resources, 21–25 November, 1966*. FAO, Rome.

FAO (1973). *Report of the Fourth FAO Expert Consultation on Animal Genetic Resources (Poultry Breeding)*. FAO, Rome.

FAO (1974). *Round Table: The Conservation of Animal Genetic Resources*. FAO, Rome.

FAO (1975). *Report of the Sixth Session of the FAO Panel of Experts on Plant Exploration and Introduction*. FAO, Rome.

Felsenstein, J. (1971). Inbreeding and variance effective numbers in populations with overlapping generations. *Genetics* **68**, 581–97.

Fenner, F. (1971). Evolution in action: myxomatosis in the Australian wild rabbit. In *Topics in the Study of Life: The Bio Source Book*, ed. A. Kramer, pp. 463–71. Harper and Row, New York.

Fisher, R. A. (1958). *The Genetical Theory of Natural Selection*, 2nd edn. Dover Publications, Inc., New York.

Flannery, K. V. (1969). Origins and ecological effects of early domestications in Iran and the Near East. In *The Domestication and Exploitation of Plants*, ed. P. J. Ucko and G. W. Dimbleby, pp. 73–100. Duckworth, London.

Flesness, N. (1977). Gene pool conservation and computer analysis. *Inter. Zoo. Yearb.* **17**, 77–81.

Foose, T. (1977). Demographic models for management of captive populations. *Inter. Zoo. Yearb.* **17**, 70–76.

Foose, T. (1978). Demographic and genetic models and management for the okapi (*Okapia johnstoni*) in captivity. *Acta Zool. Pathol. Antverpiensia* **71**, 47–52.

Foote, W. C. (1976). Feral or wild genotypes of sheep and goats and their role in meat production. Conference on the role of sheep and goats in agricultural development. Winrock International Livestock Research and Training Center, Morrilton, Arkansas.

Foote, W. H. (1966). Benton barley. *Crop Sci.* **6**, 93.

Forrester, D. J. (1971). Bighorn sheep lungworm-pneumonia complex. In *Parasitic Diseases of Wild Mammals*, ed. J. W. Davis and R. C. Anderson, pp. 158–73. Iowa State University Press, Ames, Iowa.

Fosberg, F. R. (1973). Temperate zone influence on tropical forest land use: A plea for sanity. In *Tropical Forest Ecosystems in Africa and South America: A Comparative Review*, ed. B. J. Meggers, E. S. Ayensu and W. D. Duckworth, pp. 345–50. Smithsonian Institute Press, Washington.

Foster, R. B. (1980). Heterogeneity and disturbance in tropical vegetation. In

Conservation Biology: An Evolutionary–Ecological Perspective, ed. M. E. Soulé and B. A. Wilcox, pp. 75–92. Sinauer Associates, Sunderland, Mass.

Fowler, J. A. (1964). The *Rana pipiens* problem, a proposed solution. *Amer. Natur.* **98**, 213–19.

Fox, S. F. (1975). Natural selection of morphological phenotypes of the lizard *Uta Stansburiana*. *Evolution* **29**, 95–107.

Frame, L. H. and Frame, G. W. (1976). Female African wild dogs emigrate. *Nature* **263**, 227–9.

Frankel, O. H. (1970a). Genetic conservation in perspective. In *Genetic Resources in Plants – Their Exploration and Conservation*, ed. O. H. Frankel and E. Bennett, *IBP Handbook No. 11*, pp. 469–89. Blackwell Scientific Publications, Oxford.

Frankel, O. H. (1970b). Variation, the essence of life. Sir William Macleay Memorial Lecture. *Proc. Linn. Soc. NSW* **95**, 158–69.

Frankel, O. H. (ed.) (1973). *Survey of Crop Genetic Resources in their Centres of Origin*. FAO-IBP, Rome.

Frankel, O. H. (1974). Genetic Conservation: our evolutionary responsibility. *Genetics* **78**, 53–65.

Frankel, O. H. (1975a). Genetic resources survey as a basis for exploration. In *Crop Genetic Resources for Today and Tomorrow*, ed. O. H. Frankel and J. G. Hawkes, *International Biological Programme 2*, pp. 99–109. Cambridge University Press.

Frankel, O. H. (1975b). Genetic resources centres – a co-operative global network. In *Crop Genetic Resources for Today and Tomorrow*, ed. O. H. Frankel and J. G. Hawkes, *International Biological Programme 2*, pp. 473–81. Cambridge University Press.

Frankel, O. H. (1977). Natural variation and its conservation. In *Genetic Diversity in Plants*, ed. Amir Muhammed, R. Aksel and R. C. von Borstel, pp. 21–44. Plenum Press, New York.

Frankel, O. H. (1978a). Natural resources and technology: evaluation, use and conservation of biological resources. In *Proceedings Third Inter-Congress, Pacific Science Association*, ed. C. H. Lamoureux, pp. 303–23. Indonesian Institute of Sciences.

Frankel, O. H. (1978b). Germplasm "preservation". *Plant Genet. Resour. Newsl.* **34**, 18–19.

Frankel, O. H. (1978c). Conservation of crop genetic resources and their wild relatives: an overview. In *Conservation and Agriculture*, ed. J. G. Hawkes, Duckworth, London.

Frankel, O. H. and Bennett, E. (1970a). Genetic Resources – Introduction. In *Genetic Resources in Plants – Their Exploration and Conservation*, ed. O. H. Frankel and E. Bennett, *IBP Handbook No. 11*, pp. 7–17. Blackwell Scientific Publications, Oxford.

Frankel, O. H. and Bennett, E. (eds.) (1970b). *Genetic Resources in Plants – Their Exploration and Conservation. IBP Handbook No. 11*. Blackwell Scientific Publications, Oxford.

Frankel, O. H. and Hawkes, J. G. (eds.) (1975). *Crop Genetic Resources for Today and Tomorrow. International Biological Programme 2*. Cambridge University Press.

Franklin, I. A. (1980). Evolutionary change in small populations. In *Conservation*

Biology: An Evolutionary–Ecological Perspective, ed. M E. Soulé and B. A. Wilcox, pp. 135–50. Sinauer Associates, Sunderland, Mass.

Freuchan, P. (1935). *Arctic Adventures – My Life in the Frozen North*. Farrar and Rinehart, New York.

Frisch, J. E. (1975). The relative incidence and effect of bovine infectious Keratoconjunctivitis in *Bos indicus* and *Bos taurus* cattle. *Anim. Prod.* **21,** 265–74.

Frisch, J. E. and Vercoe, J. E. (1978). Utilizing breed differences in growth of cattle in the tropics. *World Anim. Rev.* **25,** 8–12.

Fryer, R. F. (1972). Poroporo as a crop. *Proceedings of the 22nd Lincoln College Farmers' Conference*, pp. 96–103. Lincoln College, Lincoln NZ.

Fujino, K. and Kang, T. (1968). Transferrin groups of tuna. *Genetics* **59,** 79–91.

Futuyma, D. (1973). Community structure and stability in constant environments. *Amer. Natur.* **107,** 443–6.

Gale, M. D. and Law, C. N. (1977). Norin-10-based semi-dwarfism. In *Genetic Control of Diversity in Plants*, ed. Amir Muhammed, R. Aksel and R. C. von Borstel, pp. 133–51. Plenum Press, New York.

Galli, A. E., Leck, C. F. and Forman, R. T. T. (1976). Avian distribution patterns in forest islands of different sizes in central New Jersey. *Auk* **93,** 356–64.

Gallup, G. G. Jr, Boren, J. L., Gagliardi, G. J. and Wallnau, L. B. (1977). A mirror for the mind of man, or will the chimpanzee create an identity crisis for *Homo sapiens? J. Hum. Evol.* **6,** 303–13.

Garten, C. T. (1976). Relationships between aggressive behavior and genic heterozygosity in the oldfield mouse. *Peromyscus polionotus. Evolution* **30,** 59–72.

Gause, G. F. (1934). *The Struggle for Existence*. Williams and Wilkins, Baltimore.

Gersh, H. (1971). *The Animals Next Door: A Guide To Zoos and Aquariums of the Americas*. Fleet Academic Editions, New York.

Gibbons, R. W. and Mercer, P. C. (1972). Peanut disease control in Malawi, Central Africa. *J. Amer. Peanut Res. and Educ. Assoc. Inc.* **4** (1).

Gilbert, L. E. (1980). Food web organization and the conservation of Neotropical diversity. In *Conservation Biology: An Evolutionary–Ecological Perspective*, ed. M. E. Soulé and B. A. Wilcox, pp. 11–34. Sinauer Associates, Sunderland, Mass.

Gill, K. S., Sandhu, T. S., Singh, Kuldip and Brar, J. S. (1975). Evaluation of mungbean (*Vigna radiata* L. Wilczek) germplasm. *Crop Improv.* **2,** 99–104.

Gillespie, J. (1977). A general model to account for enzyme variation in natural populations. III. Multiple alleles. *Evolution* **31,** 85–90.

Gillespie, J. and Langley, C. (1974). Multilocus behavior in random environments. I. Random Levene models. *Genetics* **82,** 123–37.

Gilpin, M. E. and Diamond, J. M. (1976). Calculation of immigration and extinction curves from the species–area–distance–relation. *Proc. Natl. Acad. Sci. USA* **73,** 4130–4.

Given, D. R. (1976). A register of rare and endangered indigenous plants in New Zealand. *N.Z. J. Bot.* **14,** 135–49.

Gladstones, J. S. (1970). Lupins as crop plants. *Field Crop Abstr.* **23,** 123–48.

Goldfoot, D. A. (1977). Rearing conditions which support or inhibit later sexual potential of laboratory monkeys: hypothesis and diagnostic behaviors. *Lab. Anim. Sci.* **27,** 548–56.

Gomez-Pompa, A., Vazquez-Yanes, C. and Guevara, S. (1972). The tropical rain forest: a non renewable resource. *Science* **117**, 762–5.

Gomez-Pompa, A. and Vazquez-Yanes, C. (1976). Estudios sobre sucesion secundaria en los tropicos calidohumedos: el ciclo de vida de las especies secundarias. In *Regeneracion de Selvas*, ed. A. Gomez-Pompa *et al.*, pp. 579–93. Compania Editorial Continental, Mexico.

Goodman, D. (1980). Demographic intervention for closely managed populations. In *Conservation Biology: An Evolutionary–Ecological Perspective*, ed. M. E. Soulé and B. A. Wilcox, pp. 171–95. Sinauer Associates, Sunderland, Mass.

Goodwin, H. A. and Goodwin, J. M. (1973). List of mammals which have become extinct or are possibly extinct since 1600. *IUCN Occasional Paper No. 8*. International Union for the Conservation of Nature and Natural Resources. Morges, Switzerland.

Goodwin, H. A. and Holloway, C. W. (1972). *Red Data Book*. International Union for the Conservation of Nature and Natural Resources. Morges, Switzerland.

Gordon, I. (1975). Problems and prospects in cattle egg transfer. *Irish Vet. J.* **29**, 21–30, 29–62.

Gould, S. J. (1977). This view of life. *Nat. Hist.* **86**, 12–18.

Gould, S. J. and Eldredge, N. (1977). Punctuated equilibria: the tempo and mode of evolution reconsidered. *Paleobiology* **3**, 115–51.

Gray, J. C., Kung, S. D. and Wildman, S. G. (1975). Origin of *Nicotiana tabacum* L. detected by polypeptide composition of Fraction I protein. *Nature* **252**, 226–7.

Grayson, D. K. (1977). Pleistocene avifaunas and the overkill hypotheses. *Science* **195**, 691–3.

Greenway, J. C. Jr. (1967). *Extinction and Vanishing Birds of the World*. Dover Publications, New York.

Gregory, R. P. G. and Bradshaw, A. D. (1965). Heavy metal tolerance in populations of *Agrostis tenuis* Sibth. and other grasses. *New Phytol.* **64**, 131–43.

Grinnell, G. B. (1928). Mountain sheep. *J. Mammal.* **9**, 1–9.

Guldager, P. (1975). *Ex situ* conservation stands in the tropics. In *The Methodology of Conservation of Forest Genetic Resources*, pp. 85–92. FAO, Rome.

Guzman, E. D. de (1975). Conservation of vanishing timber species in the Philippines. In *South East Asian Plant Genetic Resources*, ed. J. T. Williams, C. H. Lamoureux and N. Wullijarni-Soetjipto, pp. 198–204. LIPI, Bogor.

Halpin, B. (1975). *Patterns of Animal Disease*. Baillière Tindall, London.

Hammons, R. O. (1977). Groundnut rust in the United States and the Carribean. *PANS* **23**, 300–4.

Hanelt, P. and Hammer, K. (1975). Bericht über eine Reise nach Ostmähren und der Slowakei 1974 zur Sammlung autochthoner Sippen von Kulturpflanzen. *Kulturpflanze* **23**, 207–15.

Harding, J., Allard, R. W. and Smeltzer, D. G. (1966). Population studies in predominantly self-pollinated species. IX. Frequency-dependent selection in *Phaseolus lunatus*. *Proc. Natl. Acad. Sci. USA* **56**, 99–104.

Hare, R. (1967). The antiquity of diseases caused by bacteria and viruses. A review of the problem from a bacteriologist's point of view. In *Diseases in Antiquity*, ed. D. Brothwell and A. T. Sandison, pp. 115–31. C. C. Thomas, Springfield, Ill.

Harlan, J. R. (1951). Anatomy of gene centers. *Am. Natur.* **85**, 97–103.

Harlan, J. R. (1956). Distribution and utilization of natural variability in cultivated plants. *Brookhaven Symp. Biol.* **9**, 191–206.

Harlan, J. R. (1975a). *Crops and Man.* American Society of Agronomy, Madison.

Harlan, J. R. (1975b). Our vanishing genetic resources. *Science* **188**, 618–21.

Harlan, J. R. (1976). Genetic resources in wild relatives of crops. *Crop Sci.* **16**, 329–33.

Harlan, J. R. and de Wet J. M. J. (1971). Toward a rational classification of cultivated plants. *Taxon* **20**, 509–17.

Harlan, J. R. and de Wet, J. M. J. (1977). Pathways of genetic transfer from *Tripsacum* to *Zea mays. Proc. Natl. Acad. Sci. USA* **74**, 3494–7.

Harlan, J. R., de Wet, J. M. J. and Price, E. G. (1973). Comparative evolution of cereals. *Evolution* **27**, 311–25.

Harper, C. W. Jr. (1975). Stability of species in geological time. *Science* **190**, 269.

Harper, J. L. (1969). The role of predation in vegetational diversity. *Brookhaven Symp. Biol.* **22**, 48–62.

Harrington, J. F. (1970). Seed and pollen storage for conservation of plant gene resources. In *Genetic Resources in Plants – Their Exploration and Conservation*, ed. O. H. Frankel and E. Bennett, *IBP Handbook No. 11*, pp. 501–21. Blackwell Scientific Publications, Oxford.

Harris, D. R. (1969). Agricultural systems, ecosystems and the origins of agriculture. In *The Domestication and Exploitation of Plants*, ed. P. J. Ucko and G. W. Dimbleby, pp. 3–15. Duckworth, London.

Hartl, D. L. and Cook, R. D. (1973). Balanced polymorphisms of quasi-neutral alleles. *Theor. Pop. Biol.* **4**, 163–72.

Hawkes, J. G. (1969). The ecological background of plant domestication. In *The Domestication and Exploitation of Plants*, ed. P. J. Ucko and G. W. Dimbleby, pp. 17–29. Duckworth, London.

Hawkes, J. G. (1970). Potatoes. In *Genetic Resources in Plants – Their Exploration and Conservation*, ed. O. H. Frankel and E. Bennett, *IBP Handbook No. 11*, pp. 311–19. Blackwell Scientific Publications, Oxford.

Hawkes, J. G. (1976). *Manual for Field Collectors (Seed Crops).* FAO, Rome.

Hayes, F. A. and Prestwood, A. K. (1969). Some considerations for diseases and parasites of white-tailed deer in the southeastern United States. In *Proceedings of a Symposium on White-Tailed Deer in the Southern Forest Habitat. Nagodoches, Texas, 25–26 March 1969*, pp. 32–6. Forest Service, US Department of Agriculture.

Hecht, M. K. (1974). Morphological transformation, the fossil record, and the mechanisms of evolution: a debate (Part I). In *Evolutionary Biology*, vol. 7, ed. T. Dobzhansky, M. K. Hecht and W. C. Steere, pp. 295–303. Plenum Press, New York.

Heck, L (1952). Über den Auerochsen und seine Rückzüchtung. *Jahrb. Nassau Ver. Naturk.* **90**, 107–24.

Hedrick, P. W. (1974). Genetic variation in a heterogeneous environment. I. Temporal heterogeneity and the absolute dominance model. *Genetics* **78**, 757–70.

Hedrick, P. W., Ginevan, M. E. and Ewing, E. P. (1976). Genetic polymorphism in heterogeneous environments. *Ann. Rev. Ecol. Syst.* **7**, 1–32.

Henshaw, G. G. (1975). Technical aspects of tissue culture storage for genetic conservation. In *Crop Genetic Resources for Today and Tomorrow*, ed. O. H. Frankel and J. G. Hawkes, *International Biological Programme 2*, pp. 349–57. Cambridge University Press.

Heron, A. C. (1972). Population ecology of a colonizing species: the pelagic tunicate *Thalia democratica*. I. Individual growth rate and generation time. *Oecologia* **10**, 269–93.

Hersh, G. N. and Rogers, D. J. (1975). Documentation and information requirements for genetic resources application. In *Crop Genetic Resources for Today and Tomorrow*, ed. O. H. Frankel and J. G. Hawkes, *International Biological Programme 2*, pp. 407–46. Cambridge University Press.

Heslop-Harrison, J. (1974). Postscript: the threatened plants committee. In *Succulents in Peril. Supplement to IOS Bulletin 3, no. 3*, ed. D. R. Hunt, pp. 30–2. Kew, Richmond, Surrey.

Heslop-Harrison, J. (1976). Introduction. In *Conservation of Threatened Plants*, ed. J. B. Simmons, R. I. Beyer, B. E. Brandham, G. Ll. Lucas, and V. T. H. Parry, pp. 3–7. Plenum Press, New York.

Hester, J. (1967). The agency of man in animal extinctions. In *Pleistocene Extinctions: The Search for a Cause*, ed. P. S. Martin and H. E. Wright. Yale University Press, New Haven and London.

Hill, W. G. (1977). Selection with overlapping generations. In *Proceedings of the International Conference on Quantitative Genetics*, ed. E. Pollak, O. Kempthorne and T. B. Bailey pp. 367–77. Iowa State University Press, Ames, Iowa.

Hinegardner, R. (1976). Evolution of genome size. In *Molecular Evolution*, ed. F. J. Ayala, pp. 179–99. Sinauer Associates, Sunderland, Mass.

Holcomb, J., Tolbert, T. M. and Jain, S. K. (1977). A diversity analysis of genetic resources in rice. *Euphytica* **26**, 441–50.

Holden, J. H. W. (1977). Potato breeding at Pentlandfield. In *Scottish Plant Breeding Station, 56th Annual Report 1976–77*, pp. 66–97.

Holt, S. B. (1961). Quantitative genetics of finger-print patterns. *Brit. Med. Bull.* **17**, 247–50.

Hopson, J. A. (1969). The origin and adaptive radiation of mammai-like reptiles and non-therian mammals. *Ann. NY Acad. Sci.* **167**, 199–216.

Hornocher, M. G. (1969). Winter territoriality in mountain lions. *J. Wildl. Manage.* **33**, 457–64.

Howard, B. H. (1975). Possible long-term cold storage of woody plant material. In *Crop Genetic Resources for Today and Tomorrow*, ed. O. H. Frankel and J. G. Hawkes, *International Biological Programme 2*, pp. 359–67. Cambridge University Press.

Howe, H. F. (1977). Bird activity and seed dispersal of a tropical wet forest tree. *Ecology,* **58**, 539–50.

Hutchinson, G. E. (1959). Homage to Santa Rosalia, or why are there so many kinds of animals? *Amer. Natur.* **93**, 145–59.

Hutchinson, G. E. (1961). The paradox of the plankton. *Amer. Natur.* **95**, 137–45.

Hutt, F. B. (1958). *Genetic Resistance to Disease in Animals*. Comstock, Ithaca, NY.

Hutt, F. B. (1965). The utilization of genetic resistance to disease in domestic animals. In *Genetics Today*, pp. 775–83. Pergamon Press, Oxford.

292   *References*

Hutt, F. B. and Crawford, R. D. (1960). On breeding chicks resistant to pullorum disease without exposure thereto. *Can. J. Genet. Cyt.* **2**, 357–70.

Huxley, J. S. (1958). Evolutionary process and taxonomy with special reference to grades. *Uppsala Univ. Arskrift*, 21–39 (not seen).

IARI (Indian Agricultural Research Institute) (1972). *Collection and Study of Cultivated and Wild Rices from North East India. USPL 480 Project, Final Report.* IARI, New Delhi.

IBP (1966). Plant gene pools. *IBP News* **5**, 48–51.

IBPGR (International Board for Plant Genetic Resources) (1976a). *Priorities Among Crops and Regions.* FAO, Rome.

IBPGR (1976b). *Report of the IBPGR Working Group on Engineering Design and Cost Aspects of Long-Term Seed Storage Facilities.* IBPGR, FAO, Rome.

IBPGR (1977a). *A Co-operative Regional Programme in South East Asia.* IBPGR, FAO, Rome.

IBPGR (1977b). *Report of the Second meeting of the Advisory Committee of the Board on the Genetic Resources Communication, Information and Documentation System (GR/CIDS).* AGPE:IBPGR/77/6. IBPGR, FAO, Rome.

IBPGR (1978a). *Annual Report 1977.* IBPGR Secretariat, Rome.

IBPGR (1978b). *IBPGR Regional Committee for Southeast Asia. Report of the First Meeting.* IBPGR Secretariat, FAO, Rome.

IBPGR (1978c). *Report of the IBPGR/GO1 Workshop on South Asian Plant Genetic Resources, New Delhi, 9–12 May 1978.* IBPGR, FAO, Rome.

IBPGR (1978d). IBPGR plans for crop collecting. In *Plant Genetic Resources. No. 33*, pp. 4–8. FAO-IBPGR, Rome.

ICRISAT (International Crops Research Institute for the Semi-Arid Tropics) (1976). *Chickpea Breeding.* ICRISAT, Hyderabad, India.

IRRI (International Rice Research Institute) (1975). *Annual Report for 1974.* IRRI, Los Baños, Philippines.

IUCN (1975). *United Nations List of National Parks and Equivalent Reserves.* International Union for Conservation of Nature and Natural Resources. Morges, Switzerland.

Jackson, J. F. and Pounds, J. A. (1979). Comments on assessing the dedifferentiating effect of gene flow. *Syst. Zool.* **28**, 78–84.

Jain, S. K. (1969). Comparative ecogenetics of two *Avena* species occurring in Central California. *Evolutionary Biology* **3**, 73–118.

Jain, S. K. (1975). Genetic reserves. In *Crop Genetic Resources for Today and Tomorrow*, ed. O. H. Frankel and J. G. Hawkes, *International Biological Programme 2*, pp. 379–96. Cambridge University Press.

Jain, S. K. and Marshall, D. R. (1967). Population studies in predominantly self-pollinating species. X. Variation in natural populations of *Avena fatua* and *Avena barbata. Amer. Natur.* **101**, 19–34.

Jain, S. K., Qualset, C. O., Bhatt, G. M. and Wu, K. K. (1975). Geographical patterns of phenotypic diversity in a world collection of durum wheats. *Crop Sci.* **15**, 700–4.

James, J. W. (1962). The spread of genes in randomly mating control populations. *Genet. Res. Camb.* **3**, 1–10.

Janzen, D. H. (1972). The uncertain future of the tropics. *Nat. Hist.* **81**, 80–94.

Janzen, D. H. (1973). Tropical agroecosystems. *Science* **182,** 1212–19.

Janzen, D. H. (1976). Additional land at what price? Responsible use of the tropics in a food-population confrontation. *Proc. Amer. Phytopathol. Soc.* **3,** 35–9.

Johnson, B. L. (1975). Identification of the apparent B-genome donor of wheat. *Can. J. Genet. Cytol.* **17,** 21–39.

Johnson, G. B. (1976). Genetic polymorphism and enzyme function. In *Molecular Evolution,* ed. F. J. Ayala, pp. 46–59. Sinauer Associates, Sunderland, Mass.

Johnson, G. B. (1977). Characterization of electrophoretically cryptic variation in the Alpine butterfly, *Colias meadii. Biochem. Genet.* **15,** 665–93.

Johnson, R. F. and Selander, R. K. (1964). House sparrows: rapid evolution of races in North America. *Science* **144,** 548–50.

Joshi, N. R. and Phillips, R. W. (1953). *Zebu cattle of India and Pakistan. FAO-Agricultural Studies No. 19.* FAO, Rome.

Karlin, S. and Lieberman, U. (1975). Random temporal variation in selection intensities – case of large population size. *Theor. Pop. Biol.* **6,** 355.

Kettlewell, H. B. D. (1957). Industrial melanism in moths and its contribution to our knowledge of evolution. *Proc. Royal Instn. (GB)* **36,** 1–14.

Kidd, K. K. (1974). Biochemical polymorphism, breed relationships, and germplasm resources in domestic cattle. *Proc. 1st World Congr. Genet. Appl. Livest. Prod.* **1,** 321–8.

Kimura, M. (1979). Model of effectively neutral mutations in which selective constraint is incorporated. *Proc. Natl. Acad. Sci. USA* **76,** 3440–4.

Kimura, M. and J. F. Crow (1963). The measurement of effective population number. *Evolution* **17,** 279–88.

Kimura, M. and Ohta, T. (1971). *Theoretical Aspects of Population Genetics.* Princeton University Press, Princeton, NJ.

King, J. L. (1972). The role of mutation in evolution. In *Proceedings of the Sixth Berkeley Symposium on Mathematics, Statistics and Probability,* ed. L. M. Le Cam, J. Neyman and E. L. Scott, pp. 69–100. University of California Press, Berkeley.

Kleiman, D. G. (1980). The sociobiology of captive propagation. In *Conservation Biology: An Evolutionary–Ecological Perspective,* ed. M. E. Soulé and B. A. Wilcox, pp. 243–62. Sinauer Associates, Sunderland, Mass.

Kleiman, D. G. and Eisenberg, J. F. (1973). Comparisons of canid and felid social systems from an evolutionary perspective. *Anim. Behav.* **21,** 637–59.

Klingel, H. (1969). The social organisation and population ecology of the plains zebra *(Equus quagga). Zool. Afr.* **4,** 249–63.

Knowles, P. F. (1969). Centers of plant diversity and conservation of germplasm: Safflower. *Econ. Bot.* **23,** 324–9.

Kobayashi, S., Ikeda, I, and Nakatani, N. (1978). Long-term storage of Citrus pollen. In *Long-term Preservation of Favourable Germplasm in Arboreal Crops,* ed. T. Akihama and K. Nakajima, pp. 8–12. Fruit Tree Research Station, M.A.F. Fujimoto, Japan.

Koehn, R. K., Milkman, R. and Mitton, J. B. (1976). Population genetics of marine pelecypods: IV. Selection, migration and genetic differentiation in the blue mussel *Mytilus edulis. Evolution* **30,** 2–32.

Koehn, R. K., Turano, F. J. and Mitton, J. B. (1973). Population genetics of marine pelecypods: II. Genetic differences in microhabitats of *Modiolus demissus*. *Evolution* **27**, 100–5.

Kühn, F., Hammer, K. and Hanelt, P. (1976). Botanische Ergebnisse einer Reise in die CSSR 1974 zur Sammlung autochthoner Landsorten von Kulturpflanzen. *Kulturpflanze* **24**, 283–347.

Kumaran, P. M., Nayar, N. M. and Murthy, K. N. (1977). A study of variation in cashew (*Anacardium occidentale* L.: Anacardiaceae) germplasm. In *3rd International Congress of the Society for the Advancement of Breeding Researches in Asia and Oceania (SABRAO)*. 1a pp. 16–19. SABRAO

Lack, D. (1954). *The Natural Regulation of Animal Numbers.* Clarendon Press, Oxford.

Lande, R. (1976). The maintenance of genetic variability by mutation in a polygenic character with linked loci. *Genet. Res. Camb.* **26**, 221–35.

Lapin, P. I. (1976). Report of the International Association of Botanic Gardens Plenary Session – Moscow 1975. In *Conservation of Threatened Plants*, ed. J. B. Simmons, R. I. Beyer, P. E. Brandham, G. Ll. Lucas and V. T. H Parry, pp. 249–257. Plenum Press, New York.

Latter, B. D. H. (1959). Genetic sampling in a random mating population of constant size and sex ratio. *Aust. J. Biol. Sci.* **12**, 500–5.

Latter, B. D. H. and Robertson, A. (1962). The effect of inbreeding and artificial selection on reproductive fitness. *Genet. Res. Camb.* **3**, 110–38.

Lauvergne, J. J. (1975). Disappearing cattle breeds in Europe and the Mediterranean basin. In *Conservation of Animal Genetic Resources*, (Pilot Study) pp. 21–41. FAO, Rome.

Law, C. N., Snape, J. W. and Worland, A. J. (1978). The genetical relationship between height and yield in wheat. *Heredity,* **40**, 133–51.

Law, C. N. and Worland, A. J. (1973). Chromosome substitutions and their use in the analysis and prediction of wheat varietal performance. In *Proceedings 4th International Wheat Genetics Symposium*, pp. 41–9. Missouri Agricultural Experimental Station, Columbia, Mo.

Leamy, L. (1974). Heritability of osteometric traits in a random bred population of mice. *J. Hered.* **65**, 109–20.

Leppik, E. E. (1970). Gene centres of plants as sources of disease resistance. *Ann. Rev. Phytopath.* **8**, 232–44.

Lerner, I. M. (1954). *Genetic Homeostasis.* Oliver & Boyd, Edinburgh.

Levene, H. (1953). Genetic equilibrium when more than one ecological niche is available. *Amer. Natur.* **87**, 331–3.

Levin, D. A. (1976). Consequences of long-term certificial selection, inbreeding and isolation in *Phlox*. II. The organization of allozymic variability. *Evolution* **30**, 463–72.

Levins, R. (1968). *Evolution in Changing Environments.* Princeton University Press, Princeton, New Jersey.

Lewontin, R. C. (1974). *The Genetic Basis of Evolutionary Change.* Columbia University Press, New York.

Lindstrom, E. W. (1941). Analysis of modern maize breeding principles and methods. In *Proceedings of the VII International Congress of Genetics*, ed. R. C. Punnett, pp. 151–6. Cambridge University Press.

Longenecker, B. M., Pazderka, F., Gavora, J. S., Spencer, J. L. and Ruth, R. F. (1976). Lymphoma induced by herpesvirus: resistance associated with a major histocompatibility gene. *Immunogenetics* **3**, 401–7.

Lowe, H. J. (1942). Rinderpest in Tangyanyika territory. *Emp. J. Expt. Agr.* **10**, 189.

Luckinbill, L. S. (1973). Coexistence in laboratory populations of *Paramecium aurelia* and its predator *Didinium nasutum*. *Ecology* **54**, 1320–7.

Lynch, C. B. (1977). Inbreeding effects upon animals derived from wild populations of *Mus musculus*. *Evolution* **31**, 525–37.

MacArthur, R. H. (1972). *Geographical Ecology*. Harper & Row, New York.

MacArthur, R. H. and Wilson, E. O. (1963). An equilibrium theory of insular zoogeography. *Evolution* **17**, 373–87.

MacArthur, R. H. and Wilson, E. O. (1967). *The Theory of Island Biogeography*. Princeton University Press, Princeton, New Jersey.

McDonald, J. F. and Ayala, F. J. (1974). Genetic response to environmental heterogeneity. *Nature* **250**, 572–4.

MacGillaury, H. J. (1968). Modes of evolution mainly among marine invertebrates. *Bijdr. Dierkd.* **38**, 69–74.

McKenzie, J. A. and Parsons, P. A. (1974). Microdifferentiation in a natural population of *Drosophila melanogaster* to alcohol in the environment. *Genetics* **77**, 385–94.

McNeill, W. H. (1976). *Plagues and Peoples*. Anchor Press, Garden City, New York.

MacNeish, R. S. (1976). Early man in the New World. *Am. Sci.* **64**, 316–27.

McPhee, H. C., Russel, E. Z. and Zeller, J. (1931). An inbreeding experiment with Poland China swine. *J. Hered.* **22**, 383–403.

Maesen, L. J. G. van der (1972). Cicer, *L. a Monograph of the Genus, with Special Reference to the Chickpea* (Cicer arietinum L.), *its Ecology and Cultivation*. *Meded. Landbouwhopesch. Wageningen*, 72–10. Veenman and Zonen, Wageningen.

Maesen, L. J. G. van der (1973). Chickpea: distribution of variability. In *Survey of Crop Genetic Resources in their Centres of Diversity*, ed. O. H. Frankel, pp. 30–4. FAO-IBP, Rome.

Maggs, D. H. (1972). Pistachios in Iran and California. *Plant Introd. Rev.* **9/1**, 12–15.

Maggs, D. H. (1973). Genetic Resources in Pistachio. *Plant Genet. Resour. Newsl.* **29**, 7–15.

Maijala, K. (1974). Conservation of animal breeds in general. In *Round Table: The Conservation of Animal Genetic Resources*, pp. 37–46. FAO, Rome.

Main, A. R. and Yadav. M. (1971). Conservation of macropods in reserves in Western Australia. *Biol. Cons.* **3**, 123–32.

Maini, J. S., Yeatman, C. W. and Teich, A. H. (1975). *In situ* and *ex situ* conservation of gene resources of *Pinus banksiana* and *Picea glauca*. In *The Methodology of Conservation of Forest Genetic Resources*, pp. 27–40. FAO, Rome.

Mangelsdorf, P. C. (1966). Genetic potentials for increasing yields of food crops and animals. *Proc. Natl. Acad. Sci. USA* **56**, 370–5.

Manwell, C. and Baker, C. M. A. (1970). *Molecular Biology and the Origin of Species*. University of Washington Press, Seattle, Washington.

Manwell, C. and Baker, C. M. A. (1976). Protein polymorphisms in domesticated species: evidence for hybrid origin? In *Population Genetics and Ecology*, ed. S. Karlin and E. Nevo, pp. 105–39. Academic Press, New York.

Manwell, C. and Baker, C. M. A. (1977). Genetic distance between the Australian Merino and the Poll Dorset sheep. *Genet. Res. Camb.* **29**, 239–53.

Marshall, D. R. (1977). The advantages and hazards of genetic homogeneity. In *The Genetic Basis of Epidemics in Agriculture*, ed. P. R. Day, pp. 1–20. New York Academy of Sciences, New York.

Marshall, D. R. and Allard, R. W. (1970). Isozyme polymorphisms in natural populations of *Avena fatua* and *A. barbata*. *Heredity* **25**, 373–82.

Marshall, D. R. and Brown, A. H. D. (1975). Optimum sampling strategies in genetic conservation. In *Crop Genetic Resources for Today and Tomorrow*, ed. O. H. Frankel and J. G. Hawkes, *International Biological Programme 2*, pp. 53–80. Cambridge University Press.

Marshall, D. R. and Brown, A. H. D. (1980). Theory of forage plant collection. (In preparation.)

Marshall, D. R. and Pryor, A. J. (1978). Multiline varieties and disease control. I. The "dirty crop" approach with each component carrying a unique single resistance gene. *Theor. Appl. Genet.* **51**, 177–84.

Martin, F. W. (1975). The storage of germplasm of tropical roots and tubers in the vegetative form. In *Crop Genetic Resources for Today and Tomorrow*, ed. O. H. Frankel and J. G. Hawkes, *International Biological Programme 2*, pp. 369–77. Cambridge University Press.

Martin, P. S. (1967). Prehistoric overkill. In *Pleistocene Extinctions: The Search for a Cause*, ed. P. S. Martin and H. E. Wright, pp. 77–120. Yale University Press, New Haven and London.

Martin, P. S. (1973). The discovery of America. *Science* **179**, 969–74.

Martin, R. D. (1975). Introduction. In *Breeding Endangered Species in Captivity*, ed. R. D. Martin, pp. xv–xxv. Academic Press, London, New York.

Mason, I. L. (1974). The conservation of animal genetic resources: Introduction to round table. *1st World Congr. Genet. Appl. Livest. Prod.* **2**, 13–21.

Mason, I. L. (1975a). Report of the mission to the Kuri cattle of Lake Chad. In *Conservation of Animal Genetic Resources*, (Pilot study), pp. 51–7. FAO, Rome.

Mason, I. L. (1975b). Preliminary survey of endangered breeds throughout the world. In *Conservation of Animal Genetic Resources* (Pilot Study), pp. 43–9. FAO, Rome.

May, R. M. (1973). *Stability and Complexity in Model Ecosystems*. Princeton University Press, Princeton, New Jersey.

Mayr, E. (1963). *Animal Species and Evolution*. Harvard University Press, Cambridge, Mass.

Mayr, E. (1967). Evolutionary challenges to the mathematical interpretation of evolution. In *Mathematical Challenges to the Neo-Darwinian Interpretation of Evolution*, ed. P. S. Moorhead and M. M. Kaplan, pp. 47–58. *The Wistar Institute Symposium, Monograph No. 5*. Wistar Institute Press. Philadelphia.

Mayr, E. (1969). *Principles of Systematic Zoology*. McGraw-Hill, New York.

Mayr, E. (1970). *Populations, Species and Evolution*. Harvard University Press, Cambridge, Mass.

Medway, L. and Wells, D. (1971). Diversity and density of birds and mammals at Kuala Lompat, Pahang. *Malay. Nat. J.* **24**, 238–47.

Melville, R. (1970). Angiospermae, a compilation. In *Red Data Book*, vol. 5, IUCN, Morges.

Mertz, D. B., Cawthorn, D. A. and Park, T. (1976). An experimental analysis of competitive indeterminacy in *Tribolium. Proc. Natl. Acad. Sci. USA*, **73**, 1368–72.

Micke, A. (1976). Introduction. In *Induced Mutations in Cross-Breeding*, pp. 1–4. IAEA, Vienna.

Miller, R. H. (1977). The need for and potential application of germplasm preservation in cattle. *J. Hered.* **68**, 365–74.

Miller, R. S. and Botkin, D. B. (1974). Endangered species: models and predictions. *Amer. Sci.* **62**, 172–81.

Mitton, J. B. (1978). Relationship between heterozygosity for enzyme loci and variation of morphological characters in natural populations. *Nature* **273**, 661–2.

Moore, N. W. and Hooper, M. D. (1975). On the number of bird species in British woods. *Biol. Conserv.* **8**, 239–50.

Morel, G. (1975). Meristem culture techniques for the long-term storage of cultivated plants. In *Crop Genetic Resources for Today and Tomorrow*, ed. O. H. Frankel and J. G. Hawkes, *International Biological Programme 2*, pp. 327–32. Cambridge University Press.

Morley, F. H. W. (1953). Selection for economic characters in Australian Merino sheep. III. Inheritance of skin-fold score in Merino sheep and problems of scale. *Aust. J. Agric. Res.* **2**, 204–12.

Morris, L. N. and Kerr, B. A. (1974). Genetic variability in macaques: correlations between biochemical and dermatoglyphic patterns. *J. Hum. Evol.* **3**, 223–35.

Mosimann, J. E. and Martin, P. S. (1975). Simulating overkill by Paleo Indians. *Amer. Sci.* **63**, 304–13.

Muller, C. H. (1966). The role of chemical inhibition (allelopathy) in vegetational composition. *Bull. Torrey Bot. Club* **93**, 332–51.

Munck, L. (1972). Improvement of nutritional value in barley. *Hereditas* **72**, 1–128.

Murphy, H. C., Wahl, I., Dinoor, A., Miller, J. D., Morey, D. D., Luke, H. H., Sechler, D. and Reyes, L. (1967). Resistance to crown rust and soil borne mosaic virus in *Avena sterilis. Pl. Dis. Reptr.* **51**, 120–4.

Myers, N. (1979). *The Sinking Ark.* Pergamon Press, Oxford.

Nag, K. K. and Street, H. E. (1973). Carrot embryogenesis from frozen cultured cells. *Nature* **245**, 270–2.

Nakagahra, M., Akihama, T. and Hayashi, K. (1975). Genetic variation and geographic cline of esterase izozymes in native rice varieties. *Jpn. J. Genet* **50**, 373–382.

Naumenko, E. V., Trut, L. N., Pavlova, S. I. and Belyaev, D. K. (1976). Genetics and phenogenetics of hormonal characteristics in animals. II. Changes in the reactivity of pituitary-adrenal system and the influence of sex glands in selection of silver foxes for behaviour. *Genetika* (Moscow) **12**, 50–5.

Nei, M. (1975). *Molecular Population Genetics and Evolution.* North Holland, New York.

Nei, M., Maruyama, T. and Chakraborty, R. (1975). The bottleneck effect and genetic variability in populations. *Evolution* **29**, 1–10.

Nevo, E. (1978). Genetic variation in natural populations: patterns and theory. *Theor. Pop. Biol.* **13**, 121–77.

Nevo, E., Kim, Y. J., Shaw, C. R. and Thaeler, C. S. Jr. (1974). Genetic variation, selection and speciation in *Thomomys talpoides* pocket gophers. *Evolution* **28**, 1–23.

Nevo, E., Zohary, D., Brown, A. H. D. and Haber, M. (1979). Allozyme–environment relationships in natural populations of wild barley in Israel. *Evolution* **33**, 815–33.

Oduye, O. O. (1975). Bovine cutaneous streptothricosis in Nigeria. *World Anim. Rev.* **16**, 13–17.

Ohta, T. and Kimura, M. (1975). Theoretical analysis of electrophoretically detectable polymorphisms: Models of very slightly deleterious mutations. *Amer. Natur.* **69**, 137–45.

Osborne, R. H. and De George, F. V. (1959). *Genetic Basis of Morphological Variation*. Harvard University Press, Cambridge, Mass.

Packard, A. (1972). Cephalopods and fish: the limits of convergence. *Biol. Rev. Camb.* **47**, 241–308.

Packer, C. (1978). Inter-group transfer and inbreeding avoidance in *Papio nubius*. *Anim. Behav.* **27**, 1–36.

Paine, R. T. (1966). Food web complexity and species diversity. *Amer. Natur.* **100**, 65–75.

Parsons, P. A. (1971). Extreme environment heterosis and genetic loads. *Heredity* **26**, 479–83.

Passmore, J. (1974). *Man's Responsibility for Nature*. Duckworth, London.

Patton, J. L., Yang, S. Y. and Myers, P. (1975). Genetic and morphologic divergence among introduced rat populations (*Rattus rattus*) of the Galápagos Archipelago, Ecuador. *Syst. Zool.* **24**, 296–310.

Perry, J., Bridgwater, D. D. and Horseman, D. L. (1972). Captive propagation: a progress report. *Zoologica* (NY) **57**, 109–17.

Perry, J. and Kibbee, P. B. (1974). The capacity of American zoos. *Inter. Zoo Yearb.* **14**, 240–7.

Peterson, P. A., Flavell, R. B. and Barratt, D. H. P. (1975). Altered mitochondrial membrane activities associated with cytoplasmically inherited disease sensitivity in maize. *Theor. Appl. Genet.* **45**, 309–14.

Pianka, E. R. (1970). On r- and K-selection. *Amer. Natur.* **104**, 592–7.

Pickett, S. T. A. and Thompson, J. N. (1978). Patch dynamics and the design of nature reserves. *Biol. Conserv.* **13**, 27–37.

Pimental, D. and Bellotti, A. C. (1976). Parasite-host population systems and genetic stability. *Amer. Natur.* **110**, 877–88.

Pinder, N. J. and Barkham, J. P. (1978). An assessment of the contribution of captive breeding to the conservation of rare mammals. *Biol. Conserv.* **13**, 187–245.

Pirchner, F. (1969). *Population Genetics in Animal Breeding*. W. H. Freeman and Co., San Francisco.

Pitman, C. R. S. (1942). *A Game Warden Takes Stock*. Nisbet, London.

Pliske, T. (1975). Attraction of lepidoptera to plants containing pyrrolizidine alkaloids. *Environ. Entomol.* **4**, 455–73.

Poppendiek, H. H. (1976). Mesembryanthemums and the problems of their cultivation. In *Conservation of Threatened Plants*, ed. J. B. Simmons, R. I. Beyer, P. E. Brandham, G. Ll. Lucas and V. T. H. Parry, pp. 55–60. Plenum Press, New York.

Powell, J. R. (1971). Genetic polymorphisms in varied environments. *Science* **174**, 1035–6.

Powell, J. R. (1975). Protein variation in natural populations of animals. In *Evolutionary Biology*, vol. 8, ed. T. Dobzhansky, M. K. Hecht and W. C. Steere, pp. 79–119. Plenum Press, New York.

Pradham, S. (1971). *Investigations on Insect Pests of Sorghum and Millets.* Indian Agricultural Research Institute, New Delhi.

Prakash, S. (1972). Origin of reproductive isolation in the absence of apparent genic differentiation in a geographic isolate of *Drosophila pseudoobscura*. *Genetics* **72**, 143–5.

Premack, D. (1971). On the assessment of language competence in the chimpanzee. In *The Behavior of Non-human Primates*, ed. A. Schrier and F. Stollnitz, pp. 185–228. Academic Press, New York.

Preston, F. W. (1962). The canonical distribution of commoness and rarity: Part I. *Ecology* **43**, 185–215; Part II. **43**, 410–32.

Price, M. V. and Waser, N. M. (1979). Pollen dispersal and optimal outcrossing in *Delphinium nelsoni*. *Nature* **277**, 294–7.

Pryor, A. J. (1977). Plant disease resistance and the acquisition of pathogen virulence. In *3rd International Congress of the Society for the Advancement of Breeding Research in Asia and Oceania (SABRAO)*, 4, pp. 15–28. SABRAO, Canberra.

Qualset, C. O. (1975). Sampling germplasm in a center of diversity: an example of disease resistance in Ethiopian barley. In *Crop Genetic Resources for Today and Tomorrow*, ed. O. H. Frankel and J. G. Hawkes, *International Biological Programme 2*, pp. 81–96. Cambridge University Press.

Qualset, C. O. and Schaller, C. W. (1969). Additional sources of resistance to the barley yellow dwarf virus in barley. *Crop Sci.* **9**, 104–5.

Ralls, K., Brugger, K. and Ballou, J. (1979). Inbreeding and juvenile mortality in small populations of ungulates. *Science* **206**, 1101–3.

Rao, A. R. (1977). Distribution patterns of indigenous wheat varieties in northern Pakistan. In *Genetic Control of Diversity in Plants*, ed. Amir Muhammed, R. Aksel and R. C. von Borstel, pp. 51–66. Plenum Press, New York.

Raup, D. M. (1976a). Species diversity in the Phanerozoic: a tabulation. *Paleobiology* **2**, 279–88.

Raup, D. M. (1976b). Species diversity in the Phanerozoic: an interpretation. *Paleobiology* **2**, 289–97.

Raven, P. R. (1976). Ethics and attitudes. In *Conservation of Threatened Plants*, ed. J. B. Simmons *et al.*, pp. 155–79. Plenum Press, New York.

Recher, H. F. (1972). The vertebrate fauna of Sydney. *Proc. Ecol. Soc. Aust.* **7**, 79–87.

Reeves, E. C. R. and Robertson, F. W. (1953). Studies in quantitative inheritance. II. Analysis of a strain of *Drosophila melanogaster* selected for long wings. *J. Genet.* **51**, 276–316.

Rensch, B. (1947). *Neuere Probleme der Abstrammungslehre, die Transspezifische Evolution*. Enke, Stuttgart.

300   References

Richardson, S. D. (1970). Gene pools in forestry. In *Genetic Resources in Plants – Their Exploration and Conservation*, ed. O. H. Frankel and E. Bennett, *IBP Handbook No. 11*, pp. 353–65. Blackwell Scientific Publications, Oxford.

Rick, C. M. (1973). Potential genetic resources in tomato species: clues from observations in native habitats. In *Genes, Enzymes, and Populations*, ed. A. M. Srb, pp. 255–69. Plenum Press, New York.

Rick, C. M. and Fobes, J. F. (1975). Allozyme variation in the cultivated tomato and closely related species. *Bull. Torrey Bot. Club* **102**, 376–84.

Rick, C. M., Fobes, J. F. and Holle, M. (1977). Genetic variation in *Lycopersicon pimpinellifolium*: evidence of evolutionary change in mating systems. *Plant. Syst. Evol.* **127**, 139–70.

Ricklefs, R. E. (1973). *Ecology*. Chiron Press, Newton, Mass.

Ricklefs, R. E. (Chrmn), Amadon, D., Conway, W., Miller, R., Nisbet, I., Schreiber, R. Seal, U., Selander, R. and Temple, S. (1978). *Report of the Advisory Panel on the California Condor*, Audubon Conservation Report No. 6. National Audubon Society, New York.

Riley, R., Chapman, V. and Johnson, R. (1968). Introduction of yellow rust resistance of *Aegilops comosa* into wheat by genetically induced homoeologous recombination. *Nature* **217**, 383–4.

Roberts, E. H. (1975). Problems of long-term storage of seed and pollen for genetic resources conservation. In *Crop Genetic Resources for Today and Tomorrow*, ed. O. H. Frankel and J. G. Hawkes, *International Biological Programme 2*, pp. 269–95. Cambridge University Press.

Roberts, E. H., Abdalla, F. H. and Owen, R. J. (1967). Nuclear damage and the ageing of seeds, with a model for seed survival curves. *Symp. Soc. Exp. Biol.* **21**, 65–100.

Robertson, A. (1955). Selection in animals: synthesis. *Cold Spring Harbor Symp. Quant. Biol.* **20**, 225–9.

Robinson, M. H. (1979). Untangling tropical biology. *New Sci.* **82**, 378–81.

Roche, L. R. (1975). Guidelines for the methodology of conservation of forest genetic resources. In *The Methodology of Conservation of Forest Genetic Resources*, pp. 107–13. FAO, Rome.

Roe, F. G. (1951). *The North American Buffalo: A Critical Study of the Species in its Wild State*. University of Toronto Press.

Rosen, M. N. (1971a). Avian cholera. In *Infectious and Parasitic Diseases of Wild Birds*, ed. J. W. Davis, R. C. Anderson, L. Karstad and D. O. Trainer, pp. 59–74. Iowa State University Press, Ames, Iowa.

Rosen, M. N. (1971b). Botulism. In *Infectious and Parasitic Diseases of Wild Birds*, ed. J. W. Davis, R. C. Anderson, L. Karstad and D. O. Trainer, pp. 100–17. Iowa State University Press, Ames, Iowa.

Rutter, R. J. and Pimlott, D. H. (1968). *The World of the Wolf.* J. B. Lippincott Co., Philadelphia.

Sakai, S., Doi, Y. and Nakayama, A. (1978). Changes in regenerative ability and carbohydrate reserve of tea root during a long-term storage for maintenance of useful germplasm. In *Long-Term Preservation of Favourable Germ Plasm in Arboreal Crops*, ed. T. Akihama and K. Nakajima, pp. 71–9. Fruit Tree Research Station, M.A.F. Fujimoto, Japan.

Sakai, A. and Noshiro, M. (1975). Some factors contributing to the survival of crop seeds cooled to the temperature of liquid nitrogen. In *Crop Genetic Resources for Today and Tomorrow*, ed. O. H. Frankel and J. G. Hawkes, *International Biological Programme 2*, pp. 317–26. Cambridge University Press.

Salisbury, G. W. and Hart, R. G. (1975). Functional integrity of spermatozoa after storage. *BioScience* **25**, 159–65.

Sastrapradja, S. (1973). Tree fruits in Java, Madura and Bali. In *Survey of Crop Genetic Resources in their Centres of Diversity*, ed. O. H. Frankel, pp. 142–59. FAO-IBP, Rome.

Sastrapradja, S. (1975). Tropical fruit germplasm in South East Asia. In *South East Asian Plant Genetic Resources*, ed. J. T. Williams, C. H. Lamoureux and N. Wullijarni Soetjipto, pp. 33–46. LIPI, Bogor.

Saville, D. B. O. (1959). Limited penetration of barriers as a factor in evolution. *Evolution* **13**, 333–43.

Scanland, T. B. (1971). Effects of Predation on Epifaunal Assemblages in a Submarine Canyon. Ph.D. dissertation. University of California, San Diego.

Schaal, B. A. and Levin, D. A. (1976). The demographic genetics of *Liatris cylindracea* Michx. (Compositae) *Amer. Natur.* **110**, 191–206.

Schaeffer, B. and Hecht, M. K. (1965). Introduction and historical background. *Syst. Zool.* **14**, 245–8.

Schaeffer, B. and Rosen, D. E. (1961). Major adaptive levels in the evolution of the actinopterygian feeding mechanism. *Amer. Zool.* **1**, 187–204.

Schaller, C. W., Rasmusson, D. C. and Qualset, C. O. (1963). Sources of resistance to the yellow-dwarf virus in barley. *Crop Sci.* **3**, 342–4.

Schaller, G. B. and Vasconcelos, J. M. C. (1978). A marsh deer in Brazil. *Oryx* **14**, 345–50.

Schaller, G. (1972). *The Serengeti Lion*. University of Chicago Press, Chicago.

Scheltema, N. (1971). Larval dispersal as a means of genetic exchange by geographically separated populations of shallow-water benthic marine gastropods. *Biol. Bull.* **140**, 284–322.

Schneider, D., Boppre, M., Schneider, H., Thompson, W. R., Borisk, C. J., Petty, R. L. and Meinwald, J. (1975). A pheromone precursor and its uptake in male *Danaus* butterflies. *J. Comp. Physiol.* **97**, 245–56.

Schoener, T. W. (1965). The evolution of bill size differences among sympatric congeneric species of birds. *Evolution* **19**, 189–213.

Schoener, T. W. (1969). Models of optimal size for solitary predators. *Amer. Natur.* **103**, 277–313.

Schopf, J. W., Haugh, L. N., Molnar, R. E. and Satterthwait, D. F. (1973). On the development of the metaphytes and metazoans. *J. Paleontol.* **47**, 1–9.

Schopf, T. J. M. (1974). Permo-Triassic extinctions; relation to sea-floor spreading. *J. Geol.* **82**, 129–43.

Schopf, T. J. M. and Gooch, J. L. (1972). A natural experiment to test the hypothesis that loss of genetic variability was responsible for mass extinctions of the fossil record. *J. Geol.* **80**, 481–3.

Scott, G. H. (1976). Foraminiferal biostratigraphy and evolutionary models. *Syst. Zool.* **25**, 78–80.

Scott, G. R. (1970). Rinderpest. In *Infectious Diseases of Wild Mammals*, ed.

J. W. Davis, L. H. Karstad and D. O. Trainer, pp. 20–35. Iowa State University Press, Ames, Iowa.

Seal, U. S. (1978). The Noah's Ark problem: multigeneration management of wild species in captivity. In *Endangered Birds: Management Techniques for Preserving Threatened Species*, ed. S. A. Temple, pp. 303–14. University of Wisconsin Press, Madison.

Sears, E. R. (1956). The transfer of leaf-rust resistance from *Aegilops umbellulata* to wheat. *Brookaven Symp. Biol.* **9**, 1–21.

Sears, E. R. (1966). Chromosome mapping with the aid of telocentrics. Proceedings of the 2nd International Wheat Genetics Symposium. *Hereditas* (suppl.) **2**, 370–80.

Seetharaman, R., Sharma, S. D. and Shastry, S. V. S. (1972). Germplasm conservation and use in India. In *Rice Breeding*, pp. 187–200. International Rice Research Institute, Los Baños, Philippines.

Seifert, G. W. (1971). Ecto- and endoparasitic effects on the growth rates of Zebu crossbred and British cattle in the field. *Aust. J. Agric. Res.* **22**, 839–50.

Selander, R. K. (1976). Genic variation in natural populations. In *Molecular Evolution*, ed. F. J. Ayala, pp. 21–45. Sinauer Associates, Sunderland, Mass.

Selander, R. K. and Kaufman, D. W. (1973a). Self fertilization and genic population structure in a colonizing land snail. *Proc. Natl. Acad. Sci USA* **70**, 1186–1190.

Selander, R. K. and Kaufman, D. W. (1973b). Genic variability and strategies of adaptations in animals. *Proc. Natl. Acad. Sci. USA* **70**, 1875–7.

Selander, R. K., Kaufman, D. W., Baker, R. J. and Williams, S. L. (1975). Genic and chromosomal differentiation in pocket gophers of the *Geomys bursarius* group. *Evolution* **28**, 557–64.

Sengbusch, R. von (1934). Die Geschichte der "Süsslupinen". *Naturwissenschaften* **22**, 278–81.

Sengbusch, R. von (1953). Ein Beitrag zur Entstehungsgeschichte unserer Nahrungs-Kulturpflanzen unter besonderer Berücksichtigung der Individualauslese. *Der Züchter* **23**, 353–64.

Senner, J. W. (1980). Inbreeding depression and the survival of zoo populations. In *Conservation Biology: An Evolutionary–Ecological Perspective*, ed. M. E. Soulé and B. A. Wilcox, pp. 209–24. Sinauer Associates, Sunderland, Mass.

Short, R. V. (1976). The introduction of new species of animals for the purpose of domestication. *Symp. Zool. Soc. Lond.* **40**, 321–33.

Siebenaller, J. F. (1978). Genetic variation in deep-sea invertebrate populations: the bathyal gastropod *Bathybembix bairdii*. *Mar. Biol.* **47**, 265–75.

Sigurbjörnsson, R. and Micke, A. (1974). Philosophy and accomplishments of mutation breeding. In *Polyploidy and Induced Mutations in Plant Breeding*, pp. 303–43. IAEA, Vienna.

Simberloff, D. S. (1974). Permo-Triassic extinctions: effects of area on biotic equilibrium. *J. Geol.* **82**, 267–74.

Simberloff, D. S. and Abele, L. G. (1976). Island biogeography theory and conservation practice. *Science* **154**, 285–6.

Simberloff, D. S. and Wilson, E. O. (1970). Experimental zoogeography of islands. A two-year record of colonization. *Ecology* **51**, 934–7.

Simmonds, N. W. (1962). Variability in crop plants, its use and conservation. *Biol. Rev.* **37**, 442–65.

Simmonds, N. W. (1969). Prospects of potato improvement. In *Scottish Plant Breeding Station, 48th Annual Report 1968–69*, pp. 18–38. Edinburgh.

Simmons, J. B., Beyer, R. I., Brandham, P. E., Lucas, G. Ll. and Parry, V. T. H. (eds.) (1976). *Conservation of Threatened Plants*. Plenum Press, New York.

Simpson, G. G. (1953). *The Major Features of Evolution*. Columbia University Press, New York.

Simpson, G. G. (1959). The nature and origin of supraspecific taxa. *Cold Spring Harbor Symp. Quant. Biol.* **24**, 255–71.

Singer, P. (1975). *Animal Liberation*. New York Review, New York.

Singh, Bhag. (1974). Current maize cultivars of north eastern Himalayan region. *SABRAO J.* **6**, 229–35.

Singh, Bhag. (1977a). *Races of Maize in India*. Indian Council of Agricultural Research, New Delhi.

Singh, Bhag. (1977b). Evaluation and use of Indian primitive cultivars in the improvement of maize. In *3rd International Congress of the Society for the Advancement of Breeding Research in Asia and Oceania (SABRAO)* pp. 1(a) 9–12. SABRAO, Canberra.

Singh, H. B., Joshi, B. S., Chandel, K. P. S., Pant, K. C. and Saxena, R. K. (1974). Genetic diversity in some Asiatic *Phaseolus* species and its conservation. *Proceedings of the 2nd General Congress of the Society for the Advancement of Breeding Researches in Asia and Oceania*, ed. S. Ramanujam and R. D. Iyer, *Indian J. Genet.* **34A**, pp. 52–7.

Singh, R. and Axtell, J. D. (1973). High lysine mutant gene (hl) that improves protein quality and biological value of grain sorghum. *Crop Sci.* **13**, 535–9.

Singh, R. S., Lewontin, R. C. and Felton, A. A. (1976). Genetic heterogeneity within electrophoretic 'alleles' of xanthine dehydrogenase in *Drosophila pseudoobscura. Genetics* **84**, 609–29.

Singh, S. M. and Zouros, E. (1978). Genetic variation associated with growth rate in the American oyster *(Crassostrea virginica). Evolution*, **32**, 342–53.

Sioli, H. (1975). Tropical rivers as expressions of their terrestrial environments. In *Tropical Ecological Systems: Trends in Terrestrial Research*, ed. F. Golley and E. Medina, pp. 275–88, Springer-Verlag, New York.

Slatis, H. M. (1960). An analysis of inbreeding in the European bison. *Genetics* **45**, 275–87.

Slobodkin, L. B. (1968). How to be a predator. *Amer. Zool.* **8**, 43–51.

Smith, F. E. (1954). Quantitative aspects of population growth. In *Dynamics of Growth Processes*, ed. E. Boell, pp. 277–94. Princeton University Press, Princeton, New Jersey.

Smith, M. R. (1936). Distribution of the Argentine ant in the United States and suggestions for its control and eradication. *Circ. US Dep. Agric.* **387**, 1–39.

Sokal, R. R. and Crovello, T. J. (1970). The biological species concept: a critical evaluation. *Amer. Natur.* **104**, 122–53.

Sommer, A. (1976). Attempt at an assessment of the world's tropical forests. *Unasylva* **28**, 5–25.

Soulé, M. (1966). Trends in the insular radiation of a lizard. *Amer. Natur.* **100**, 47–64.

Soulé, M. E. (1972). Phenetics of natural populations. III. Variation in insular populations of a lizard. *Amer. Natur.* **106**, 429–46.

Soulé, M. E. (1973). The epistasis cycle: a theory of marginal populations. *Annu. Rev. Ecol. Syst.* **4**, 165–87.

Soulé, M. E. (1976). Allozyme variation: its determinants in space and time. In *Molecular Evolution*, ed. F. J. Ayala, pp. 60–77. Sinauer Associates, Sunderland, Mass.

Soulé, M. E. (1979). Heterozygosity and developmental stability: another look. *Evolution* **33**, 396–401.

Soulé, M. E. (1980). Thresholds for survival: maintaining fitness and evolutionary potential. In *Conservation: An Evolutionary–Ecological Perspective*, ed. M. E. Soulé and B. A. Wilcox, pp. 151–70. Sinauer Associates, Sunderland, Mass.

Soulé, M. E. (1981). Phenetics of natural populations. V. Allomeric variation. Submitted.

Soulé, M. E. and Wilcox, B. A. (eds.) (1980). *Conservation Biology: An Evolutionary–Ecological Perspective*. Sinauer Associates, Sunderland, Mass.

Soulé, M. E., Wilcox, B. A. and Holtby, C. (1979). Benign neglect: a model of faunal collapse in the game reserves of East Africa. *Biol. Conserv.* **15**, 259–72.

Soulé, M. E., Yang, S. Y., Weiler, M. G. W. and Gorman, G. C. (1973). Island lizards: the genetic–phenetic variation correlation. *Nature* **242**, 190–2.

Specht, R. L., Roe, E. M. and Boughton, V. H. (1974). Conservation of major plant communities in Australia and Papua New Guinea. *Aust. J. Bot.* Suppl. No. 7.

Spiess, E. B. (1977). *Genes in Populations*. Wiley, New York.

Spillett, J. J., Bunch, T. D. and Foote, W. C. (1975). The use of wild and domestic animals and the development of new genotypes. *J. Anim. Sci.* **40**, 1009–15.

Spooner, R. L., Bradley, J. S. and Young, G. B. (1975). Genetics and disease in domestic animals with particular reference to dairy cattle. *Vet. Rec.* **97**, 125–30.

Spurway, H. (1952). Can wild animals be bred in captivity? *New Biol.* **13**, 11–30.

Stanley, S. M. (1975). Stability of species in geologic time. *Science* **190**, 267–8.

Stephenson, A. B., Wyatt, A. J. and Nordskog, A. W. (1953). Influence of inbreeding on egg production in the domestic fowl. *Poult. Sci.* **32**, 510–17.

Stewart, J. L. (1951). The West African Shorthorn cattle. Their value as Trypanosomiasis-resistant animals. *Vet. Rec.* **27**, 454–7.

Stølen, O. (1965). Investigations on the tolerance of barley varieties to high hydrogen-ion concentration in soil. *Royal Veterinary and Agricultural College Yearbook 1965*, Copenhagen.

Strong, L. (1978). Inbred mice in science. In *Origins of Inbred Mice*, ed. H. C. Morse, pp. 45–67. Academic Press, New York.

Sugawara, Y. and Sakai, S. (1978). Some factors contributing to survival of suspension-cultured cells cooled to temperature of liquid nitrogen. In *Long Term Preservation of Favourable Germ Plasm in Arboreal Crops*, ed. T. Akihama and K. Nakajima, pp. 121–9. Fruit Tree Research Station, M. A. F. Fujimoto, Japan.

Sved, J. A., Reed, T. E. and Bodmer, W. F. (1967). The number of balanced polymorphisms that can be maintained in a natural population. *Genetics* **55**, 469–81.

Svihla, A. (1936). The Hawaiian rat. *Murrelet* **17**, 3–14.

Sykes, J. T. (1972). Tree fruit resources in Turkey. *Plant Gen. Res. Newsl.* **27**, 17–21.

Syme, P. D. (1977). Observations on the longevity and fecundity of *Orgilus obscurator* (Hymenoptera: Brachonidae) and the effects of certain foods on longevity. *Can. Ent.* **109**, 995–1000.

Temin, R. G., Meyer, H. D., Dawson, P. S. and Crow, J. F. (1969). The influence of epistasis on homozygous viability depression in *Drosophila melanogaster*. *Genetics* **61**, 497–515.

Terborgh, J. (1974). Preservation of natural diversity: the problem of extinction prone species. *BioScience* **24**, 715–22.

Terborgh, J. (1975). Faunal equilibria and the design of wildlife preserves. In *Tropical Ecological Systems: Trends in Terrestrial and Aquatic Research*, ed. F. Golley and E. Medina, pp. 369–80. Springer-Verlag, New York.

Terborgh, J. W. (1976). Island biogeography and conservation: strategy and limitations. *Science* **193**, 1029–30.

Terborgh, J. W. and Winter, B. (1980). Some causes of extinction. In *Conservation Biology: An Evolutionary–Ecological Perspective*, ed. M. E. Soulé and B. A. Wilcox, pp. 119–34. Sinauer Associates, Sunderland, Mass.

Thoday, J. M. (1953). Components of fitness. *Symp. Soc. Exp. Biol.* **7**, 96–113.

Thompson, P. A. (1975). The collection, maintenance and environmental importance of the genetic resources of wild plants. *Environ. Conserv.* **2**, 223–8.

Thompson, P. A. (1976). Factors involved in the selection of plant resources for conservation as seed in gene banks. *Biol. Conserv.* **10**, 159–67.

Thompson, P. A. and Brown, G. E. (1972). The seed unit at the Royal Botanic Gardens, Kew, *Kew Bull.* **26**, 445–56.

Tinbergen, L. (1960). The natural control of insects in pinewoods. I: Factors influencing the intensity of predation by songbirds. *Arch. Neerl. Zool.* **13**, 266–336.

Tinkle, D. W. and Selander, R. K. (1973). Age-dependent allozymic variation in a natural population of lizards. *Biochem. Genet.* **8**, 231–7.

Tracey, M. W., Bellet, N. F. and Graven, C. B. (1975). Excess of allozyme homozygosity and breeding population structure in the mussel *Mytilus californianus*. *Mar. Biol.* **32**, 303–11.

Treus, V. D. and Lobanov, N. V. (1971). Acclimatisation and domestication of the eland at Askenya-Nova Zoo. *Zoo Yearb.* **11**, 147–56.

Turner, H. G. and Short, A. J. (1972). Effects of field infestations of gastrointestinal helminths and of the cattle tick (*Boophilus microplus*) on growth of three breeds of cattle. *Aust. J. Agric. Res.* **23**, 177–93.

Turner, H. N. (1977). International collaboration in animal breeding. *SABRAO J.* **9**, 1–8.

Turner, H. N. (1978). Selection for reproduction rate in Australian Merino sheep: direct responses. *Aust. J. Agric. Res.* **29**, 327–50.

Turton, I. D. (1974). The collection, storage and dissemination of information on breeds of livestock. In *Round Table: The Conservation of Animal Genetic Resources*, pp. 61–74. FAO, Rome.

Tyler, W. J., Chapman, A. B. and Dickerson, G. E. (1949). Growth and production of inbred and outbred Holstein-Friesian cattle. *J. Dairy Sci.* **32**, 247–56.

Underwood, J. H. (1979). *Human Variation and Human Evolution*. Prentice-Hall, Englewood Cliffs, New Jersey.

UNESCO (1973). Expert Panel on Project 8: Conservation of natural areas and of the genetic material they contain. In *MAB Report Series* No. 12. UNESCO, Paris.

UNESCO (1974). MAB Task Force on criteria and guidelines for the choice and establishment of biosphere reserves. In *MAB Report Series* No. 22. UNESCO, Paris.

Vairavan, E., Siddiq, E. A., Arunachalam, V. and Swaminathan, S. M. (1973). A study of the nature of genetic divergence in rice from Assam and North East Himalayas. *Theor. Appl. Genet.* **43**, 213–21.

Valentine, J. W. (1976). Genetic strategies of adaptation. In *Molecular Evolution*, ed. F. J. Ayala, pp. 78–94. Sinauer Associates, Sunderland, Mass.

Valentine, J. W. and Ayala, F. J. (1976). Genetic variation in krill. *Proc. Natl. Acad. Sci. USA* **73**, 658–60.

Van der Plank, J. E. (1975). *Principles of Plant Infection*. Academic Press, New York.

Van Valen, L. (1974). Molecular evolution as predicted by natural selection. *J. Mol. Evol.* **3**, 89–101.

Van Valen, L. and Sloan, R. E. (1977). Ecology and the extinction of the dinosaurs. *Evol. Theory* **2**, 37–64.

Vavilov, N. I. (1949–50). The origin, variation, immunity, and breeding of cultivated plants. *Chron. Bot.* **13**, 1–366.

Vermeej, G. J. (1973). Adaptation, versatility and evolution. *Syst. Zool.* **22**, 466–77.

Villiers, T. A. (1975). Genetic maintenance of seeds in imbibed storage. In *Crop Genetic Resources for Today and Tomorrow*, ed. O. H. Frankel and J. G. Hawkes, *International Biological Programme 2*, pp. 297–315. Cambridge University Press.

Viswanathan, T. V. (1976). Cytogenetic studies on some medicinal plants of the genus *Solanum* utilized by the tribals of Kerala and Tamil Nadu. Ph.D. Thesis, University of Calicut, India.

Vogel, O. A., Allan, R. E. and Peterson, C. J. (1963). Plant and performance characteristics of semidwarf winter wheats producing most efficiently in Eastern Washington. *Agron. Jour.* **55**, 397–8.

Volf, J. (1975). Breeding of Przewalski wild horses. In *Breeding of Endangered Species in Captivity*, ed. R. D. Martin, pp. 263–70. Academic Press, New York.

Wallace, B. (1970). *Genetic load: Its Biological and Conceptual Aspects*. Prentice-Hall, Inc., Englewood Cliffs, NJ.

Wallace, B. and Madden, C. (1965). Studies on inbred strains of *Drosophila melanogaster*. *Amer. Natur.* **99**, 495–509.

Wark, D. C. (1963). *Nicotiana* species as sources of resistance to blue mould (*Peronospora tabacina* Adam) for cultivated tobacco. In *Proceedings of the 3rd World Tobacco Scientific Congress*, pp. 252–9. Salisbury.

Wark, D. C. (1970). Development of flue-cured tobacco cultivars resistant to a common strain of blue mould. *Tob. Sci.* **171**, 147–50.

Warland, M. A. G. (1975). A cautionary note on breeding endangered species in captivity. In *Breeding Endangered Species in Captivity*, ed. R. D. Martin, pp. 373–7. Academic Press, New York.

Watson, I. A. (1970a). The utilization of wild species in the breeding of cultivated crops resistant to plant pathogens. In *Plant Genetic Resources – Their Exploration and Conservation*, ed. O. H. Frankel and E. Bennett, *IBP Handbook No. 11*, pp. 441–57. Blackwell Scientific Publications, Oxford.

Watson, I. A. (1970b). Changes in virulence and population shifts in plant pathogens. *Ann. Rev. Phytopath.* **8**, 209–30.

Watson, I. A. and Singh, D. (1952). The future of rust resistant wheat in Australia. *J. Aust. Inst. Agric. Sci.* **18**, 190–7.

Watt, W. B. (1977). Adaptation of specific loci: I. Natural selection on phosphoglucose isomerase of *Colias* butterflies: biochemical and population aspects. *Genetics* **87**, 177–94.

Webb, S. D. (1969). Extinction–origination equilibria in late Cenozoic land mammals of North America. *Evolution* **23**, 688–702.

Webb, S. D. (1976). Mammalian faunal dynamics of the great American interchange. *Paleobiology* **2**, 220–34.

Weir, B. S., Allard, R. W. and Kahler, A. I. (1972). Analysis of complex allozyme polymorphisms in a barley population. *Genetics* **72**, 505–23.

Went, F. W. (1957). *The Experimental Control of Plant Growth*. Chronica Botanica. Waltham, Mass.

Whitcomb, R. F., Lynch, J. F., Opler, P. A. and Robbins, C. S. (1976). Island biogeography and conservation: strategy and limitations. *Science* **193**, 1030–2.

White, M. J. D. (1959). Speciation in animals. *Aust. J. Sci.* **22**, 32–9.

White, M. J. D. (1978). *Modes of Speciation*. W. H. Freeman and Co., San Francisco.

Whitehead, G. K. (1978). Captive breeding as a practical aid to preventing extinction and providing animals for reintroduction. In *Threatened Deer*, pp. 353–63. IUCN, Morges, Paris.

Whitmore, T. C. (1975a). South East Asian forests as an unexploited source of fast growing timber. In *South East Asian Plant Genetic Resources*, ed. J. T. Williams, C. H. Lamoureux and N. Wullijarni-Soetjipto, pp. 205–12. LIPI, Bogor.

Whitmore, T. C. (1975b). *Tropical Rain Forests of the Far East*. Clarendon Press, Oxford.

Whitmore, T. C. (1980). The conservation of tropical rain forest. In *Conservation Biology: An Evolutionary–Ecological Perspective*, ed. M. E. Soulé and B. A. Wilcox, pp. 303–18. Sinauer Associates, Sunderland, Mass.

Whittingham, D. G. (1975). Low temperature storage of mammalian embryos. In *Basic Aspects of Freeze Preservation of Mouse Strains*, (Bar Harbor Workshop) ed. O. Muhlbock and A. Spiegel. Gustav Fischer Verlag, New York.

Whyte, R. O. (1958). *Plant Exploration, Collection and Introduction*. FAO, Rome.

Whyte, R. O. and Julén, G. (1963). *Proceedings of a Technical Meeting on Plant Exploration and Introduction. Genetica Agraria* **17**, 573 pp.

Wilcox, B. A. (1980). Insular ecology and conservation. In *Conservation Biology: An Evolutionary–Ecological Perspective*, ed. M. E. Soulé and B. A. Wilcox, pp. 95–118. Sinauer Associates, Sunderland, Mass.

Williams, E. E. (1969). The ecology of colonization seen in the zoogeography of anoline lizards on small islands. *Quant. Rev. Biol.* **44**, 345–89.

Williams, E. E. (1972). The origin of Faunas. Evolution of lizard congeners in a

308   *References*

complex island fauna: a trial analysis. In *Evolutionary Biology*, ed. T. Dobzhansky, M. Hecht and W. Steere, pp. 47–88. Appleton-Century-Crofts, New York.

Williams, J. T., Lamoureux, C. H. and Wullijarni-Soetjipto, N. (eds.) (1975). *South East Asian Plant Genetic Resources*. LIPI, Bogor.

Willis, E. O. (1974). Populations and local extinctions of birds on Barro Colorado Island, Panama. *Ecol. Monogr.* **44**, 153–69.

Willis, E. O. (1980). The composition of avian communities in remanescent woodlots in southern Brazil. *Papels Avulsos Museu Paulisto*, (in press).

Wills, C. (1978). Rank-order selection is capable of maintaining all genetic polymorphisms. *Genetics* **89**, 403–17.

Wilson, A. C. (1976). Gene regulation in evolution. In *Molecular Evolution*, ed. F. J. Ayala, pp. 225–36. Sinauer Associates, Sunderland, Mass.

Wilson, A. C., Bush, G. L., Case, S. M. and King, M. C. (1975). Social structuring of mammalian populations and rate of chromosomal evolution. *Proc. Natl. Acad. Sci. USA* **72**, 5061–5.

Wilson, E. O. (1961). The nature of the taxon cycle in the Melanesian ant fauna. *Amer. Natur.* **95**, 169–93.

Wilson, E. O. (1975). *Sociobiology: The New Synthesis*. Belknap Press, Cambridge, Mass.

Wilson, E. O. and Bossert, W. H. (1971). *A Primer of Population Biology*. Sinauer Associates, Sunderland, Mass.

Wilson, E. O. and Brown, W. L. (1953). The subspecies concept and its taxonomic application. *Syst. Zool.* **2**, 97–111.

Wilson, E. O. and Willis, E. O. (1975). Applied biogeography. In *Ecology and Evolution of Communities*, ed. M. L. Cody and J. M. Diamond, pp. 522–34. Harvard University Press, Cambridge, Mass.

Witcombe, J. R. and Rao, A. R. (1976). The genecology of wheat in a Nepalese centre of diversity. *J. Appl. Ecol.* **13**, 915–24.

Withers, L. A. (1978). Freeze preservation of cultured cells and tissues. In *Frontiers of Plant Tissue Culture 1978*, ed. T. A. Thorpe, pp. 297–306. International Association of Plant Tissue Culture, Calgary, Alberta.

Wolliams, K. R. (1976). The propagation of Hawaiian endangered species. In *Conservation of Threatened Plants*, ed. J. B. Simmons, R. I. Beyer, P. E. Brandham, G. Ll. Lucas and V. T. H. Parry, pp. 73–83. Plenum Press, New York.

Wood-Jones, F. (1951–52). A contribution to the history and anatomy of Père Davids Deer (*Elaphurus davidianus*). *Proc. Zool. Soc. Lond.* **121**, 319–70.

Wright, S. (1921). Systems of mating. *Genetics* **6**, 111–78.

Wright, S. (1931). Evolution in Mendelian populations. *Genetics* **16**, 97–159.

Wright, S. (1937). The distribution of gene frequencies in populations. *Proc. Natl. Acad. Sci. USA*, **23**, 307–20.

Wright, S. (1969). *Evolution and the Genetics of Populations*, vol. 2, *The Theory of Gene Frequencies*. University of Chicago Press, Chicago.

Wright, S. (1977). *Evolution and the Genetics of Populations*, vol. 3, *Experimental Results and Evolutionary Deductions*. University of Chicago Press, Chicago.

Wright, S. (1978). *Evolution and the Genetics of Populations*, vol. 4, *Variability Within and Among Natural Populations*. University of Chicago Press, Chicago.

Wu, K. K. and Jain, S. K. (1977). A note on germplasm diversity in the world collections of safflower. *Econ. Bot.* **31**, 72–5.

Young, E. G. Jr. and Murray, J. C. (1966). Heterosis and inbreeding depression in the diploid and tetraploid cottons. *Crop Sci.* **6**, 436–8.

Zagaja, S. W. (1970). Temperate zone tree fruits. In *Plant Genetic Resources – Their Exploration and Conservation*, ed. O. H. Frankel and E. Bennett, *IBP Handbook No. 11*, pp. 327–33. Blackwell Scientific Publications, Oxford.

Zeuner, F. E. (1963). *A History of Domesticated Animals*. Hutchinson, London.

Zeven, A. C. and Zhukovsky, P. M. (1975). *Dictionary of Cultivated Plants and their Centres of Diversity*. Centre for Agricultural Publishing and Documentation, Wageningen.

Zimmer, D. E. and Kinman, M. L. (1972). Downy mildew resistance in cultivated sunflower and its inheritance. *Crop Sci.* **12**, 749–51.

Ziswiler, V. (1967). *Extinct and Vanishing Animals*. Springer-Verlag Inc., New York.

Zohary, D. and Feldman, M. (1962). Hybridization between amplidiploids and the evolution of polyploids in the wheat (Aegilops-Triticum) group. *Evolution* **16**, 44–61.

Zohary, D., Harlan, J. R. and Vardi, J. (1969). The wild diploid progenitors of wheat and their breeding value. *Euphytica* **18**, 58–65.

Zohary, D. and Spiegel-Roy, P. (1975). Beginnings of fruit growing in the Old World. *Science* **187**, 319–27.

Index

aborigines, no record of extinction of a
species by, 17
Aconitum heterophyllum (source of
aconite), endangered by
over-exploitation, 210
adaptation
to environmental change, 84
to extreme habitats, 164, 165
to favourable conditions, 200–1
of plants under domestication, 176;
ecotypic, 178–9; species-specific,
177–8
time, space, and fitness as components
of, 1
Addax nasomaculatus (addax), breeding in
captivity, 139
Aegilops spp.
in ancestry of modern wheats, 179
examined for protein characteristics, 241
Aegilops umbellulata: resistance of, to rusts,
transferred to wheat, 205
aggressiveness
associated with heterozygosity in
Peromyscus, 54, 71
in inbred domesticated and in wild rats,
150
silver foxes selected for lack of, 253–4
Alca impennis (great auk), extinction of,
16
allelopathy, 11
amniocentesis, for detection of chromosome
abnormalities in foetus, 155
amphibians
crosses between remote populations of,
often sterile, 154
estimates of genetic variability in, 43
Anacardium occidentale (cashew nut), low
variability of, 218
Andropogon hallii, selection for
non-shattering of seed in, 177
antelopes, large: inbreeding of, in nature
reserves, 130

anthrax: mortality of animals from, in the
wild and in nature reserves, 99
antibiotics: evolution of resistance to, in
bacteria, 84
Antilope cervicapra (black buck): successful
return of, to nature, 141
ants
Argentine fire ant (*Iridomyrmex
humilis*), in competition with North
American species, 11–12
hosts of larva of *Maculinia* butterfly,
losing their habitat because of scarcity
of rabbits, 115
on islands, 12
Aotus trivirgatus (owl monkey), variation of
chromosome number in, 154
apes, great: trisomics in, 155
Arachis hypogaea (groundnut), pathogen
resistance in wild relatives of 192–
193
aroids of tropical forests, bees pollinating,
115
Ashkania Nova Zoo, eland in, 144
Asia
South-East: disappearance of wild fruits
in, 218–19; endangered forest species
in, 224; five countries in, combine to
collect and preserve genetic resources
of crops, and to set aside nature
reserves, 5, 247; multi-purpose forest
reserves suggested for, 228
asymmetry of structure (developmental
instability), negatively correlated with
heterozygosity in side-blotched lizard,
54–5
Australia
extinction of species in, 15, 106
introduction of pasture grasses and
legumes into, 210
Western: salvage of wild flowers of, by
commercial planting, and protection in
the wild, 167

fitness—*contd*
relation of, to genetic variability in natural populations, 47–9, 57, 119; an asymptotic relation? 58, 76
relation of, to individual's deviation from average character, 53
fitness characters, 65
Food and Agriculture Organization of United Nations (FAO), 214, 242, 245, 266
Panel of Experts on Forest Gene Resources, 246
Panel of Experts on Plant Exploration and Introduction, 215, 233, 235, 245, 246; disbanded, 246
forest reserves, 228, 229, 231
forests
acres of, per person, and extinction of bird species in West Indies, 25–6
bird species in isolated patches of, increase in number with size of patch, 102
controlled artificial disturbance of, to maintain successional stages, 119; must avoid destruction of rare species, 128, 131–2
multiple use of temperate coniferous, not applicable to tropics, 118
natural disturbances affecting, 116
as reservoirs of economic plants as yet unused, 209, 210, 212
as sources of genetic diversity, 210, 224
time scale for conservation of, 2
tropical: chain reactions from extinction of species in, 111, 112–13; fragility of soil and ecosystems in, 111, 118; not renewable, 111; present rate of destruction of, 3, 25, 164, 209
tropical rain, canopy species in, 118
foxes
in extinction of Australian marsupials, 15
silver, bred in captivity: selection of, for lack of aggressiveness, accompanied by neuro-endocrinal changes, 253–4
Franklinia alatamahu, conserved by introduction into horticultural trade, 167
fruit-eating animals
dependent on successional species, 115
migration of, on failure of fruit crop, 116, 118, 123
fruit-trees
difficulty of maintaining species of, because of lack of or recalcitrant seeds, 255
disappearance of old varieties of, in Near East, 217–18
need for germplasm collections of, 239
Fundulus heteroclitus (killifish), superiority of relatively heterozygous individuals of, 53
future, man's responsibility towards, 4–5, 7–9

Gazella gazella arabica (Arabian gazelle), breeding in captivity, 139
gene flow
can natural selection over-ride effects of?, 84, 87–8
irrelevant in dealing with remnant stocks, 88, 96
gene pools
conservation of livestock genes by grouping related local breeds to form, 271, 273, 277
of livestock, almost entirely domestic, 254
of plant crops, enriched from wild and weedy relatives, 178
tend to resist change, 88
generation time, and intrinsic growth rate of population, 84–5
generations: overlapping of, and inbreeding, 39
genes
assumption of correlation of variation in different sets of (genetic uniformitarianism), 91, 96
deleterious dominant, always exposed to selection and readily eliminated, 63, 73
deleterious recessive: load of, and maximum permissible rate of inbreeding, 73; may be removed or fixed by inbreeding, 68–9; responsible for inbreeding depression, 44, 62–5
loss of, in population bottlenecks, 33, 34–5, 37, 75
genetic diversity, heterozygosity, variability
advantages of, in crop plants, 201
'cornucopian' assumption concerning, 79–80, 82, 83–5, 119; not applicable to large plants and animals, 85–7, 89–90, 95
and fitness, in natural populations, 47–9, 57, 58, 76, 119
loss of; accompanies progress towards increased productivity of livestock, 256; in breeding animals in captivity, 145, 146; in conservation of plants in botanic gardens, 172
minimum population size to maintain useful level of, 91
in natural populations, 42, 82–3; estimated from electrophoretic data, 42–4, 49; models for maintenance of, 44–7; studies of, between populations, 54–7, and within populations, 50–3
in plants, 55–6, 76, 223; decreased in modern cultivars, 181; dynamics of, 179–81; estimates of extent of, 43; evolutionary continuum of, 175; few data on, in outcrossing plants, 95; levels